Enological Repercussions of Non-*Saccharomyces* Species

Enological Repercussions of Non-*Saccharomyces* Species

Special Issue Editor

Antonio Morata

MDPI • Basel • Beijing • Wuhan • Barcelona • Belgrade

MDPI

Special Issue Editor
Antonio Morata
Universidad Politécnica de Madrid (UPM)
Spain

Editorial Office
MDPI
St. Alban-Anlage 66
4052 Basel, Switzerland

This is a reprint of articles from the Special Issue published online in the open access journal *Fermentation* (ISSN 2311-5637) from 2018 to 2019 (available at: https://www.mdpi.com/journal/fermentation/special_issues/non-saccharomyces)

For citation purposes, cite each article independently as indicated on the article page online and as indicated below:

LastName, A.A.; LastName, B.B.; LastName, C.C. Article Title. *Journal Name* **Year**, *Article Number, Page Range.*

ISBN 978-3-03921-558-4 (Pbk)
ISBN 978-3-03921-559-1 (PDF)

Cover image courtesy of Maria Antonia Bañuelos.

Contents

About the Special Issue Editor

Antonio Morata is Professor in Food Science and Technology at the Universidad Politécnica de Madrid (UPM), Spain, specializing in Wine Technology, where he is also Coordinator of the Master in Food Engineering in addition to being Professor of Enology and Wine Technology at the European Master of Viticulture and Enology, Euromaster Vinifera-Erasmus+. Morata is the Spanish delegate of a group of experts in Wine Microbiology and Wine Technology in the International Organisation of Vine and Wine (OIV). Morata has authored over 70 research articles, 3 books, 13 book chapters, and 3 patents, in addition to serving as Editor of 3 books. Morata is an Editorial Board member of the MDPI journals *cursive* and *Beverages*.

fermentation

MDPI

Editorial

Enological Repercussions of Non-*Saccharomyces* Species in Wine Biotechnology

Antonio Morata

Department of Chemistry and Food Technology, Universidad Politécnica de Madrid (UPM), 28040 Madrid, Spain; antonio.morata@upm.es

Received: 30 July 2019; Accepted: 1 August 2019; Published: 5 August 2019

The use of non-*Saccharomyces* yeasts in enology has increased since the beginning of the current century because of the potential improvements they can produce in wine sensory quality. Several review articles have described the potential of some non-*Saccharomyces* species [1–3] and the suitable criteria to select them [4,5] according to the effects of the species on wine color, aroma, body or structure. Most non-*Saccharomyces* species have low fermentative power, which makes it necessary to use them in sequential fermentations with *S. cerevisiae* to completely deplete the sugars. Moreover, some of them have slow fermentation kinetics, which is a drawback for a competitive implantation in must containing *S. cerevisiae* indigenous populations. Emerging technologies to control wild indigenous yeasts can facilitate the development, growth and fermentative activity of the inoculated non-*Saccharomyces* yeasts and, therefore, the suitable expression of their metabolic properties [6]. This special issue is focused on the description and review of several non-*Saccharomyces* species with great potential in wine biotechnology, some of which are frequently used at the winery scale, but also produced industrially as dried yeast or liquid inoculant [7].

Wine acidity, especially the pH, is a key parameter in wine that controls microbial development and chemical stability. Traditional pH control is driven by acidification processes with tartaric acid or modern ion exchanger techniques, which unfortunately affect sensory quality. The biological modulation of wine acidity can be done efficiently by several non-*Saccharomyces* species, by the production of lactic acid by *Lachancea thermotolerans* or succinic acid by *Candida stellata*, the demalication by *Schizosaccharomyces pombe* or *Pichia kudriavzevii*, and the control of volatile acidity in sequential fermentations with *Torulaspora delbrueckii* or *Zygosaccharomyces florentinus* highlight the possibilities of non-*Saccharomyces* in the improvement of wine acidity [8].

Biological acidification by *L. thermotolerans* is a powerful tool to control pH in warm areas [9]. The production of acidity is performed from sugars and the product lactic acid is a stable metabolite during winemaking but also through stabilization and aging. The formation of several metabolites with sensory repercussions has also been described in this species. Acidification by *L. thermotolerans* is a natural biotechnology that helps to keep lower and more effective levels of molecular and free SO_2. Currently, in our laboratory we have selected strains of this species able to ferment at more than 12% potential alcohol, which opens the door to single fermentations with single inoculums of *L. thermotolerans*.

Wine deacidification by metabolization of malic acid is an essential step in red winemaking. This acid is unstable during stabilization and aging, and can produce microbial hazes if not eliminated previously. Usually, malic acid is transformed into lactic acid by malolactic fermentation produced by lactic acid bacteria, mainly *Oenococcus oeni*, due to the specific composition of wine. Alternatively, *S. pombe* is able to metabolize malic acid by the maloalcoholic fermentation pathway. The advantages are the fast and efficient degradation of malic acid and at the same time *S. pombe* can produce the alcoholic fermentation. Moreover, its use reduces the formation of biogenic amines. Also, the peculiar metabolism of *S. pombe* facilitates the formation of vitisin A pyranoanthocyanin pigments, with positive effects on color stability [10].

Among the pioneer species used in enology is *T. delbrueckii*, with medium fermentative power, some strains reach 9%–10% in alcohol with a high fermentation purity. The production of acetic esters and other specific aromas makes this yeast a key option to improve wine aroma, but it also has interesting effects on the body and structure [11]. Recently, it has been used in sparkling wines to make more complex base wines, whilst also increasing the structure during bottle aging [12].

The production of acetic esters is an interesting strategy to improve a wine's aromatic profile. The use of *Wickerhamomyces anomalus* helps to increase the contents of several esters, specifically 2-phenyl-ethyl acetate, with positive floral profiles [13]. The main drawback of this species is the high production of acetic acid, which can be partially controlled with suitable strain selection, but also through its use in sequential fermentation with *S. cerevisiae*. Apiculate species, such as the *Hanseniaspora/Kloeckera* genera, are also described as strong producers of acetate esters, and many species enhance the formation of 2-phenyl-ethyl acetate; some also produce benzenoids or nor-isoprenoids. Moreover, they tend to have an interesting effect on structure by producing full bodied wines [14]. Some of these species, as well as *Metschnikowia pulcherrima* and *C. stellata*, are able to release extracellular hydrolytic enzymes, such as β-glucosidases or c-lyases, that help improve the varietal aroma by releasing free terpenes or thiols [15,16]. A wide pool of enzymatic activities can also be found in saprophytic *Aureobasidium pullulans*, several of these enzymes can be purified with useful applications in enology [17]. *A. pullulans* is a typical yeast-fungus that can be found in the indigenous microbiota of the berry together with the apiculate genera *Hanseniaspora/Kloeckera*.

Spoilage yeasts such as *Zygosaccharomyces rouxii*, *Saccharomycodes ludwigii* or *Brettanomyces bruxellensis* may be difficult to handle at specific winemaking stages. Usually, the main concern of the enologist is their control and elimination from musts and wines, but also the analysis of their populations and their main marker metabolites. However, these non-*Saccharomyces* species sometimes have interesting applications in fermentative industries. *Zygosaccharomyces rouxii* is a frequent osmophilic spoilage species that causes re-fermentations in sweet wines and other drinks, such as fruit juices and soft beverages. Its control can be done using additives as DMDC, emerging antimicrobials as LfcinB, or cold pasteurization processes as DBD, US, UHPH or PEFs [18]. *Saccharomycodes ludwigii* is a strong fermenting yeast able to completely finish grape sugars; it also shows a strong resistance to high SO_2 levels. Some interesting applications are now being described, such as the use of this species in the reduction of the alcoholic degree of beers or in the production of ciders. In enology, the production of off-flavors reduces a lot the potential use of *S. ludwigii* in wine fermentation. The control measures used to reduce its prevalence in wines are the use of emerging physical technologies, chemical additives such as DMDC, but also natural products such as chitosan or biological control with killer yeasts [19]. The use of biological control with yeasts able to produce antimicrobial peptides is a novelty in the elimination of *Brettanomyces* spp. [20]. This spoilage yeast degrades the sensory quality of the wine as it develops during barrel aging, usually affecting more expensive wines by producing several unpleasant molecules [21]. Conventional control is based on the use of SO_2 and hygiene measures, however both parameters are difficult to control and maintain during long periods in difficult materials such as barrel wood. The use of *C. intermedia* as a selective bio-controller is a natural way to reduce the damages produced by *Brettanomyces*. Bio-protection and biological management of spoilage and undesired yeast can be also done by using *M. pulcherrima*, the production of the pigment pulcherrimin and their effect on iron chelation helps to eliminate competitive yeasts in grapes or at the beginning of fermentation [15].

If the twentieth century saw the explosion of *S. cerevisiae* applications, non-*Saccharomyces* yeasts open up a world of new biotechnologies in the twenty-first century, including improved fermentations, with more complex and differentiated sensory profiles in wines, bioprotection applications, enzymatic activities, acidity modulation, improvement of aging processes, reduction of toxic molecules and additives, and many other possibilities to discover. Some of these potentials contribute to the adaptation of wine to regions and terroirs, even to the ecological changes produced by global warming.

Funding: This research received no external funding.

Fermentation **2019**, *5*, 72

Conflicts of Interest: The authors declare no conflict of interest.

References

1. Ciani, M.; Maccarelli, F. Oenological properties of non-*Saccharomyces* yeasts associated with wine-making. *World J. Microbiol. Biotechnol.* **1997**, *14*, 199–203. [CrossRef]
2. Jolly, N.P.; Augustyn, O.P.H.; Pretorius, I.S. The role and use of non-*Saccharomyces* yeasts in wine production. *South Afr. J. Enol. Vitic.* **2006**, *27*, 15–39. [CrossRef]
3. Jolly, N.P.; Varela, C.; Pretorius, I.S. Not your ordinary yeast: Non-*Saccharomyces* yeasts in wine production uncovered. *FEMS Yeast Res.* **2014**, *14*, 215–237. [CrossRef] [PubMed]
4. Comitini, F.; Gobbi, M.; Domizio, P.; Romani, C.; Lencioni, L.; Mannazzu, I.; Ciani, M. Selected non-*Saccharomyces* wine yeasts in controlled multistarter fermentations with *Saccharomyces cerevisiae*. *Food Microbiol.* **2011**, *28*, 873–882. [CrossRef] [PubMed]
5. Suárez-Lepe, J.A.; Morata, A. New trends in yeast selection for winemaking. *Trends Food Sci. Technol.* **2012**, *23*, 39–50. [CrossRef]
6. Morata, A.; Loira, I.; Vejarano, R.; González, C.; Callejo, M.J.; Suárez-Lepe, J.A. Emerging preservation technologies in grapes for winemaking. *Trends Food Sci. Technol.* **2017**, *67*, 36–43. [CrossRef]
7. Morata, A.; Suárez Lepe, J.A. New biotechnologies for wine fermentation and ageing. In *Advances in Food Biotechnology*; Ravishankar Rai, P.V., Ed.; John Wiley & Sons, Ltd.: West Sussex, UK, 2016; pp. 293–295.
8. Vilela, A. Use of Non-conventional Yeasts for Modulating Wine Acidity. *Fermentation* **2019**, *5*, 27. [CrossRef]
9. Morata, A.; Loira, I.; Tesfaye, W.; Bañuelos, M.A.; González, C.; Suárez Lepe, J.A. *Lachancea thermotolerans* Applications in Wine Technology. *Fermentation* **2018**, *4*, 53. [CrossRef]
10. Loira, I.; Morata, A.; Palomero, F.; González, C.; Suárez-Lepe, J.A. *Schizosaccharomyces pombe*: A Promising Biotechnology for Modulating Wine Composition. *Fermentation* **2018**, *4*, 70. [CrossRef]
11. Ramírez, M.; Velázquez, R. The Yeast *Torulaspora delbrueckii*: An Interesting But Difficult-To-Use Tool for Winemaking. *Fermentation* **2018**, *4*, 94. [CrossRef]
12. Ivit, N.N.; Kemp, B. The Impact of Non-*Saccharomyces* Yeast on Traditional Method Sparkling Wine. *Fermentation* **2018**, *4*, 73. [CrossRef]
13. Padilla, B.; Gil, J.V.; Manzanares, P. Challenges of the Non-Conventional Yeast *Wickerhamomyces anomalus* in Winemaking. *Fermentation* **2018**, *4*, 68. [CrossRef]
14. Martin, V.; Valera, M.J.; Medina, K.; Boido, E.; Carrau, F. Oenological Impact of the *Hanseniaspora*/Kloeckera Yeast Genus on Wines—A Review. *Fermentation* **2018**, *4*, 76. [CrossRef]
15. Morata, A.; Loira, I.; Escott, C.; del Fresno, J.M.; Bañuelos, M.A.; Suárez-Lepe, J.A. Applications of *Metschnikowia pulcherrima* in Wine Biotechnology. *Fermentation* **2019**, *5*, 63. [CrossRef]
16. García, M.; Esteve-Zarzoso, B.; Cabellos, J.M.; Arroyo, T. Advances in the Study of *Candida stellata*. *Fermentation* **2018**, *4*, 74. [CrossRef]
17. Bozoudi, D.; Tsaltas, D. The Multiple and Versatile Roles of *Aureobasidium pullulans* in the Vitivinicultural Sector. *Fermentation* **2018**, *4*, 85. [CrossRef]
18. Escott, C.; Del Fresno, J.M.; Loira, I.; Morata, A.; Suárez-Lepe, J.A. *Zygosaccharomyces rouxii*: Control Strategies and Applications in Food and Winemaking. *Fermentation* **2018**, *4*, 69. [CrossRef]
19. Vejarano, R. *Saccharomycodes ludwigii*, Control and Potential Uses in Winemaking Processes. *Fermentation* **2018**, *4*, 71. [CrossRef]
20. Peña, R.; Chávez, R.; Rodríguez, A.; Ganga, M.A. A Control Alternative for the Hidden Enemy in the Wine Cellar. *Fermentation* **2019**, *5*, 25. [CrossRef]
21. Suárez, R.; Suárez-Lepe, J.A.; Morata, A.; Calderón, F. The production of ethylphenols in wine by yeasts of the genera *Brettanomyces* and *Dekkera*: A review. *Food Chem.* **2007**, *102*, 10–21. [CrossRef]

fermentation

MDPI

Review

Use of Nonconventional Yeasts for Modulating Wine Acidity

Alice Vilela

CQ-VR, Chemistry Research Centre, School of Life Sciences and Environment, Department of Biology and Environment, University of Trás-os-Montes and Alto Douro (UTAD), Enology Building, 5000-801 Vila Real, Portugal; avimoura@utad.pt

Received: 12 February 2019; Accepted: 12 March 2019; Published: 18 March 2019

Abstract: In recent years, in line with consumer preferences and due to the effects of global climate change, new trends have emerged in wine fermentation and wine technology. Consumers are looking for wines with less ethanol and fruitier aromas, but also with a good balance in terms of acidity and mouthfeel. Nonconventional yeasts contain a wide range of different genera of non-*Saccharomyces*. If in the past they were considered spoilage yeasts, now they are used to enhance the aroma profile of wine or to modulate wine composition. Recent publications highlight the role of non-*Saccharomyces* as selected strains for controlling fermentations mostly in cofermentation with *Saccharomyces*. In this article, I have reviewed the ability of some bacteria and non-*Saccharomyces* strains to modulate wine acidity.

Keywords: wine acidity; volatile acidity; malolactic bacteria; *Lactobacillus plantarum*; *Lachancea thermotolerans*; *Schizosaccharomyces pombe*; *Candida stellate*; *Torulaspora delbrueckii*; *Zygotorulaspora florentina*; *Pichia kudriavzevii*; *Stermerella bacillaris*

---

1. Acids Present in Grapes and Wines and Their Perceived Taste

Organic acids, next to sugars, are the most abundant solids present in grape juice. They are a significant constituent of juice and wine. Responsible for the sour/acid taste, they also influence wine stability, color, and pH. The quality and quantity of organic acids in conjunction with the sugars has a significant effect on the mouthfeel quality of wines [1].

Acid composition and concentration within the grape-must or wine are influenced by many factors, such as grape variety, soil composition, and climatic conditions. Accumulation of grape acids, namely tartaric acid, usually occurs at the beginning of grape berry development and is, to a large extent, completed at the beginning of ripening [2].

Amerine [3] reported that in berries, tartaric, malic, citric, ascorbic, phosphoric, and tannic acids were present, and soon after, Stafford [4] confirmed the occurrence of all but ascorbic and tannic acids in grapevine leaves, and included oxalic acid, in the form of idioblast crystals of calcium oxalate. Kliewer [5] identified 23 acids in berries, although most of these were found only in trace amounts. Nowadays we know that, by far, the predominant acids are tartaric and malic, which together may account for over 90% of the total acidity in the berry, existing at crudely a 1:1 to 1:3 ratio of tartaric to malic acid [6], both contributing to the pH of the juice, must, and wine during vinification and subsequent aging (Figure 1) [7].

Tartaric and malic acids are diprotic, with two dissociable protons per molecule. It is the first proton dissociation, with pKa values of around 2.98 (tartaric) and 3.46 (malic) that are meaningful properties in a winemaking context. At a typical wine pH (3.4), tartaric acid will be three times as acidic as malic acid [7].

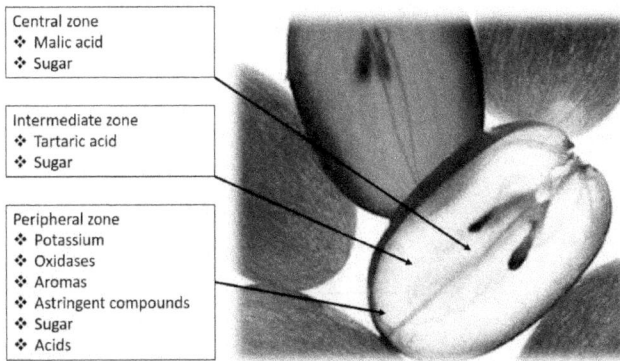

Figure 1. Grape-berry flavor zones and distribution of tartaric and malic acids.

The bitartrate and bimalate monoanions have important sensory roles to play in wine taste. Malic acid presents a harsh metallic taste (Table 1), sometimes correlated with the taste of green-apples, while the taste attributed to tartaric acid is frequently referred to as being 'mineral' or citrus-like. To compensate malic acid 'lost' in the late stages of berry ripening, the addition of tartaric acid at crush, or thereafter, can be performed, providing, in this way, control of must/wine pH. However, tartaric acid, unlike malic acid, is not a metabolic substrate for lactic acid bacteria or even yeasts.

Table 1. Organic acids present in grapes and wines and major acids' sensory descriptors. Adapted from Boulton et al. [8].

Fixed Acids		Volatile Acids	
Major Acids	**Minor Acids**	**Major Acids**	**Minor Acids**
L-tartaric (citrus-like taste)	Amino-acids	Acetic (vinegar-like)	Formic
L-malic (metallic, green-apples taste)	Pyruvic		Propionic
L-lactic (sour and spicy)	α-Ketoglutaric		2-Methylpropionic
Citric (fresh and citrus-like)	Isocitric		Butyric
Succinic (sour, salty, and bitter)	2-Oxoglutaric		2-Methylbutyric
	Dimethyl glyceric		3-Methylbutyric
	Citramalic		Hexanoic
	Gluconic acid [(1)]		Octanoic
	Galacturonic		Decanoic
	Glucuronic, Mucic, Coumaric, and Ascorbic		

[(1)] Present in wine made with grapes infected with *Botrytis cinerea*.

Citric acid, that presents a pleasant citrus-like taste (Table 1), has many uses in wine production. Citric acid is a weak organic acid that presents antimicrobial activity against molds and bacteria. It can create a relationship with antioxidants by chelating metal ions, thus helping in browning prevention. Citric acid occurs in the metabolism of almost every organism because it is an important intermediate in the tricarboxylic acid cycle (TCA cycle) [9], Figure 2.

During the winemaking process, it is advisable to monitor the concentration of organic acids in order to ensure the quality of the wine, and a distinction is made between acids directly produced in grapes (tartaric, malic, and citric) and those originating during the fermentation—alcoholic and malolactic—succinic, lactic, and acetic acids, among others [11], Table 1.

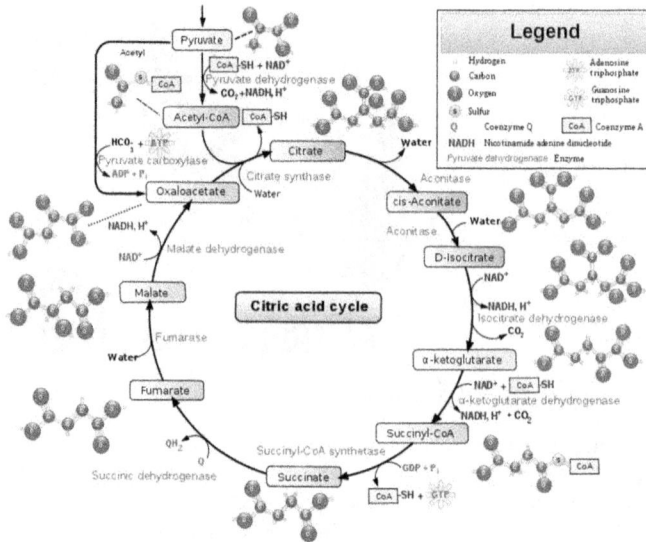

Figure 2. Schematic representation of the main steps, intermediated compounds, and enzymes of the TCA cycle [10].

Acetic acid, in quantities higher than 0.8–0.90 gL^{-1} [11], is immediately recognizable due to the vinegar smell an acrid taste, causing the wine to be considered spoiled. The maximum acceptable limit for volatile acidity in most wines is 1.2 gL^{-1} of acetic acid [12]. Acetic acid can appear on the grapes or the grape-must due to the presence of yeasts like *Hansenula* spp. and *Brettanomyces* spp., filamentous fungi (*Aspergillus niger*, *Aspergillus tenuis*, *Cladosporium herbarum*, *Penicillium* spp., and *Rhizopus arrhizus*), and bacteria (LAB-like indigenous *Lactobacilli*, and acetic acid bacteria). During alcoholic fermentation, acetic acid usually is formed in small quantities (0.2–0.5 gL^{-1} acetic acid) as a byproduct of *S. cerevisiae* metabolism. If the amounts are higher, some contamination spoilage yeasts and bacteria can be present: *Candida krusei*, *Candida stellate*, *Hansaniaspora uvarum/Kloeckera apiculate*, *Pichia anomala*, *Saccharomycodes ludwigii*, *Acetobacter pasteurianus*, and *Acetobacter liquefaciens*; after malolactic fermentation, heterofermentative species of *Oenococcus* and *Lactobacillus* also have the potential to produce acetic acid through the metabolism of residual sugar [13].

Succinic acid, with a sour, salty, and bitter taste, is the major acid produced by yeast during fermentation. This acid is resistant to microbial metabolism under fermentative conditions. During a period from 1991 to 2003, Coulter and coworkers [14] studied 93 red and 45 white Australian wines and found that the concentration of succinic acid in red wines reached from "none" (detection limit of 0.1 gL^{-1}) to 2.6 gL^{-1}, with a mean value of 1.2 gL^{-1}, while the concentration in white wines was between 0.1 gL^{-1} to 1.6 gL^{-1}, with a mean value of 0.6 gL^{-1}. Thus, succinic acid plays an important role in wine acidity [14].

Lactic acid, that usually is perceived as sour and spicy, is mainly produced by lactic acid bacteria during malolactic fermentation. However, small amounts can also be synthesized by yeast.

Today, the range of wines on the market is huge. On the other hand, wine companies tend to develop a style. Last year's tendency was to attribute "medals" to balanced flavor profile wines with bordering notes of vegetal-green, chemical, earthy, or sulfur characters, aromas of fruit and oak, hot/full mouthfeel (generally related to the alcohol content), low bitterness, and high sweetness [15]. Consumers and winemakers, giving more importance to flavors and sweetness, tend to "despise" acidity, even if it is one of the most important components of the wine. So, it is important for the wine industry to be able to modulate wine acidity, having in mind the concept of "healthy" and "biological", without the addition of enological products.

In this article, a review is made of some microorganisms, namely non-*Saccharomyces* yeasts, that, due to their physiological and genetic traits, are able to modulate wine acidity, either by increasing the wine's acid content (biological acidification) or by decreasing it—biological deacidification.

2. Wine Biological Acidity Modulation by Bacteria via Malolactic Fermentation

Physicochemical deacidification of wines is time-consuming, requires labor, capital input, and may reduce wine quality [16]. Biological deacidification of wine with malolactic bacteria (MLB), most often strains of *Oenococcus oeni*, previously known as *Leuconostoc oenos* [17], is the traditional method used for removing excess wine acidity. However, one important thing must be taken into account; of wine's total acidity, biological deacidification only affects the malic acid portion, it does not reduce tartaric acid.

While during alcoholic fermentation, wine yeast strains convert the grape sugars into ethanol and other flavors/mouthfeel compounds, after sugar depletion and the decline of yeasts population, LAB proliferates by utilizing the remaining sugars and thereafter performs malolactic fermentation (MLF). Despite its name "malolactic fermentation", this biological process is not a fermentation, but an enzymatic reaction in which malic acid (L (−) malic acid) is decarboxylated to lactic acid (L (+) lactic acid) and CO_2, Figure 3, [18,19]. This process also reduces the potential carbon source for spoilage microorganisms and leads to wine microbial stabilization [20].

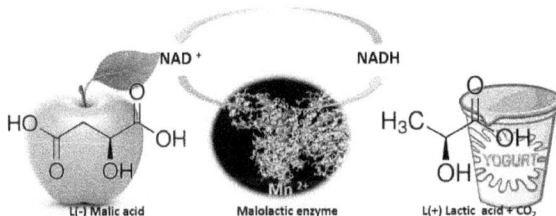

Figure 3. Schematic representation of malolactic enzyme action. Malolactic bacteria convert sharp green-apple-like malic acid into softer, much less tart lactic acid, releasing CO_2 along the way.

During MLF, the metabolism of *O. oeni* can improve the wine's sensory characteristics by producing a myriad of secondary metabolites [21]. However, the success of MLF is influenced by oenological parameters, such as temperature, pH, alcohol content, SO_2 concentration [22], and yeast inhibitory metabolites, such as medium chain fatty acids [23] or peptic fractions [24].

Several working groups are focused on alternative LAB, such as *Lactobacillus plantarum*, to perform MLF in wine [25,26]. *Lactobacillus plantarum* can survive under winemaking stress conditions, and during the fermentation process they are also able to produce a huge number of secondary metabolites important for the wine's aroma and flavor, including β-glucosidases, esterases, phenolic acid decarboxylases, and citrate lyases [27–29] once they contain genes encoding important enzymes that are active under winemaking conditions [30,31]; and can even improve red wine's color and solve problems associated with wine filtration due to tannase activities [29].

More specific studies have found that depending on the stress conditions in the wine, the gene coding for the malolactic enzyme works differently for *O. oeni* [32] and *L. plantarum* [33]. Miller et al. [33] found that the expression of mle (malolactic enzyme) *L. plantarum* gene presented an increased expression in the middle of MLF and was inducible by the presence of malic acid and low pH wine values, decreasing, nevertheless, in the presence of ethanol. Later, Iorizzo and coworkers [34] reported that some strains of *L. plantarum* were able to grow at pH values ranging from 3.2 to 3.5 and in the presence of 13% (*v*/*v*) ethanol. Several strains of *L. plantarum* were also found to be able to tolerate the presence of sulfite and in the concentrations used in winemaking [35].

Moreover, *L. plantarum* strains produce high concentrations of lactic acid, which may contribute to "biological acidification" in low acidity wines, and thus improving wine mouthfeel [36].

3. Wines Biological Acidity Modulation by Nonconventional Yeasts

The malolactic fermentation process is not free from collateral effects (production of off-flavors, wine quality loss, and human health issues due to the production of biogenic amines). Benito et al. [37] developed a new red winemaking methodology by combining the use of two non-*Saccharomyces* yeast strains as an alternative to the traditional MLF. According to the authors, malic acid is consumed by *Schizosaccharomyces pombe*, while *Lachancea thermotolerans* produces lactic acid in order to increase the acidity of wines produced from low acidity musts. The main fermentative properties of interesting non-*Saccharomyces* yeasts reported as advantageous for fermented beverages and that can modulate wine acidity are described in Table 2.

Table 2. Percentage of ethanol formed during fermentation, sugars fermented, main volatile compounds formed and effect on wine acidity of seven non-*Saccharomyces* yeasts.

Yeast Species	Ethanol Formation (%, v/v)	Sugars Fermented	Volatile Compounds	Effect on Wine Acidity	Ref.
Lachancea thermotolerans	<9	Glucose Fructose Maltose Galactose	2-phenylethyl acetate Ethyl lactate	Acidity enrichment (lactic acid)/Acidity reduction (acetic acid)	[37–40]
Schizosaccharomyces pombe	12–14	Glucose Fructose Sucrose Maltose	Higher alcohols Esters	Maloalcoholic deacidification	[37,41]
Candida stellate	10.6 + 9.81 gL^{-1} glycerol (in co-culture with *S. cerevisiae*)	Glucose Sucrose Raffinose (slow fermentation)	Esters Acetoin	Acidity enrichment (Succinic acid)	[42]
Torulaspora delbrueckii	11 (table wine) 13-14 [i] (in co-culture with *S. cerevisiae*)	Glucose Galactose [ii] Maltose [ii] Sucrose [ii] a,a-Trehalose [ii] Melibiose [ii]	Long-chain alcohols, esters, aldehydes, and glycerol	Low production of acetic acid	[43–45]
Z. florentinus/Z. Florentina	>13 [iv] (in co-culture with *S. cerevisiae*)	Frutose [iii] Glucose Galactose Sucrose Maltose Raffinose Trehalose	higher alcohols and esters	Low production of acetic acid. Some species are able to consume acetic acid [v]	[46–50]
Pichia kudriavzevii/Issatchenkia orientalis	>7 [vi] (in microvinifications with chemically defined grape juice)	Glucose Fructose Sucrose, Maltose, Raffinose Xylose [vii]	Esters and Higher alcohols	Consume L-malic acid	[51,52]
Starmerella bacillaris/Candida zemplinina	11.7–12.1 [viii]	Glucose Fructose [xix]	Higher level of some terpenes, lactones and thiols [x]	Malic acid degradation; Reduction of acetic acid in sweet wines; Production of pyruvic acid.	[53–57]

[i] The musts were obtained from botrytized Semillon grapes with initial sugar concentrations of 360 gL^{-1} [43]. [ii] Variable according to strain. [iii] Some *Zygosaccharomyces* are fructophilic. *Z. rouxii* and *Z. bailii* possess genes (FFZ) that encode specific fructose facilitators and proteins [49]. [iv] In white grape juice, not added with SO$_2$, with 231 gL^{-1} sugar content [48]. [v] *Z. bailii* is known to consume acetic acid [50]. [vi] In microvinifications with chemically defined grape juice with similar nitrogen and acidic fraction composition to Patagonian Pinot noir juice (gL^{-1}: glucose 100 gL^{-1}, fructose 100 gL^{-1}, potassium tartrate 5 gL^{-1}, L-malic acid 3 gL^{-1}, citric acid 0.2 gL^{-1}, easily assimilable nitrogen 0.208 gL^{-1} and pH 3.5) [51]. [vii] *Pichia kudriavzevii* presents the ability to produce ethanol from xylose. Xylose is a sugar found in wood, meaning this can used as an alternative for ethanol production, which is particularly useful in the biofuel industry [52]. [viii] A decrease up to 0.7% (v/v) of ethanol when *S. cerevisiae* was inoculated with a delay of 48 h with respect to the inoculation of *Starmerella bacillaris* [53]. [xix] *S. bacillaris* show fructophilic, cryotellerant, and osmophylic characters of interest for the winemakers [55]. [x] Sauvignon blanc wines fermented by mixed cultures (*S. bacillaris* and *S. cerevisiae*) contained significantly higher levels of thiols [57].

Schizosaccharomyces pombe, *Lachancea thermotolerans*, and *Torulaspora delbrueckii* are presently produced at the industrial level by biotechnological companies [38]. *Torulaspora delbrueckii* is commercialized in the form of a pure culture, selected for its properties to increase aromatic complexity, mouthfeel, low production of volatile acidity, and high resistance to initial osmotic shock and it is highly recommended for the fermentation of late harvest wines in sequential culture with *S. cerevisiae*.

3.1. Lachancea thermotolerans: Wine Acidification/Deacetification

L. thermotolerans cells are rather similar in both shape and size to *S. cerevisiae* and impossible to distinguish by optical microscopy. They also reproduce asexually by multipolar budding. In fermentation conditions, an alcohol degree of 9% (*v/v*) is the limit of ethanol produced and tolerated [38]. *L. thermotolerans* can produce lactic acid during fermentation, up to 9.6 gL^{-1} [58], and glycerol [59].

All these interesting features can be a way to address the problems of increased alcohol content/reduction in the total acidity of wines associated with global climate changes [60]. Since 2013 [59], studies have been made in several wines and wine-regions that elucidate *L. thermotolerans* wine-making features—Sangiovese and Cabernet-Sauvignon wines where a significant increase in the spicy notes was found [59]; Airén wines, an increased lactic acid concentration up to 3.18 gL^{-1} and a pH reduction of 0.22 were accomplished [61], Emir wines, where an increase in final total acidity of 5.40–6.28 gL^{-1} was achieved [60].

However, *L. thermotolerans* is also capable of another interesting wine-making feature. This yeast can be used to develop a controlled biological deacetification process of wines with high volatile acidity, with the process being oxygen-dependent, which means that its metabolism must shift more towards respiration than fermentation [40,50].

To verify the potential application of *L. thermotolerans* wine deacetification, the strain was inoculated in two wine-supplemented mineral media, (I) simulating the refermentation of a wine with freshly crushed grapes (130 gL^{-1} of glucose and 4% ethanol (*v/v*)) and (II) simulating the refermentation of a wine with the residual marc from a finished wine fermentation (33 gL^{-1} glucose and 10% ethanol (*v/v*)). The volatile acidity of both mixtures was 1.13 gL^{-1} of acetic acid. *L. thermotolerans* was able to consume 94.6% of the initial acetic acid, in the high-glucose medium, under aerobic conditions and the final "wine" was left with a volatile acidity of 0.06 gL^{-1} [50].

3.2. Schizosaccharomyces pombe: Biological Acidity Modulation via Maloalcoholic Fermentation

Usually, wine is produced by using *Saccharomyces* yeast to transform sugars into alcohol, followed by *Oenococus* bacteria to complete malolactic fermentation and thus rendering the wine with its pleasantness and microbial stability. This methodology has some unsolved problems: (i) the management of highly acidic musts; (ii) the production of potentially toxic products, including biogenic amines and ethyl carbamate [62]. To overcome these issues, the use of non-*Saccharomyces* yeast strains able to perform alcoholic fermentation and malic acid degradation are being studied. *S. cerevisiae* has long been known as a poor metabolizer of extracellular malate, due to the lack of a mediated transport system for the acid [63]. One example is the fission yeast from the genus *Schizosaccharomyces*. These yeasts are able to consume malic acid by converting it to ethanol and CO_2 [64], Figure 4.

Schizosaccharomyces pombe is highly appreciated in colder regions due to its particular metabolism of maloalcoholic fermentation [64], significantly reducing the levels of ethyl carbamate precursors and biogenic amines without the need for any bacterial MLF [62]. *Schizosaccharomyces* is also able, during fermentation, to increase the formation of vitisins and vinylphenolic pyranoanthocyanin [65]; these pigments intensify the color of the finished wine [66].

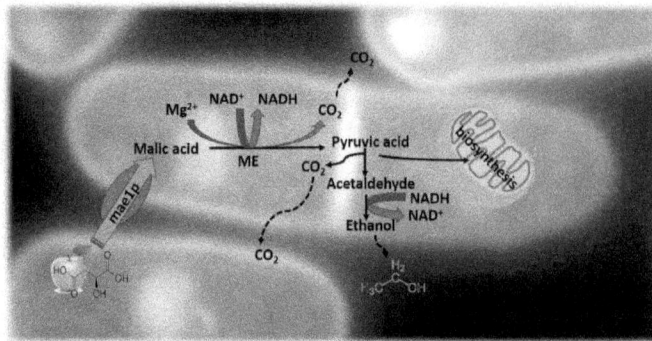

Figure 4. Schematic representation of the maloalcoholic pathway. Malic acid is transported into the yeast cell by mae1p carboxylic acid transporter. The malic enzyme (ME) converts malate into pyruvate, supplying pyruvic acid to the mitochondria for biosynthesis.

3.3. Candida stellata: Biological Acidity Enrichment

Candida stellata is an imperfect yeast of the genus *Candida*, order Saccharomycetales, phylum *Ascomycota*. *C. stellata* was originally isolated from an overripe grape must in Germany, but we can find it widespread in natural and artificial habitats [67].

During the beginning of alcoholic fermentation, many species can grow simultaneously in the grape must; *C. stellata* species have been described in this stage of fermentation [68]. *C. stellata* cells are spherical/ovoid; usually found as single cells. This yeast ferments glucose, sucrose, and raffinose and uses lysine as its sole N source. It is also able to grow at higher pH values and it is not sensitive to ethanol [67]. These features make it a good candidate for co-inoculation with *S. cerevisiae*. This yeast is also frequently associated with musts proceeding from botrytized grapes [69].

An interesting feature is the ability of *C. stellata* to form succinic acid. Maurizio and Ferraro [70] found that *C. stellata* yeast strains produced more succinic acid than a *Saccharomyces* control strain in synthetic grape juice. The higher production of the acid was also associated with higher glycerol levels. The yeasts also presented high ethanol production and ethanol tolerance [42,71]. Succinic acid could positively influence the sensory/mouthfeel profile of wines with insufficient acidity. Nevertheless, due to its 'salt-bitter-sour' taste, excessive levels would negatively influence wine quality.

Moreover, Magyar and Tóth [72] found that *C. stellata* is similar to *C. zemplinina* in its strong fructophilic character. However, *C. stellata* produces more glycerol and more ethanol, which is comparable with that produced by *S. uvarum*. The latter species is known to produce low acetic acid and low ethanol when compared with *S. cerevisiae*.

3.4. Torulaspora delbrueckii: Volatile Acidity Modulation in Very Sweet Musts

Torulaspora delbrueckii (formerly *Saccharomyces rosei*) is reported to have a positive effect on the flavor of alcoholic beverages, and exhibits low production of acetic acid, acetaldehyde, ethyl acetate, and acetoin [71].

T. delbrueckii was studied by Lafon-Lafourcade et al. [73] in order to evaluate the possibility of using it to improve the quality of botrytized wine (wine made with *Botrytis cinerea*-infected grapes, commonly known as "noble rot"). These wines are made with grapes with sugar concentrations up to 350–450 gL^{-1} and present a challenge to *S. cerevisiae* in terms of acetic acid production. It is well known that *S. cerevisiae* produces acetic acid as a stress response to the high-sugar grape must [13]. The volatile acidity of these wines may be over 1.8 gL^{-1} (acetic acid), a value above the human nose detection threshold and over the EEC (European Economic Community) legal limit (1.5 gL^{-1}) [12].

Bely et al. [43] used mixed cultures of *T. delbrueckii* and *S. cerevisiae* in high-sugar fermentation musts with a higher concentration of *T. delbrueckii* to promote its growth. This mixed inoculum produced lower levels of acetic acid and acetaldehyde, up to 55% and 68% less, respectively, when compared with the pure cultures of *S. cerevisiae*. An indication of the positive impact of *T. delbrueckii* activity on wine quality was also demonstrated by Azzolini et al. [74] in Vino Santo, a sweet wine, due to its low production of acetic acid.

Azzolini et al. [75] studied the impact of *T. delbrueckii* in the production of Amarone wine (officially named, *Amarone della Valpolicella*), a high-alcohol dry red wine obtained from withered grapes. Winery trials were inoculated by a selected strain of *T. delbrueckii* in co-inoculation and/or sequentially with an *S. cerevisiae* strain. *T. delbrueckii* was able to promote the formation of alcohols, fermentative esters, fatty acids, and lactones, which are important in the Amarone wine flavor. In terms of aromatic capacity, in white table wines, *T. delbrueckii* was able to produce dried fruit/pastry aromas, pleasant to the tasters [44,74].

3.5. Zygosaccharomyces Florentinus/Zygotorulaspora Florentina: Volatile Acidity Modulation

The genus *Zygosaccharomyces* is involved in food and beverage spoilage since it shows high tolerance to osmotic stress, and consequently, the *Zygosaccharomyces* species can grow in severe environments with high sugar concentration, pH values closer to 2 [76], low 'water activity' (a_w), and presence of organic acids and preservatives such as SO_2 and ethanol [77]. *Zygosaccharomyces* spp. produces high ethanol and acetoin content in wines and may play an important role as non-*Saccharomyces* yeasts in differentiated wine products [78]. The odors and flavors produced by *Zygosaccharomyces* are generally described as being wine-like, so the main problem with this yeast in wines, is the formation of a haze/deposit after bottling.

Kurtzman et al. [79] and Kurtzman & Robnett [80], reported that the genus *Zygosaccharomyces* comprised 11 species, divided into two groups. After, by means of multigene sequence analysis, Kurtzman [81] proposed a new division: the *Zygosaccharomyces* clade should be divided into four phylogenetic groups: (1) *Zygosaccharomyces*, which included *Z. bailii*, *Z. bisporus*, *Z. kombuchaensis*, *Z. lentus*, *Z. mellis*, and *Z. rouxii*; (2) *Zygotorulaspora*, comprising two species, *Zygotorulaspora florentinus* (florentina) and *Zt. mrakii*; (3) *Torulaspora*, with *Torulaspora microellipsoides* and (4) the *Lachancea* clade, in which *Lachancea cidri* and *L. fermentati* were included [82].

Zygosaccharomyces are well known for their ability to spoil food and beverages [83]. Nevertheless, some strains are used in industrial production of balsamic vinegar, miso paste, and soy sauce—*Z. rouxii* [84] and *Z. kombuchaensis* in the production of kombucha [79].

Domizio and co-workers in 2011 [48] used Zy42 *Zygosaccharomyces florentinus* strain to perform grape juice fermentation in mixed cultures with a well-known commercial *Saccharomyces* strain—EC1118 (Lalvin EC-1118 it was originally named "prise de mousse"). It was isolated in Champagne and its use is validated by the Comité Interprofessionnel du Vin de Champagne (CIVC)). The EC1118/Zy42 association increased the production of propanol, higher alcohols, and esters and produced lower levels of acetic acid than the comparative trial with a pure culture of EC1118 (0.3 and 0.4 gL^{-1}, respectively).

Later, Lencioni and coworkers in 2016 [46] evaluated the strain *Zygotorulaspora florentina* (formerly *Zygosaccharomyces florentinus*) in mixed culture fermentations with *S. cerevisiae*, from the laboratory scale to the winery scale. At the lab scale, the resulting mixed fermentations showed a reduction of volatile acidity and an improvement of polysaccharides and 2-phenylethanol (rose-like aroma) concentrations. At the winery scale, they used red grape-must from the Sangiovese grape variety. The resulting wine presented a higher concentration of glycerol and esters; sensorially, the wine astringency was lower and the aromatic character was defined by floral notes.

Recently, the same authors [47], studied the possibility to decrease wine volatile acidity in mixed fermentations with *Zygotorulaspora florentina* and *Starmerella bacillaris* (syn., *Candida zemplinina*). Independent of fermentation temperature, the mixed fermentations with *Z. florentina* performed best to reduce volatile acidity, thus being a valuable tool for performing fermentation of high-sugar musts.

3.6. Pichia kudriavzevii/Issatchenkia orientalis: Malic Acid Consumption

Since 1966 [85] that there have been several reports on yeast strains degrading extracellular malic acid, including *Schizosaccharomyces pombe* [64,85]. Among the yeast strains degrading malic acid, *S. malidevorans* and *S. pombe* are the strains studied most intensively as a means of reducing wine acidity [64,85,86]. However, it has been reported that *Schizosaccharomyces* sp. may produce off-flavors/aromas in the wines [87].

Degradation of malic acid by the strain *Issatchenkia orientalis* KMBL 5774, an acidophilic yeast strain, isolated from Korean grape wine pomace, was investigated by Seo et al. [88]. In the mentioned work, degradation of malic acid by *I. orientalis* KMBL 5774 was investigated in YNB-malic acid liquid media under various culture conditions (malic acid concentration, pH, temperature, ... etc.). The maximal growth was obtained when 3% malic acid was used and maximal malic acid degradation ratio was obtained at 1–2% of the malic acid concentration (94.4 and 94.6%, respectively) [89].

A few years ago, in 2014, Mónaco and coworkers [51] studied fifty-seven Patagonian non-*Saccharomyces* yeasts of oenological origin, and tested their ability to consume L-malic acid as a carbon source. Only four isolates belonging to *Pichia kudriavzevii* (also known as *Issatchenkia orientalis*) species showed this ability, and one was selected and studied further—*P. kudriavzevii* ÑNI15. This isolate was able to degrade L-malic acid in microvinification assays (38% of L-malic acid reduction when compared with *S. cerevisiae* (22%)), increasing the pH by 0.2–0.3 units. Furthermore, *P. kudriavzevii* ÑNI15 produced low levels of ethanol and significant levels of glycerol (10.41 ± 0.48 gL^{-1}). The final wines presented fruity and cooked pears aromas, once *Pichia kudriavzevii* was unable to synthesize ethyl acetate but it showed good production of ethyl esters from fatty acids when compared with the *Saccharomyces* yeast [51].

However, under the names *Pichia kudriavzevii*, *Issatchenkia orientalis*, and *Candida glycerinogenes*, the same yeast, including genetically modified strains, is used for industrial-scale production of glycerol and succinate. In 2018, Douglass and coworkers [89] investigated the genomic diversity of a yeast species that is both an opportunistic pathogen and an important industrial yeast. Under the name *Candida krusei*, it is responsible for about 2% of yeast infections caused by Candida species in humans. Bloodstream infections with *C. krusei* are problematic because most isolates are fluconazole-resistant. In their work, Douglass et al. [89] sequenced the strains of *C. krusei* (CBS573T) and *P. kudriavzevii* (CBS5147T), and the results showed, conclusively, that they are the same species, with collinear genomes 99.6% identical in DNA sequence. Phylogenetic analysis of SNPs does not segregate clinical and environmental isolates into separate clades, suggesting that *C. krusei* infections are frequently acquired from the environment. More studies are needed to ensure public safety when using these non-*Saccharomyces* strains in food/wine production.

3.7. Starmerella bacillaris/Candida zemplinina: Wine Biological Acidification and Deacidification

One approach to reducing the ethanol content of wines is by co-inoculation of *Saccharomyces* and non-*Saccharomyces* yeast during must fermentation. The selection and use of non-*Saccharomyces* wine yeasts can potentially lead to a reduction of the overall sugar–ethanol yield [53]. The yeast *Starmerella bacillaris* (syn. *Candida zemplinina*) is often isolated from grape and winery environments specifically associated with overripe and botrytized grapes [90]. Its enological use in mixed fermentation with *S. cerevisiae* has been investigated, and several interesting features, such as low ethanol and high glycerol production, and a fructophilic aptitude, have been found [54]. This yeast also presents acidogenic, psychrotolerant, and osmotolerant properties, and therefore seems to be well adapted to sweet wine fermentations [55].

It was in 2002 that David Mills and coworkers [90] isolated the yeast *Starmerella bacillaris* for the first time from sweet wines made with botrytized grapes in Napa Valley (California, USA), and due to their enological features, it has been studied since then. Englezos and coworker in 2016 [53] investigated the potential application of *Starmerella bacillaris* in combination with *S. cerevisiae*, in co-inoculated and

Fermentation **2019**, *5*, 27

sequential cultures to reduce the ethanol in wines. Lab scale fermentations showed a decrease of up to 0.7% (v/v) of ethanol and an increase of about 4.2 gL^{-1} of glycerol.

Continuing their work on *Starmerella bacillaris* and the application of this yeast to the wine industry and winemaking, Englezos and coworker in 2017 [55] wrote an interesting review. In that report, they concluded that among other *Starmerella bacillaris* enological features, this yeast species could contribute to malic acid degradation due to its ability to produce a wide spectrum of extracellular hydrolytic enzymes [91]. It is also capable of producing pyruvic acid, acting as a natural acidification agent by reducing the wine's pH [92], which may have an impact on wine color, due to the reaction of pyruvic acid with anthocyanins producing stable colored pigments such as Vitisin A [93].

Moreover, in 2012, Rantsiou and coworkers [94] showed that the use of mixed fermentations with *Starmerella bacillaris* and *S. cerevisiae* could be a biological method to reduce acetic acid in sweet wines. According to the mentioned authors, the co-inoculation produced wines with a decrease of 0.3 gL^{-1} of acetic acid; sequential inoculation produced wines with about half of the acetic acid content, compared to wines produced with pure cultures of *S. cerevisiae* [94].

Additionally, sequential inoculations of *Starmerella bacillaris* and *S. cerevisiae* possess great potential in affecting and modulating the chemical and aromatic profile of white wines, especially those produced from Sauvignon blanc grapes [57]. The volatile profile of Chardonnay, Muscat, Riesling, and Sauvignon blanc white wines, fermented with sequential inoculation of *Starmerella bacillaris* and *S. cerevisiae*, were studied by Englezos and coworker in 2018 [57]. Mixed fermentations with both these strains affected the chemical composition of wines by modulating various metabolites of oenological interest. For volatile compounds, mixed fermentations led to a reduction of ethyl acetate, which may be a compound responsible for wine deterioration above certain limits. Interestingly, Sauvignon blanc wines, fermented by mixed cultures, contained significantly higher levels of esters and thiols, both considered positive aromatic attributes [57].

4. Final Remarks

In conclusion, many non-*Saccharomyces* yeasts present interesting oenological properties in terms of fermentation purity and production of secondary metabolites or even ethanol. When used in single or mixed cultures with *S. cerevisiae*, these yeasts strains can modulate wine acidity and increase production of some interesting compounds, such as polysaccharides, glycerol, and volatile compounds, such as 2-phenyl ethanol and 2-methyl 1-butanol.

Thus, this review confirms that some non-*Saccharomyces* yeasts that are most often considered as spoilage yeasts are actually tools of great value for the winemaking industry.

A deeper study of the oenological traits of these yeasts will provide new data for consideration in the control of fermentation, with special regard to warm-climate wine and botrytized sweet wines, where they are commonly found in mixed populations.

Acknowledgments: We appreciate the financial support provided to the Research Unit in Vila Real (PEst-OE/QUI/UI0616/2014) by FCT—Portugal and COMPETE. Additional thanks to the Project NORTE-01-0145-FEDER-000038 (I&D INNOVINE&WINE—Innovation Platform of Vine & Wine).

Conflicts of Interest: The author declares no conflict of interest.

References

1. Liu, H.F.; Wu, B.H.; Fan, P.G.; Xu, H.Y.; Li, S.H. Inheritance of sugars and acids in berries of grape (*Vitis vinifera* L.). *Euphytica* **2007**, *153*, 99–107. [CrossRef]
2. Cosme, F.; Gonçalves, B.; Inês, A.; Jordão, A.M.; Vilela, A. Grape and wine metabolites: Biotechnological approaches to improve wine quality. In *Grape and Wine Biotechnology*; Morata, A., Loira, I., Eds.; InTechOpen: London, UK, 2016; pp. 187–214.
3. Amerine, M.A. The Maturation of Wine Grapes. *Wines Vines* **1956**, *37*, 53–55.
4. Stafford, H. Distribution of Tartaric Acid in the Leaves of Certain Angiosperms. *Am. J. Bot.* **1959**, *46*, 347–352. [CrossRef]

5. Kliewer, W. Sugars and Organic Acids of *Vitis vinifera*. *Plant Physiol.* **1966**, *41*, 923–931. [CrossRef] [PubMed]
6. Lamikanra, O.; Inyang, I.; Leong, S. Distribution and effect of grape maturity on organic acid content of red Muscadine grapes. *J. Agric. Food Chem.* **1995**, *43*, 3026–3028. [CrossRef]
7. Ford, C.M. The Biochemistry of Organic Acids in the Grape. In *The Biochemistry of the Grape Berry*; e-Book, Gerós, H., Chaves, M., Delrot, S., Eds.; Bentham Science Publishers: Sharjah, UAE, 2012; pp. 67–88. ISBN 978-1-60805-360-5.
8. Boulton, R.B.; Singleton, V.L.; Bisson, L.F.; Kunkee, R.E. *Principles and Practices of Winemaking*; Chapman and Hall: New York, NY, USA, 1996; pp. 102–181.
9. Sharma, R.K. Citric Acid. In *Natural Food Antimicrobial Systems*; Naidu, A.S., Ed.; CRC Press LLC: New York, NY, USA, 2000.
10. Citric Acid Cycle. Available online: https://commons.wikimedia.org/wiki/File:Citric_acid_cycle_with_aconitate_2.svg#/media/File:Citric_acid_cycle_with_aconitate_2.svg (accessed on 30 January 2019).
11. Ribéreau-Gayon, P.; Glories, Y.; Maujean, A.; Dubourdieu, D. Alcohols, and other volatile compounds. The chemistry of wine stabilization and treatments. In *Handbook of Enology*, 2nd ed.; John Wiley & Sons Ltd.: Chichester, UK, 2006; Volume 2, pp. 51–64.
12. Office Internationale de la Vigne et du Vin. *International Code of Oenological Practices*; OIV: Paris, France, 2010.
13. Vilela-Moura, A.; Schuller, D.; Mendes-Faia, A.; Silva, R.F.; Chaves, S.R.; Sousa, M.J.; Côrte-Real, M. The impact of acetate metabolism on yeast fermentative performance and wine quality: Reduction of volatile acidity of grape musts and wines—Minireview. *Appl. Microbiol. Biotechnol.* **2011**, *89*, 271–280. [CrossRef] [PubMed]
14. Coulter, A.D.; Godden, P.W.; Pretorius, I.S. Succinic acid-how is it formed, what is its effect on titratable acidity, and what factors influence its concentration in wine? *Wine Ind. J.* **2004**, *19*, 16–24.
15. Hopfer, H.; Heymann, H. Judging wine quality: Do we need experts, consumers or trained panelists? *Food Qual. Prefer.* **2014**, *32*, 221–233. [CrossRef]
16. Pretorius, I.S. Tailoring wine yeast for the new millennium: Novel approaches to the ancient art of winemaking. *Yeast* **2000**, *16*, 675–729. [CrossRef]
17. Dicks, L.M.T.; Dellaglio, F.; Collins, M.D. Proposal to reclassify *Leuconostoc oenos* as *Oenococcus oeni* [corrig.] gen. nov., comb. nov. *Int. J. Syst. Bacteriol.* **1995**, *45*, 395–397. [CrossRef] [PubMed]
18. Henick-Kling, T. Malolactic fermentation. In *Wine Microbiology and Biotechnology*; Fleet, G.H., Ed.; Harwood Academic Publishers: Reading, UK, 1993; pp. 289–327.
19. Volschenk, H.; Van Vuuren, H.J.J.; Viljoen-Bloom, M. Malic acid in wine: Origin, function and metabolism during vinification. *S. Afr. J. Enol. Vitic.* **2006**, *27*, 123–136. [CrossRef]
20. Lasik-Kurdyś, M.; Majcher, M.; Nowak, J. Effects of Different Techniques of Malolactic Fermentation Induction on Diacetyl Metabolism and Biosynthesis of Selected Aromatic Esters in Cool-Climate Grape Wines. *Molecules* **2018**, *23*, 2549. [CrossRef] [PubMed]
21. Bartowsky, E.; Borneman, A. Genomic variations of *Oenococcus oeni* strains and the potential to impact on malolactic fermentation and aroma compounds in wine. *Appl. Microbiol. Biotechnol.* **2011**, *92*, 441–447. [CrossRef] [PubMed]
22. Lerm, E.; Engelbrecht, L.; du Toit, M. Malolactic fermentation: The ABC's of MLF. *S. Afr. J. Enol. Vitic.* **2010**, *31*, 186–212. [CrossRef]
23. Alexandre, H.; Costello, P.J.; Remize, F.; Guzzo, J.; Guilloux-Benatier, M. *Saccharomyces cerevisiae-Oenococcus oeni* interactions in wine: Current knowledge and perspectives. *Int. J. Food Microbiol.* **2004**, *93*, 141–154. [CrossRef]
24. Nehme, N.; Mathieu, F.; Taillandier, P. Impact of the co-culture of *Saccharomyces cerevisiae-Oenococcus oeni* on malolactic fermentation and partial characterization of a yeast-derived inhibitory peptidic fraction. *Food Microbiol.* **2010**, *27*, 150–157. [CrossRef] [PubMed]
25. Berbegal, C.; Peña, N.; Russo, P.; Grieco, F.; Pardo, I.; Ferrer, S.; Spano, G.; Capozzi, V. Technological properties of *Lactobacillus plantarum* strains isolated from grape must fermentation. *Food Microbiol.* **2016**, *57*, 187–194. [CrossRef] [PubMed]
26. Lucio, O.; Pardo, I.; Krieger-Weber, S.; Heras, J.M.; Ferrer, S. Selection of Lactobacillus strains to induce biological acidification in low acidity wines. *LWT Food Sci. Technol.* **2016**, *73*, 334–341. [CrossRef]
27. Matthews, A.; Grimaldi, A.; Walker, M.; Bartowsky, E.; Grbin, P.; Jiranek, V. Lactic acid bacteria as a potential source of enzymes for use in vinification. *Appl. Environ. Microbiol.* **2004**, *70*, 5715–5731. [CrossRef]

28. Grimaldi, A.; Bartowsky, E.; Jiranek, V. Screening of *Lactobacillus* spp. and *Pediococcus* spp. for glycosidase activities that are important in enology. *J. Appl. Microbiol.* **2005**, *99*, 1061–1069. [CrossRef]
29. Brizuela, N.; Tymczyszyn, E.E.; Semorile, L.C.; La Hens, D.V.; Delfederico, L.; Hollmann, A.; Bravo-Ferrada, B. *Lactobacillus plantarum* as a malolactic starter culture in winemaking: A new (old) player? *Electron. J. Biotechnol.* **2019**, *38*, 10–18. [CrossRef]
30. Du Toit, M.; Engelbrecht, L.; Lerm, E.; Krieger-Weber, S. *Lactobacillus*: The next generation of malolactic fermentation starter cultures—An overview. *Food Bioprocess Techmol.* **2011**, *4*, 876–906. [CrossRef]
31. Lerm, E.; Engelbrecht, L.; Toit, M.D. Selection and Characterisation of *Oenococcus oeni* and *Lactobacillus plantarum* South African Wine Isolates for Use as Malolactic Fermentation Starter Cultures. *S. Afr. J. Enol. Vitic.* **2011**, *32*, 280–295. [CrossRef]
32. Olguín, N.; Bordons, A.; Reguant, C. Multigenic expression analysis as an approach to understanding the behavior of *Oenococcus oeni* in wine-like conditions. *Int. J. Food Microbiol.* **2010**, *144*, 88–95. [CrossRef]
33. Miller, B.J.; Franz, C.M.; Cho, G.S.; du Toit, M. Expression of the malolactic enzyme gene (mle) from *Lactobacillus plantarum* under winemaking conditions. *Curr. Microbiol.* **2011**, *62*, 1682–1688. [CrossRef]
34. Iorizzo, M.; Testa, B.; Lombardi, S.J.; García-Ruiz, A.; Muñoz-González, C.; Bartolomé, B.; Moreno-Arribas, M.V. Selection and technological potential of *Lactobacillus plantarum* bacteria suitable for wine malolactic fermentation and grape aroma release. *LWT Food Sci. Technol.* **2016**, *73*, 557–566. [CrossRef]
35. Bravo-Ferrada, B.M.; Hollmann, A.; Delfederico, L.; La Hens, D.V.; Caballero, A.; Semorile, L. Patagonian red wines: Selection of *Lactobacillus plantarum* isolates as potential starter cultures for malolactic fermentation. *World J. Microbiol. Biotechnol.* **2013**, *29*, 1537–1549. [CrossRef]
36. Berbegal, C.; Spano, G.; Tristezza, M.; Grieco, F.; Capozzi, V. Microbial Resources and Innovation in the Wine Production Sector. *S. Afr. J. Enol. Vitic.* **2017**, *38*, 156–166. [CrossRef]
37. Benito, A.; Calderón, F.; Palomero, F.; Benito, S. Combined use of selected *Schizosaccharomyces pombe* and *Lachancea thermotolerans* yeast strains as an alternative to the traditional malolactic fermentation in red wine production. *Molecules* **2015**, *20*, 9510–9523. [CrossRef]
38. Morata, A.; Suárez-Lepe, J.A. New biotechnologies for wine fermentation and ageing. In *Advances in Food Biotechnology*, 1st ed.; Ravishankar Rai, V., Ed.; John Wiley & Sons, Ltd.: West Sussex, UK, 2016; pp. 287–301.
39. Vilela, A. Targeting Demalication and Deacetification Methods: The Role of Carboxylic Acids Transporters. *Biochem. Physiol.* **2017**, *6*, 224. [CrossRef]
40. Vilela, A. *Lachancea thermotolerans*, the Non-*Saccharomyces* Yeast that Reduces the Volatile Acidity of Wines. *Fermentation* **2018**, *4*, 56. [CrossRef]
41. Suárez-Lepe, J.A.; Morata, A. New trends in yeast selection for winemaking. *Trends Food Sci. Technol.* **2012**, *23*, 39–50. [CrossRef]
42. Ferraro, L.; Fatichenti, F.; Ciani, M. Pilot scale vinification process using immobilized *Candida stellata* cells and *Saccharomyces cerevisiae*. *Process. Biochem.* **2000**, *35*, 1125–1129. [CrossRef]
43. Bely, M.; Stoeckle, P.; Masneuf Pomarède, I.; Dubourdieu, D. Impact of mixed *Torulaspora delbrueckii Saccharomyces cerevisiae* culture on high sugar fermentation. *Int. J. Food Microbiol.* **2008**, *122*, 312–320. [CrossRef] [PubMed]
44. Velázquez, R.; Zamora, E.; Álvarez, M.L.; Hernández, L.M.; Ramírez, M. Effects of new *Torulaspora delbrueckii* killer yeasts on the must fermentation kinetics and aroma compounds of white table wine. *Front. Microbiol.* **2015**, *6*, 1222. [CrossRef] [PubMed]
45. Pacheco, A.; Santos, J.; Chaves, S.; Almeida, J.; Leão, C.; Sousa, M.J. The Emerging Role of the Yeast *Torulaspora delbrueckii* in Bread and Wine Production: Using Genetic Manipulation to Study Molecular Basis of Physiological Responses. In *Structure and Function of Food Engineering*; Eissa, A.A., Ed.; IntechOpen: London, UK, 2012; Chapter 13; pp. 339–370.
46. Lencioni, L.; Romani, C.; Gobbi, M.; Comitini, F.; Ciani, M.; Domizio, P. Controlled mixed fermentation at winery scale using *Zygotorulaspora florentina* and *Saccharomyces cerevisiae*. *Int. J. Food Microbiol.* **2016**, *234*, 36–44. [CrossRef] [PubMed]
47. Lencioni, L.; Taccari, M.; Ciani, M.; Domizio, P. *Zygotorulaspora florentina* and *Starmerella bacillaris* in multistarter fermentation with *Saccharomyces cerevisiae* to reduce volatile acidity of high sugar musts. *Aust. J. Grape Wine Res.* **2018**, *24*, 368–372. [CrossRef]

48. Domizio, P.; Romani, C.; Lencioni, L.; Comitini, F.; Gobbi, M.; Mannazzu, I.; Ciani, M. Outlining a future for non-*Saccharomyces* yeasts: Selection of putative spoilage wine strains to be used in association with *Saccharomyces cerevisiae* for grape juice fermentation. *Int. J. Food Microbiol.* **2011**, *147*, 170–180. [CrossRef]

49. Pina, C.; Gonçalves, P.; Prista, C.; Loureiro-Dias, M.C. Ffz1, a New Transporter Specific for Fructose from *Zygosaccharomyces bailii*. *Microbiology* **2004**, *150*, 2429–2433. [CrossRef]

50. Vilela-Moura, A.; Schuller, D.; Mendes-Faia, A.; Côrte-Real, M. Reduction of volatile acidity of wines by selected yeast strains. *Appl. Microbiol. Biotechnol.* **2008**, *80*, 881–890. [CrossRef]

51. Mónaco, S.; Barda, N.; Rubio, N.; Caballero, A. Selection and characterization of a Patagonian *Pichia kudriavzevii* for wine deacidification. *J. Appl. Microbiol.* **2014**, *117*, 451–464. [CrossRef] [PubMed]

52. Hello, My Name is Pichia kudriavzevii; Eureka Brewing. Available online: https://eurekabrewing. wordpress.com/2014/02/16/hello-my-name-is-pichia-kudriavzevii (accessed on 16 February 2019).

53. Englezos, V.; Rantsiou, K.; Cravero, F.; Torchio, F.; Ortiz-Julien, A.; Gerbi, V.; Rolle, L.; Cocolin, L. *Starmerella bacillaris* and *Saccharomyces cerevisiae* mixed fermentations to reduce ethanol content in wine. *Appl. Microbiol. Biotechnol.* **2016**, *100*, 5515–5526. [CrossRef]

54. Masneuf-Pomarede, I.; Juquin, E.; Miot-Sertier, C.; Renault, P.; Laizet, Y.; Salin, F.; Alexandre, H.; Capozzi, V.; Cocolin, L.; Colonna-Ceccaldi, B.; et al. The yeast *Starmerella bacillaris* (synonym *Candida zemplinina*) shows high genetic diversity in winemaking environments. *FEMS Yeast Res.* **2015**, *5*, fov045. [CrossRef] [PubMed]

55. Englezos, V.; Giacosa, S.; Rantsiou, K.; Rolle, L.; Cocolin, L. *Starmerella bacillaris* in winemaking: Opportunities and risks. *Curr. Opin. Food Sci.* **2017**, *17*, 30–35. [CrossRef]

56. Sadoudi, M.; Tourdot-Maréchal, R.; Rousseaux, S.; Steyer, D.; Gallardo-Chacón, J.-J.; Ballester, J.; Vichi, S.; Guérin-Schneider, R.; Caixach, J.; Alexandre, H. Yeast–yeast interactions revealed by aromatic profile analysis of Sauvignon Blanc wine fermented by single or co-culture of non-*Saccharomyces* and *Saccharomyces* yeasts. *Food Microbiol.* **2012**, *2*, 243–253. [CrossRef] [PubMed]

57. Englezos, V.; Rantsiou, K.; Cravero, F.; Torchio, F.; Pollon, M.; Fracassetti, D.; Ortiz-Julien, S.; Gerbi, V.; Rolle, L.; Cocolin, L. Volatile profile of white wines fermented with sequential inoculation of *Starmerella bacillaris* and *Saccharomyces cerevisiae*. *Food Chem.* **2018**, *257*, 350–360. [CrossRef]

58. Kapsopoulou, K.; Kapaklis, A.; Spyropoulos, H. Growth and fermentation characteristics of a strain of the wine yeast *Kluyveromyces thermotolerans* isolated in Greece. *World J. Microbiol. Biotechnol.* **2005**, *21*, 1599–1602. [CrossRef]

59. Gobbi, M.; Comitini, F.; Domizio, P.; Romani, C.; Lencioni, L.; Mannazzu, I.; Ciani, M. *Lachancea thermotolerans* and *Saccharomyces cerevisiae* in simultaneous and sequential co-fermentation: A strategy to enhance acidity and improve the overall quality of wine. *Food Microbiol.* **2013**, *33*, 271–281. [CrossRef]

60. Balikci, E.K.; Tanguler, H.; Jolly, N.P.; Erten, H. Influence of *Lachancea thermotolerans* on cv. Emir wine fermentation. *Yeast* **2016**, *33*, 313–321. [CrossRef]

61. Benito, Á.; Calderón, F.; Palomero, F.; Benito, S. Quality and composition of Airén wines fermented by sequential inoculation of *Lachancea thermotolerans* and *Saccharomyces cerevisiae*. *Food Technol. Biotechnol.* **2016**, *54*, 135–144. [CrossRef]

62. Benito, Á.; Jeffares, D.; Palomero, F.; Calderón, F.; Bai, F.Y.; Bähler, J.; Benito, S. Selected *Schizosaccharomyces pombe* Strains Have Characteristics That Are Beneficial for Winemaking. *PLoS ONE* **2016**, *11*, e0151102. [CrossRef] [PubMed]

63. Salmon, J.M. L-Malic acid permeation in resting cells of anaerobically grown *Saccharomyces cerevisiae*. *Biochim. Biophys. Acta* **1987**, *901*, 30–34. [CrossRef]

64. Volschenk, H.; Vuuren, H.J.; Viljoen–Bloom, M. Malo-ethanolic fermentation in *Saccharomyces* and *Schizosaccharomyces*. *Curr. Genet.* **2003**, *43*, 379–391. [CrossRef] [PubMed]

65. Loira, I.; Morata, A.; Comuzzo, P.; Callejo, M.J.; González, C.; Calderón, F.; Suárez-Lepe, J.A. Use of *Schizosaccharomyces pombe* and *Torulaspora delbrueckii* strains in mixed and sequential fermentations to improve red wine sensory quality. *Food Res. Int.* **2015**, *76*, 325–333. [CrossRef]

66. Morata, A.; González, C.; Suárez-Lepe, J.A. Formation of vinylphenolic pyranoanthocyanins by selected yeasts fermenting red grape musts supplemented with hydroxycinnamic acids. *Int. J. Food Microbiol.* **2007**, *116*, 144–152. [CrossRef] [PubMed]

67. García, M.; Esteve-Zarzoso, B.; Cabellos, J.M.; Arroyo, T. Advances in the Study of *Candida stellata*. *Fermentation* **2018**, *4*, 74. [CrossRef]

68. Mora, J.; Mulet, A. Effects of some treatments of grape juice on the population and growth of yeast species during fermentation. *Am. J. Enol. Vitic.* **1991**, *42*, 133–136.

69. Jackson, R.S. *Wine Science-Principles, Practice, Perception*, 2nd ed.; Academic Press: San Diego, CA, USA, 2000.

70. Maurizio, C.; Ferraro, L. Enhanced glycerol content in wines made with immobilized *Candida stellata* cells. *Appl. Environ. Microbiol.* **1996**, *62*, 128–132.

71. Ciani, M.; Maccarelli, F. Oenological properties of non-*Saccharomyces* yeasts associated with wine-making. *World J. Microbiol. Biotechnol.* **1998**, *14*, 199–203. [CrossRef]

72. Magyar, I.; Tóth, T. Comparative evaluation of some oenological properties in wine strains of *Candida stellata, Candida zemplinina, Saccharomyces uvarum* and *Saccharomyces cerevisiae*. *Food Microbiol.* **2011**, *28*, 94–100. [CrossRef]

73. Lafon-Lafourcade, S.; Lucmaret, V.; Joyeaux, A.; Ribereau-Gayon, P. Utilisation de levains mixtes dans l élaboration'des vins de pourriture noble en vue de réduire lácidité volatile. *Comptes Rendues de L'Académie d'Agriculture* **1981**, *67*, 616–622.

74. Azzolini, M.; Tosi, E.; Lorenzini, M.; Finato, F.; Zapparoli, G. Contribution to the aroma of white wines by controlled *Torulaspora delbrueckii* cultures in association with *Saccharomyces cerevisiae*. *World J. Microbiol. Biotechnol.* **2015**, *31*, 277–293. [CrossRef] [PubMed]

75. Azzolini, M.; Fedrizzi, B.; Tosi, E.; Finato, F.; Vagnoli, P.; Scrinzi, C.; Zapparoli, G. Effects of *Torulaspora delbrueckii* and *Saccharomyces cerevisiae* mixed cultures on fermentation and aroma of Amarone wine. *Eur. Food Res. Technol.* **2012**, *235*, 303–313. [CrossRef]

76. Gordon, J.L.; Wolfe, K.H. Recent allopolyploid origin of *Zygosaccharomyces rouxii* strain ATCC 42981. *Yeast* **2008**, *25*, 449–456. [CrossRef] [PubMed]

77. Steels, H.; James, S.A.; Roberts, I.N.; Stratford, M. *Zygosaccharomyces lentus*: A significant new osmophilic, preservative-resistant spoilage yeast, capable of growth at low temperature. *J. Appl. Microbiol.* **1999**, *87*, 520–527. [CrossRef] [PubMed]

78. Romano, P.; Suzzi, G. Higher Alcohol and Acetoin Production by *Zygosaccharomyces* Wine Yeasts. *J. Appl. Bacteriol.* **1993**, *75*, 541–545. [CrossRef]

79. Kurtzman, C.P.; Robnett, C.J.; Basehoar-Powers, E. *Zygosaccharomyces kombuchaensis*, a new ascosporogenous yeast from 'Kombucha tea'. *FEMS Yeast Res.* **2001**, *1*, 133–138. [CrossRef] [PubMed]

80. Kurtzman, C.P.; Robnett, C.J. Phylogenetic relationships among yeasts of the 'Saccharomyces complex' determined from multigene sequence analyses. *FEMS Yeast Res.* **2003**, *3*, 417–432. [CrossRef]

81. Kurtzman, C.P. Phylogenetic circumscription of *Saccharomyces Kluyveromyces* and other members of the *Saccharomycetaceae*, and the proposal of the new genera *Lachancea, Nakaseomyces, Naumovia, Vanderwaltozyma* and *Zygotorulaspora*. *FEMS Yeast Res.* **2003**, *4*, 233–245. [CrossRef]

82. Cabral, S.; Prista, C.; Loureiro-Dias, M.; Leandro, M. Occurrence of FFZ genes in yeasts and correlation with fructophilic behavior. *Microbiology* **2015**, *161*, 2008–2018. [CrossRef]

83. Martorell, P.; Stratford, M.; Steels, H.; Fernández-Espinar, M.T.; Querol, A. Physiological characterization of spoilage strains of *Zygosaccharomyces bailii, Zygosaccharomyces rouxii* isolated from high sugar environments. *Int. J. Food Microbiol.* **2007**, *114*, 234–242. [CrossRef] [PubMed]

84. Sá-Correia, I.; Guerreiro, J.F.; Loureiro-Dias, M.C.; Leão, C.; Côrte-Real, M. Zygosaccharomyces. In *Encyclopedia of Food Microbiology*, 2nd ed.; Batt, C.A., Tortorello, M.-L., Eds.; Academic Press: New York, NY, USA, 2014; Volume 3, pp. 849–855.

85. Rankine, B.C. Decomposition of L-malic acid by wine yeasts. *J. Sci. Food Agric.* **1966**, *17*, 312–316. [CrossRef] [PubMed]

86. Snow, P.G.; Gallander, J.F. Deacidification of white table wines through partial fermentation with *Schizosaccharomyces pombe*. *Am. J. Enol. Vitic.* **1979**, *30*, 45–48.

87. Rodriquez, S.B.; Thornton, R.J. Factors influencing the utilization of L-malate by yeasts. *FEMS Microbiol. Lett.* **1990**, *72*, 17–22. [CrossRef]

88. Seo, S.-H.; Rhee, C.-H.; Park, H.-D. Degradation of Malic Acid by *Issatchenkia orientalis* KMBL 5774, an Acidophilic Yeast Strain Isolated from Korean Grape Wine Pomace. *J. Microbiol.* **2007**, *45*, 521–527.

89. Douglass, A.P.; Offei, B.; Braun-Galleani, S.; Coughlan, A.Y.; Martos, A.A.; Ortiz-Merino, R.A.; Byrne, K.P.; Wolfe, K.H. Population genomics shows no distinction between pathogenic *Candida krusei* and environmental *Pichia kudriavzevii*: One species, four names. *PLoS Pathog.* **2018**, *14*, e1007138. [CrossRef]

90.	Mills, D.A.; Johannsen, E.A.; Cocolin, L. Yeast diversity and persistence in botrytis-affected wine fermentations. *Appl. Environ. Microbiol.* **2002**, *68*, 4884–4893. [CrossRef]
91.	Englezos, V.; Rantsiou, K.; Torchio, F.; Rolle, L.; Gerbi, V.; Cocolin, L. Exploitation of the non-*Saccharomyces* yeast *Starmerella bacillaris* (synonym *Candida zemplinina*) in wine fermentation: Physiological and molecular characterizations. *Int. J. Food Microbiol.* **2015**, *199*, 33–40. [CrossRef]
92.	Magyar, I.; Nyitrai-Sárdy, D.; Leskó, A.; Pomázi, A.; Kállay, M. Anaerobic organic acid metabolism of *Candida zemplinina* in comparison with *Saccharomyces* wine yeasts. *Int. J. Food Microbiol.* **2014**, *178*, 1–6. [CrossRef]
93.	Romboli, Y.; Mangani, S.; Buscioni, G.; Granchi, L.; Vincenzini, M. Effect of *Saccharomyces cerevisiae* and *Candida zemplinina* on quercetin, vitisin A and hydroxytyrosol contents in Sangiovese wines. *World J. Microbiol. Biotechnol.* **2015**, *31*, 1137–1145. [CrossRef]
94.	Rantsiou, K.; Dolci, P.; Giacosa, S.; Torchio, F.; Tofalo, R.; Torriani, S.; Suzzi, G.; Rolle, L.; Cocolin, L. *Candida zemplinina* can reduce acetic acid production by *Saccharomyces cerevisiae* in sweet wine fermentations. *Appl. Environ. Microbiol.* **2012**, *78*, 1987–1994. [CrossRef] [PubMed]

fermentation

MDPI

Review

Lachancea thermotolerans Applications in Wine Technology

Antonio Morata [1],*, Iris Loira [1], Wendu Tesfaye [1], María Antonia Bañuelos [2], Carmen González [1] and José Antonio Suárez Lepe [1]

[1] Department of Chemistry and Food Technology, ETSIAAB, Technical University of Madrid, 28040 Madrid, Spain; iris.loira@upm.es (I.L.); wendu.tesfaye@upm.es (W.T.); carmen.gchamorro@upm.es (C.G.); joseantonio.suarez.lepe@upm.es (J.A.S.L.)
[2] Department of Biotechnology-Plant Biology, ETSIAAB, Technical University of Madrid, 28040 Madrid, Spain; mantonia.banuelos@upm.es
* Correspondence: antonio.morata@upm.es

Received: 20 June 2018; Accepted: 6 July 2018; Published: 11 July 2018

Abstract: *Lachancea (kluyveromyces) thermotolerans* is a ubiquitous yeast that can be naturally found in grapes but also in other habitats as soil, insects and plants, extensively distributed around the world. In a 3-day culture, it shows spherical to ellipsoidal morphology appearing in single, paired cells or short clusters. It is a teleomorph yeast with 1–4 spherical ascospores and it is characterized by a low production of volatile acidity that helps to control global acetic acid levels in mixed or sequential inoculations with either *S. cerevisiae* or other non-*Saccharomyces* species. It has a medium fermentative power, so it must be used in sequential or mixed inoculations with *S. cerevisiae* to get dry wines. It shows a high production of lactic acid able to affect strongly wine pH, sometimes decreasing wine pH by 0.5 units or more during fermentation. Most of the acidification is produced at the beginning of fermentation facilitating the effect in sequential fermentations because it is more competitive at low alcoholic degree. This application is especially useful in warm areas affected by climatic change. pH reduction is produced in a natural way during fermentation and prevents the addition of tartaric acid, that produces tartrate precipitations, or the use of cation exchangers resins highly efficient reducing pH but with undesirable effects on wine quality. Production of lactic acid is done from sugars thus reducing slightly the alcoholic degree, especially in strains with high production of lactic acid. Also, an improvement in the production of 2-phenylethanol and glycerol has been described.

Keywords: *Lachancea thermotolerans*; *Kluyveromyces thermotolerans*; acidification; wines; sequential fermentations; non-*Saccharomyces*

1. Introduction

Lachancea thermotolerans was formerly known as *Kluyveromyces thermotolerans*, but it was recently reassigned in the genera *Lachancea* according to multigene sequence analysis [1]. *L. thermotolerans* (LT) is a global yeast species that can be usually found in grapes but also in other habitats as soil, insects and plants [2] and extensively distributed around the world [3]. It can be found in natural spontaneous wine fermentations with a low prevalence on days 2–4 of fermentation [4]. Morphologically, it is globous or ellipsoidal, undistinguishable from *S. cerevisiae* (Figure 1) and can be found as single cells in liquid media or in small groups. It is a teleomorph yeast presenting sexual reproduction with the formation of spherical ascospores (1–4). Asexual reproduction is produced with multilateral budding. LT forms creamy colonies with butyrous texture in solid media.

LT is able to ferment glucose and sucrose [5] and weakly galactose. It shows variable capacity to ferment maltose, trehalose and raffinose [6]. Nitrogen nutrition is similar to *S. cerevisiae* being necessary a minimum of 200 mg/L of YAN (yeast assimilable nitrogen) to avoid sluggish or stuck fermentations [7]. Serine as N-source also has shown an improvement in the fermentation performance of LT [8]. Strains of LT can express the following extracellular enzymatic activities with effect in wine aroma or phenol extraction: Esterase, Esterase-Lipase, ß-glucosidase, Pectinase, Cellulase, Xylanase, Glucanase [9].

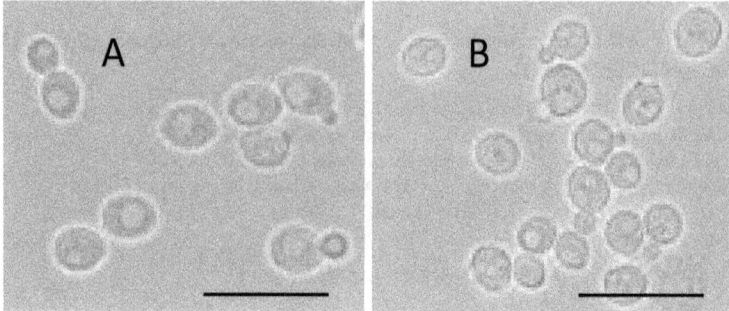

Figure 1. Optical microscopy (**A**) *Saccharomyces cerevisiae* (**B**) *Lachancea thermotolerans*. Scale 10 μm.

LT has a moderate fermentative power, and an ethanol tolerance around 5–9% v/v has been published [10–14]. Some effect can be observed in the reduction of the alcoholic degree of wines (0.7% v/v; [15]). Concerning resistance in the time, it was observed that it is able to survive several days in the presence of 9% v/v of ethanol [11,16], and it also has a good persistence even when the fermentation is dominated by *S. cerevisiae* [17]. These metabolic properties make it appear during the intermediate phase of the fermentative process before the full prevalence of high fermentative *S. cerevisiae* strains. The use of LT in sequential or mixed fermentations has some tendency to produce sluggish fermentations with more difficulties fermenting the fructose fraction of the grape sugars [7]. Moreover, an oxygen availability requirement for LT persistence seems to be higher than for *S. cerevisiae* [18]. The tolerance to temperature is similar to average strains of *S. cerevisiae* showing a good growth at 25–30 °C, but slower growth below 20 °C [5].

Comitini et al. [12] identified 5 isolates that are able to resist 10–20 mg/L of free SO_2, but it is possible to find strains resistant to more than 100 mg/L of total SO_2 [14]. The production of H_2S is variable from medium to high (25 isolates). Comitini et al. [12] also observed in 5 strains of LT a production of SH_2 ranging from 3–5 in a 0–5 scale. Resistance to DMDC is low, from 25 to 100 mg/L for populations ranging log2–log6 CFU/mL, while typical values for *S. cerevisiae* are 100–300 mg/L with same population [19].

LT has been produced at the commercial level as dry yeasts since 2012 (CONCERTO™ and MELODY™) and recommended to increase flavor complexity and intensity, to improve total acidity and to reduce volatile acidity [20,21]. It is well established that LT is capable of producing wines with higher 'spicy' and acidic notes, thus improving the overall quality of wine [13,22]. It is also described as producer of ethyl isobutyrate (strawberry nuances). Improvements in fruitiness, probably favored by the increase in acidity, are typical sensory descriptors when LT is used to ferment neutral varieties [23].

Some LT strains have been used as fungal biocontrol agents in grapes and vines to inhibit the growth of *Aspergillus* [24]. These strains do not affect the metabolic properties and performance of *S. cerevisiae* during alcoholic fermentation [25].

2. Isolation and Selection

As most of other non-*Saccharomyces* yeasts, LT can be distinguished from *S. cerevisiae* using lysine media (Figure 2; [26] and culturing temperatures in the range 25–28 °C. The use of chromogenic media is quite useful (Figure 2) for the initial isolation of yeasts belonging to this species. In CHROMagar®

they show a characteristic red-brown color that can be distinguished easily from the purple colonies of *Saccharomyces* or the creamy colors of most of the other yeasts.

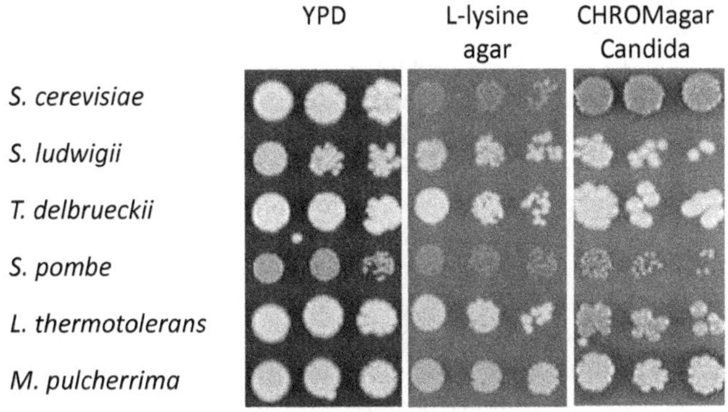

Figure 2. Colony shape and color in YPD, L-lysine specific to isolate non-*Saccharomyces* and Chromogenic media (Adapted from Loira et al. [26]).

The use of PCR-denaturing gradient gel electrophoresis (PCR-DGGE) has been proposed to identify LT during the study of the ecology of wine grapes, but with a low sensibility, needing a population of at least log2 CFU/mL [27]. Microsatellite markers and a multilocus SSR analysis have been developed to assess the genetic diversity of LT isolates [28]. Identification of LT isolates can be performed by sequencing of the D1/D2 region of the 26S rRNA gene and RAPD fingerprinting what allows yeast identification at species level [29,30]. Moreover, restriction patterns of amplified regions of 26S rDNA can be used as a routine methodology to identify non-*Saccharomyces* yeast species during red wine fermentation [31] (Figure 3). Finally, specific PCR primer pairs for the intron 2 of the mitochondrial COX1 gene, allow detect *L. thermotolerans* in wine at 10^4 cells/mL and with a *S. cerevisiae/L. termotholerans* ratio of 1000/1 [32].

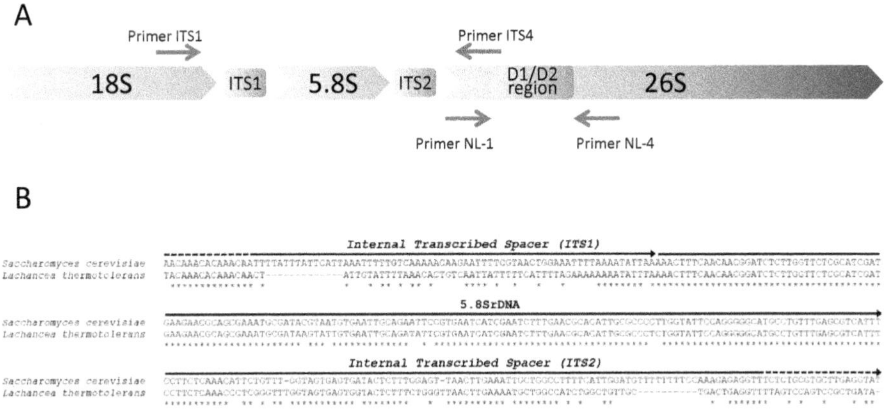

Figure 3. (**A**) Schematic diagram of the yeast rDNA gene cluster. The 18S, 5.8S and 25–28S rDNA genes are separated by the internal transcribed spacers 1 (ITS1) and 2 (ITS2). Primers for routine sequencing are shown; (**B**) Alignments of complete 5.8S rDNA sequence and partial sequences of Internal Transcribed Spacer 1 and 2. The marked sequences are identical regions. GenBank accession numbers: for *S. cerevisiae*, KT958553.1 and for *L. thermotolerans*, CU928180.1.

3. Biotechnological Application: Wine Acidification

Selected LT strains have been used for acidification of fermented beverages as wines [13,33,34] and beers [35–37]. LT is a strong producer of lactic acid [33] with significant influence in wine pH, also in other fermented beverages (Table 1). Lactic acid is stable after fermentation and ageing because it is neither chemically degraded nor microbiologically metabolized under enological conditions. Concentrations can range 1–16.8 g/L [28]. Most of the acidification is produced at the beginning of fermentation (Figure 4A), which facilitates the acidification in sequential fermentations with *S. cerevisiae* at low-medium ethanol levels (<6% *v/v*) when LT is competitive with *S. cerevisiae*. The acidification strongly influences pH being possible to reduce more than 0.5 pH units from an initial pH of 3.8–4 (Figure 4B). This has interesting implications in wines of warm areas because at pH near 4 the wine is unprotected and many spoilage microorganisms can grow even in absence of residual sugars and high alcoholic degree. Molecular SO_2 levels at pH 4 are very low (<0.5 mg/L) even when free SO_2 concentration can be higher than 50 mg/L. These values are unsuitable for conflictive spoilage yeasts like *Brettanomyces/Dekkera* that needs 0.8 mg/L of molecular SO_2 to be controlled [38]. The use of LT in mixed or sequential fermentations with *S. cerevisiae* when pH is decreased to 3.5–3.7 promotes better levels of molecular SO_2 with low contents of total sulfites, making the fermentations and especially the ageing process safer.

Table 1. Acidification biotechnologies using *Lachancea thermotolerans* in wine fermentation suitable to be used in high pH musts from warm regions.

Biotechnology	LT Fermentation Time	Initial pH	Final pH	Comments
Sequential fermentation LT → *S. Cerevisiae* LT → *S. pombe*	0–4 days Most of the acidification is performed in the 3 first days	3.9–4.2	3.5–3.7 depending on LT strains and implantation success	Population inoculated of LT must be >log6 CFU/mL
Mixed fermentation LT + *S. Cerevisiae* LT + *S. pombe*	0–6 days	3.9–4.2	3.5–3.7 depending on LT strains and implantation success	Population inoculated of LT must be >log6 CFU/mL Ratio between LT + *S. cerevisiae* (or *S. pombe*) must be log6/log2 including wild Sacch.
Coinoculation LT + LAB (*O. oeni*) and subsequent inoculation of *S. cerevisiae*	0–6 days	3.9–4.2	3.3–3.5 depending on LT strains, implantation success and lactic acid production by LAB	Strong pH reduction. Light alcohol degree reduction

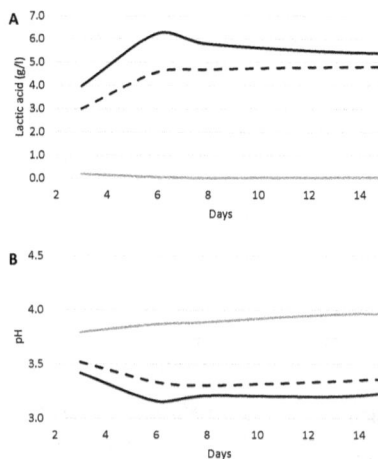

Figure 4. Lactic acid production (**A**) and pH (**B**) evolution during the fermentation of a red grape must with 240 g/L of sugar by *Lachancea thermotolerans* strain L3.1 (black line), *L. thermotolerans* and sequentially *S. cerevisiae* (dashed line) and *S. cerevisiae* (grey line).

LT is unable to entirely ferment a grape must as said, reaching maximum ethanol concentrations around 9% *v/v*, even when especially selected yeasts are used. So, it is necessary to use sequential or mixed fermentations with *S. cerevisiae* [7,12] or *S. pombe* [39,40] to get completely dry wines without residual sugars. The use of an inoculation ratio log7:log3 cfu/mL (LT/*S. cerevisiae*) is suitable to see a significant effect in pH reduction [12]. Suitable implantation to reach an effective acidification has also been tested according to inoculum size in mixed fermentations with *S. cerevisiae* [13]. However, the best results were reached in sequential inoculation after 48 h, decreasing the pH values from 3.53 in *S. cerevisiae* control fermentation to 3.33 [13]. When the inoculation is performed after 24 h, pH reduction is lower than 0.1 units, statistically significant, but probably without enological repercussions.

The production of lactic acid during fermentations with variable glucose contents in model media (8, 12, 16% *w/w*) is stable and independent of sugar levels, ranging from 2.48 ± 0.8–2.66 ± 0.9 g/L for LT strain L3.1 [41]. However, when the same sugar concentrations were tested in the presence of 2 g/L of lactic acid, the production by LT was affected, ranging from 1.25 ± 0.7 g/L (in model media with 8% *w/w* glucose) to 3.68 ± 0.3 g/L (with 16% *w/w* glucose). Nitrogen contents also affect the production of lactic acid by LT, being reduced below 150 mg/L of yeast assimilable nitrogen (YAN), but also at high concentration (>500 mg/L) (Figure 5). Highest lactic acid production is correlated with suitable YAN levels for yeast nutrition (150–200 mg/L) and subsequently with higher yeast populations (Figure 5).

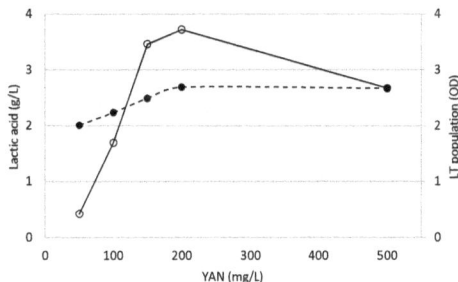

Figure 5. Lactic acid production during fermentation by *L. thermotolerans* strain L3.1 at variable yeast assimilable nitrogen (YAN) concentrations (black line). LT population estimated by OD (dashed line). Adapted from Hernández [41].

The use of LT fermentations together with *S. pombe* has been described as an alternative biotechnological tool to emulate malolactic fermentation (MLF) [42]. However, even when it can be similar in terms of acidification, the sensory profile reached in a typical MLF is more complex [43].

Initial evidence suggests that the use of LT is compatible with MLF [44]. However, lactic acid at high concentration (>4 g/L), as can be produced when LT is used during fermentation, can behave as an inhibitor of lactic acid bacteria (LAB), thus hindering MLF. It is possible to use yeast-bacteria co-inoculations to facilitate the MLF by promoting the simultaneous development of alcoholic and malolactic fermentations. The use of LT-LAB co-inoculations with subsequent *S. cerevisiae* sequential fermentation (LT-LAB-SC) is quite effective to degrade malic acid and at the same time reach lower pH values. When this biotechnology has been used at industrial level in fermentations of 1 ton of crushed grapes, the final pH reached 3.3 while the control only fermented by *S. cerevisiae* remained at pH 4 that was the initial grape value. In this case, some acidification is produced by LT and a complementary amount of lactic acid is produced by fermentation of sugars by LAB. Both microorganisms work promoting the formation of stable acidity in natural enological conditions. Achieving these pH values is only possible by using ion exchange resins, but affecting strongly wine composition and quality.

Traditionally, MLF is a way to obtain microbiologically stable red wines with a better sensory profile, but it can also affect the freshness in wines of warm areas. However, the use of LT with the production of high contents of lactic acid (>4 g/L) because of its inhibitory effect on LAB, can be an approach to obtain fresh wines in warm regions protecting malic acidity but also increasing lactic acidity and with a suitable stability.

4. Metabolic Profile and Influence on Wine Aroma and Flavor

The use of LT in wine has been described with a low production of volatile acidity (0.3–0.5 g/L [12,16]). LT was reported as a useful biotechnological tool to decrease volatile acidity [45]. Acetaldehyde levels similarly can be diminished by using LT during fermentation [22,46]. LT has been described as a moderated producer of higher alcohols [22]. The production of ethyl acetate is quite moderate (40–60 mg/L, [13]). Sequential fermentations of LT with *S. pombe* produce similar concentrations than with *S. cerevisiae* (40–50 mg/L, [40]). It is an interesting yeast to control volatile acidity produced by *S. pombe* to make full fermentations in absence of *S. cerevisiae* [39]. The production of ethyl lactate is moderate in sequential fermentations with *S. cerevisiae* (7–8 mg/L) and in mixed fermentations with *S. pombe* (8–32 mg/L) [39,40].

LT have been described as producer of β-D-glucosidase (βDG) [47] and carbon-sulfur lyase (CSL) [48], enzymes involved in the release of aroma compounds from must varietal precursors (Table 2). Usually non-*Saccharomyces* are more effective producing 3-mercaptohexan-1-ol (3MH) than 4-mercapto-4-methylpentan-2-one (4MMP) [49]. However, when LT have been used in must fermentation, significant amounts of 4MMP also moderate amounts of 3MH were released. The production of significant concentrations of 4-methyl-4-sulfanylpentan-2-one (4MSP; box-tree aroma) and 3-sulfanylhexan-1-ol (3SH; grapefruit and passion fruit hints) has also been described [48].

Table 2. Effect of *Lachancea thermotolerans* in wine fermentation.

Acidification pH or Lactic Acid (g/L)	Fermentative Power (Ethanol % v/v)	Aroma, Flavor, Polysaccharides and Color	Molecules	Reference
3.5 → 3.2; 5.1 g/L lactic acid	9			[16]
3.2 → 2.9 in coinoculation	4–8	Acidity	Lactic acid	[12]
3.53 → 3.33 sequential 48 h 0.1 units reduction sequential 24 h	10.5			[13]
1–16.6 g/L				[28]
1.2–2.6 g/	9.5–10.4			[14]
		Esters	2-phenylethanol, phenethyl propionate, ethyl salicylate, methyl salicylate, 3-methylthio-1-propanol	[50]
		Enhanced formation of terpenes & Thiols	Nerol, terpinen-4-ol 4MSP & 3SH	[48]
		β-D-glucosidase Carbon-sulfur lyase	Free terpenes and thiols	[47,48]
	7.7	Polysaccharides/ mannoproteins	*N*-acetyl hexosamines	[51]
		Polyalcohols	Glycerol	[12,16]
		Polymeric pigments	malvidin-3-glucoside-ethyl-catechin dimer	[40]

The use of LT in the fermentation of Syrah and Sauvignon blanc musts increased the formation of 2-phenylethanol, phenethyl propionate, ethyl salicylate, methyl salicylate, 3-methylthio-1-propanol [50]. The contents of terpenes nerol and terpinen-4-ol were also positively affected (Table 2). LT general effect in aroma profile is the production of several acetate esters and certain terpenes [52].

One of the most highlighted roles of *L. thermotolerans* is the production of glycerol during wine fermentation. This increase in glycerol is observed during spontaneous fermentation [12,53,54], and sequential inoculations between *L. thermotolerans* and *S. cerevisiae* [12,13,16]. However, in the case of sequential inoculations the main advantage is that glycerol is generated with a decreased volatile acidity and acetic acid concentration [12,55]. The production of glycerol is also highly related to the fermentation temperature [13] and it increases with oxygenation [56]. Generally speaking, the extent of influence that *L. thermotolerans* can exert on a given fermentation is relative to the amount of time it spends alone in contact with the grape must [13,16]. Glycerol, the next major yeast metabolite after ethanol, is associated with the smoothness (mouth-feel), sweetness and complexity in wines [57]. However, the sensory impact of glycerol is also intimately related to the grape variety and wine style [58].

Other relevant application of LT is the sensory improvement of typical regional wines [59,60]. Recently, it has been observed the key contribution of microbioma influence in terroir finger print of regional wines [61].

5. Effect on Wine Color

Yeasts can affect wine color by pH reduction [62], favoring the formation of stable pigments such as pyranoanthocyanins [63–66] or polymeric pigments [40,67], reducing the adsorption of grape anthocyanins in cell walls [66,68,69], or protecting the anthocyanins from oxidative damage by releasing reductive compounds like glutathione during fermentation and ageing on lees [62].

As for acidification, some yeasts are able to reduce wine pH during fermentation through the release or transformation of certain organic acids. Color intensity of anthocyanins is pH dependent. The production of lactic acid by LT during fermentation can affect strongly the acidity with reductions of 0.3–0.5 pH units in some cases, so it can affect wine color in a significant way. Moreover, as the lactic acid is stable during ageing, this effect can be permanent in wine. Additionally, it should be considered that acidity also helps to protect wine color by producing higher levels of molecular SO_2.

The effect of the formation of pyranoanthocyanins like vitisins during fermentation is promoted by the release of the precursors: Pyruvic acid for vitisin A and acetaldehyde for vitisin B. The correlation between the excretion of these metabolites by *S. cerevisiae* strains and the subsequent condensation with malvidin-3-*O*-glucoside has been previously reported [63]. Not significant effects on the formation of vitisins have been observed when LT has been used in sequential fermentations with *S. cerevisiae* [39,40]. Some improvements can be seen when LT is used sequentially with *S. pombe*, but this is due to the contribution of this last yeast. The selection of appropriate strains can increase the formation of vitisins during fermentation, but it does not seem that LT is a good promoter for fermentative formation of vitisins. Similar results have been published for the formation of vinylphenolic pyranoanthocyanins [39,40], so probably most of the LT strains do not express hydroxycinnamate decarboxylase activity. Concerning the formation of polymeric pigments, sequential fermentations of LT with *S. cerevisiae* and especially with *S. pombe* favors the formation of malvidin-3-glucoside-ethyl-catechin dimer [40] (Table 2).

The absorption of anthocyanins in yeast cell walls can be between 3 and 6% of total content in wines [69]. Adsorption is strain dependent in *S. cerevisiae* [68] but also variable among non-*Saccharomyces* species [62]. Compared with other species, LT has shown a medium-high adsorption capacity.

6. Special Wines

The use of LT could be also interesting in special sweet wines to better balance sweetness and acidity. The fermentative production of lactic acid by LT can help to make the strongly sweet wines like ice wines more pleasant, increasing the freshness when the grape acidity levels are unsuitable. Using LT, it could be possible to ferment at around 8–10% *v*/*v* of ethanol, remaining natural residual sugars together with a well-balanced acidity.

The quality of natural sparkling wines with second fermentation in bottle is strongly dependent on freshness, and therefore, in acidity. When the base wine is produced in warm areas frequently it lacks of enough acidity what affects sensory quality in mouth, but also the stability during second fermentation and ageing. The use of LT in the production of base wines is an interesting biotechnological tool to provide them of enough stable acidity to make safer fermentations in bottle, but also reaching better sensory profiles.

The complementary use of fructophilic non-*Saccharomyces* yeasts, such as *Candida zemplinina* [70], helps to increase sugar consumption enhancing at the same time flavor, what can be useful in some wine types. *C. zemplinina* also improve mouthfeel and roundness due to its high production of glycerol [70].

7. Conclusions

LT is a really interesting non-*Saccharomyces* yeast species to improve the quality of wine fermentation. The natural acidification by production of stable lactic acid facilitates safer wines, less prone to spoilage, with a higher freshness and with lower levels of SO_2. Also, it opens the possibility of new biotechnologies as the simultaneous co-fermentation with lactic acid bacteria reaching even lower pHs and also consuming sugars what can reduce slightly the alcohol content. It would be interesting to go deeper into the selection of LT strains to get a stronger fermenter able to surpass 10% v/v in ethanol content.

Author Contributions: A.M. Revision of the articles and writing; I.L. Revision of the articles and writing; W.T. Revision, writing and critical reading; M.A.B. Figures and molecular biology tests; C.G. Revision and ritical Reading; J.A.S.L. Critical Reading.

Funding: This research received no external funding.

Conflicts of Interest: The authors declare no conflicts of interest.

References

1. Kurtzman, C.P. Phylogenetic circumscription of *Saccharomyces*, *Kluyveromyces* and other members of the *Saccharomycetaceae*, and the proposal of the new genera *Lachancea*, *Nakaseomyces*, *Naumovia*, *Vanderwaltozyma* and *Zygotorulaspora*. *FEMS Yeast Res.* **2003**, *4*, 233–245. [CrossRef]
2. Ganter, P.F. Yeast and invertebrate associations. In *Biodiversity and Ecophysiology of Yeasts*; Rosa, C.A., Gabor, P., Eds.; Springer: Berlin, Germany, 2006; pp. 303–370.
3. Hranilovic, A.; Bely, M.; Masneuf-Pomarede, I.; Jiranek, V.; Albertin, W. The evolution of *Lachancea thermotolerans* is driven by geographical determination, anthropisation and flux between different ecosystems. *PLoS ONE* **2017**, *12*, e0184652. [CrossRef] [PubMed]
4. Combina, M.; Elía, A.; Mercado, L.; Catania, C.; Ganga, A.; Martinez, C. Dynamics of indigenous yeast populations during spontaneous fermentation of wines from Mendoza, Argentina. *Int. J. Food Microbiol.* **2005**, *99*, 237–243. [CrossRef] [PubMed]
5. Schnierda, T.; Bauer, F.F.; Divol, B.; van Rensburg, E.; Görgens, J.F. Optimization of carbon and nitrogen medium components for biomass production using non-*Saccharomyces* wine yeasts. *Lett. Appl. Microbiol.* **2014**, *58*, 478–485. [CrossRef] [PubMed]
6. Lachance, M.-A.; Kurtzman, C.P. Chapter 41—*Lachancea* Kurtzman. In *The Yeasts*, 5th ed.; Elsevier: New York, NY, USA, 2011; pp. 511–519. ISBN 9780444521491.
7. Ciani, M.; Beco, L.; Comitini, F. Fermentation behaviour and metabolic interactions of multistarter wine yeast fermentations. *Int. J. Food Microbiol.* **2006**, *108*, 239–245. [CrossRef] [PubMed]
8. Kemsawasd, V.; Viana, T.; Ardö, Y.; Arneborg, N. Influence of nitrogen sources on growth and fermentation performance of different wine yeast species during alcoholic fermentation. *Appl. Microbiol. Biotechnol.* **2015**, *99*, 10191–10207. [CrossRef] [PubMed]
9. Escribano, R.; González-Arenzana, L.; Garijo, P.; Berlanas, C.; López-Alfaro, I.; López, R.; Gutiérrez, A.R.; Santamaría, P. Screening of enzymatic activities within different enological non-*Saccharomyces* yeasts. *J. Food Sci. Technol.* **2017**, *54*, 1555–1564. [CrossRef] [PubMed]

10. Fleet, G.H. Yeast interactions and wine flavor. *Int. J. Food Microbiol.* **2003**, *86*, 11–22. [CrossRef]

11. Kapsopoulou, K.; Kapaklis, A.; Spyropoulos, H. Growth and fermentation characteristics of a strain of the wine yeast *Kluyveromyces thermotolerans* isolated in Greece. *World J. Microbiol. Biotechnol.* **2005**, *21*, 1599–1602. [CrossRef]

12. Comitini, F.; Gobbi, M.; Domizio, P.; Romani, C.; Lencioni, L.; Mannazzu, I.; Ciani, M. Selected non-*Saccharomyces* wine yeasts in controlled multistarter fermentations with *Saccharomyces cerevisiae*. *Food Microbiol.* **2011**, *28*, 873–882. [CrossRef] [PubMed]

13. Gobbi, M.; Comitini, F.; Domizio, P.; Romani, C.; Lencioni, L.; Mannazzu, I.; Ciani, M. *Lachancea thermotolerans* and *Saccharomyces cerevisiae* in simultaneous and sequential co-fermentation: A strategy to enhance acidity and improve the overall quality of wine. *Food Microbiol.* **2013**, *33*, 271–281. [CrossRef] [PubMed]

14. Aponte, M.; Blaiotta, G. Potential role of yeast strains isolated from grapes in the production of Taurasi DOCG. *Front. Microbiol.* **2016**, *7*, 809. [CrossRef] [PubMed]

15. Ciani, M.; Morales, P.; Comitini, F.; Tronchoni, J.; Canonico, L.; Curiel, J.A.; Oro, L.; Rodrigues Alda, J.; Gonzalez, R. Non-conventional yeast species for lowering ethanol content of wines. *Front. Microbiol.* **2016**, *7*, 642. [CrossRef] [PubMed]

16. Kapsopoulou, K.; Mourtzini, A.; Anthoulas, M.; Nerantzis, E. Biological acidification during grape must fermentation using mixed cultures of *Kluyveromyces thermotolerans* and *Saccharomyces cerevisiae*. *World J. Microbiol. Biotechnol.* **2007**, *23*, 735–739. [CrossRef]

17. Mills, D.A.; Johannsen, E.A.; Cocolin, L. Yeast diversity and persistence in *Botrytis*-affected wine fermentations. *Appl. Environ. Microbiol.* **2002**, *68*, 4884–4893. [CrossRef] [PubMed]

18. Holm Hansen, E.; Nissen, P.; Sommer, P.; Nielsen, J.C.; Arneborg, N. The effect of oxygen on the survival of non-*Saccharomyces* yeasts during mixed culture fermentations of grape juice with *Saccharomyces cerevisiae*. *J. Appl. Microbiol.* **2001**, *91*, 541–547. [CrossRef] [PubMed]

19. Costa, A.; Barata, A.; Malfeito-Ferreira, M.; Loureiro, V. Evaluation of the inhibitory effect of dimethyl dicarbonate (DMDC) against wine microorganisms. *Food Microbiol.* **2008**, *25*, 422–427. [CrossRef] [PubMed]

20. CHR-Hansen. CONCERTO™. Available online: https://www.chr-hansen.com/en/food-cultures-and-enzymes/wine/cards/product-cards/concerto?countryreset=1 (accessed on 30 June 2018).

21. Petruzzi, L.; Capozzi, V.; Berbegal, C.; Corbo, M.R.; Bevilacqua, A.; Spano, G.; Sinigaglia, M. Microbial resources and enological significance: Opportunities and benefits. *Front. Microbiol.* **2017**, *8*, 995. [CrossRef] [PubMed]

22. Balikci, E.K.; Tanguler, H.; Jolly, N.P.; Erten, H. Influence of *Lachancea thermotolerans* on cv. Emir wine fermentation. *Yeast* **2016**, *33*, 313–321. [CrossRef] [PubMed]

23. García, M.; Esteve-Zarzoso, B.; Crespo, J.; Cabellos, J.M.; Arroyo, T. Yeast monitoring of wine mixed or sequential fermentations made by native strains from D.O. "Vinos de Madrid" using Real-Time Quantitative PCR. *Front. Microbiol.* **2017**, *8*, 2520. [CrossRef] [PubMed]

24. Fiori, S.; Urgeghe, P.P.; Hammami, W.; Razzu, S.; Jaoua, S.; Migheli, Q. Biocontrol activity of four non- and low-fermenting yeast strains against *Aspergillus carbonarius* and their ability to remove ochratoxin A from grape juice. *Int. J. Food Microbiol.* **2014**, *189*, 45–50. [CrossRef] [PubMed]

25. Nally, M.C.; Ponsone, M.L.; Pesce, V.M.; Toro, M.E.; Vazquez, F.; Chulze, S. Evaluation of behaviour of *Lachancea thermotolerans* biocontrol agents on grape fermentations. *Lett. Appl. Microbiol.* **2018**, *67*, 89–96. [CrossRef] [PubMed]

26. Loira, I.; Morata, A.; Bañuelos, M.A.; Suárez-Lepe, J.A. Isolation, selection and identification techniques for non-*Saccharomyces* yeasts of oenological interest. In *Biotechnological Progress and Beverage Consumption*; Beverage Series; Academic Press-Elsevier: Cambridge, MA, USA, 2018; Volume 19, in press.

27. Prakitchaiwattana, C.J.; Fleet, G.H.; Heard, G.M. Application and evaluation of denaturing gradient gel electrophoresis to analyse the yeast ecology of wine grapes. *FEMS Yeast Res.* **2004**, *4*, 865–877. [CrossRef] [PubMed]

28. Banilas, G.; Sgouros, G.; Nisiotou, A. Development of microsatellite markers for *Lachancea thermotolerans* typing and population structure of wine-associated isolates. *Microbiol. Res.* **2016**, *193*, 1–10. [CrossRef] [PubMed]

29. Kurtzman, C.P.; Robnett, C.J. Identification and phylogeny of ascomycetous yeasts from analysis of nuclear large subunit (26S) ribosomal DNA partial sequences. *Antonie Van Leeuwenhoek* **1998**, *73*, 331–371. [CrossRef] [PubMed]

30. Lopandic, K.; Tiefenbrunner, W.; Gangl, H.; Mandl, K.; Berger, S.; Leitner, G.; Abd-Ellah, G.A.; Querol, A.; Gardner, R.C.; Sterflinger, K.; et al. Molecular profiling of yeasts isolated during spontaneous fermentations of Austrian wines. *FEMS Yeast Res.* **2008**, *8*, 1063–1075. [CrossRef] [PubMed]

31. Baleiras Couto, M.M.; Reizinho, R.G.; Duarte, F.L. Partial 26S rDNA restriction analysis as a tool to characterise non-*Saccharomyces* yeasts present during red wine fermentations. *Int. J. Food Microbiol.* **2005**, *102*, 49–56. [CrossRef] [PubMed]

32. Zara, G.; Ciani, M.; Domizio, P.; Zara, S.; Budroni, M.; Carboni, A.; Mannazzu, I. A culture-independent PCR-based method for the detection of *Lachancea thermotolerans* in wine. *Ann. Microbiol.* **2014**, *64*, 403–406. [CrossRef]

33. Mora, J.; Barbas, J.I.; Mulet, A. Growth of yeast species during the fermentation of musts Inoculated with *Kluyveromyces thermotolerans* and *Saccharomyces cerevisiae*. *Am. J. Enol. Vitic.* **1990**, *41*, 156–159.

34. Jolly, N.P.; Varela, C.; Pretorius, I.S. Not your ordinary yeast: Non-*Saccharomyces* yeasts in wine production uncovered. *FEMS Yeast Res.* **2014**, *14*, 215–237. [CrossRef] [PubMed]

35. Domizio, P.; House, J.F.; Joseph, C.M.L.; Bisson, L.F.; Bamforth, C.W. *Lachancea thermotolerans* as an alternative yeast for the production of beer. *J. Inst. Brew.* **2016**, *122*, 599–604. [CrossRef]

36. Callejo, M.J.; González, C.; Morata, A. Use of non-*Saccharomyces* yeasts in bottle fermentation of aged beers. In *Brewing Technology*; Kanauchi, M., Ed.; IntechOpen: London, UK, 2017; Available online: https://www.intechopen.com/books/brewing-technology/use-of-non-saccharomyces-yeasts-in-bottle-fermentation-of-aged-beers (accessed on 30 June 2018).

37. Osburn, K.; Amaral, J.; Metcalf, S.R.; Nickens, D.M.; Rogers, C.M.; Sausen, C.; Caputo, R.; Miller, J.; Li, H.; Tennessen, J.M.; et al. A novel bacteria-free method for sour beer production. *Food Microbiol.* **2018**, *70*, 76–84. [CrossRef] [PubMed]

38. Suárez, R.; Suárez-Lepe, J.A.; Morata, A.; Calderón, F. The production of ethylphenols in wine by yeasts of the genera *Brettanomyces* and *Dekkera*: A review. *Food Chem.* **2007**, *102*, 10–21. [CrossRef]

39. Del Fresno, J.M.; Morata, A.; Loira, I.; Bañuelos, M.A.; Escott, C.; Benito, S.; González Chamorro, C.; Suárez-Lepe, J.A. Use of non-*Saccharomyces* in single-culture, mixed and sequential fermentation to improve red wine quality. *Eur. Food Res. Technol.* **2017**, *243*, 2175–2185. [CrossRef]

40. Escott, C.; Del Fresno, J.M.; Loira, I.; Morata, A.; Tesfaye, W.; González, M.C.; Suárez-Lepe, J.A. Formation of polymeric pigments in red wines through sequential fermentation of flavanol-enriched musts with non-*Saccharomyces* yeasts. *Food Chem.* **2018**, *239*, 975–983. [CrossRef] [PubMed]

41. Hernández, P. Use of Lachancea thermotolerans to Improve pH in Red Wines. Effect of the Coinoculation with Oenococcus oeni. Master's Thesis, Universidad Politécnica de Madrid, Madrid, Spain, 2018.

42. Benito, Á.; Calderón, F.; Palomero, F.; Benito, S. Combine use of selected *Schizosaccharomyces pombe* and *Lachancea thermotolerans* yeast strains as an alternative to the traditional malolactic fermentation in red wine production. *Molecules* **2015**, *20*, 9510–9523. [CrossRef] [PubMed]

43. Lerm, E.; Engelbrecht, L.; du Toit, M. Malolactic Fermentation: The ABC's of MLF. *S. Afr. J. Enol. Vitic.* **2010**, *31*, 186–212. [CrossRef]

44. Du Plessis, H.W.; du Toit, M.; Hoff, J.W.; Hart, R.S.; Ndimba, B.K.; Jolly, N.P. Characterisation of non-*Saccharomyces* yeasts using different methodologies and evaluation of their compatibility with malolactic fermentation. *S. Afr. J. Enol. Vitic.* **2017**, *38*, 46–63. [CrossRef]

45. Vilela-Moura, A.; Schuller, D.; Mendes-Faia, A.; Côrte-Real, M. Reduction of volatile acidity of wines by selected yeast strains. *Appl. Microbiol. Biotechnol.* **2008**, *80*, 881–890. [CrossRef] [PubMed]

46. Ciani, M.; Comitini, F. Non-*Saccharomyces* wine yeasts have a promising role in biotechnological approaches to winemaking. *Ann. Microbiol.* **2011**, *61*, 25–32. [CrossRef]

47. Rosi, I.; Vinella, M.; Domizio, P. Characterization of β-glucosidase activity in yeasts of oenological origin. *J. Appl. Bacteriol.* **1994**, *77*, 519–527. [CrossRef] [PubMed]

48. Zott, K.; Thibon, C.; Bely, M.; Lonvaud-Funel, A.; Dubourdieu, D.; Masneuf-Pomarede, I. The grape must non-*Saccharomyces* microbial community: Impact on volatile thiol release. *Int. J. Food Microbiol.* **2011**, *151*, 210–215. [CrossRef] [PubMed]

49. Padilla, B.; Gil, J.V.; Manzanares, P. Past and future of non-*Saccharomyces* yeasts: From spoilage microorganisms to biotechnological tools for improving wine aroma complexity. *Front. Microbiol.* **2016**, *7*, 411. [CrossRef] [PubMed]

50. Beckner Whitener, M.E.; Carlina, S.; Jacobson, D.; Weighill, D.; Divol, B.; Conterno, L.; Du Toit, M.; Vrhovsek, U. Early fermentation volatile metabolite profile of non-*Saccharomyces* yeasts in red and white grape must: A targeted approach. *LWT Food Sci. Technol.* **2015**, *64*, 412–422. [CrossRef]

51. Domizio, P.; Liu, Y.; Bisson, L.F.; Barile, D. Use of non-*Saccharomyces* wine yeasts as novel sources of mannoproteins in wine. *Food Microbiol.* **2014**, *43*, 5–15. [CrossRef] [PubMed]

52. Beckner Whitener, M.E.; Stanstrup, J.; Panzeri, V.; Carlin, S.; Divol, B.; Du Toit, M.; Vrhovsek, U. Untangling the wine metabolome by combining untargeted SPME–GCxGC-TOF-MS and sensory analysis to profile Sauvignon blanc co-fermented with seven different yeasts. *Metabolomics* **2016**, *12*, 53. [CrossRef]

53. Romano, P.; Suzzi, G.; Comi, G.; Zironi, R.; Maifreni, M. Glycerol and other fermentation products of apiculate wine yeasts. *J. Appl. Microbiol.* **1997**, *82*, 615–618. [CrossRef] [PubMed]

54. Henick-Kling, T.; Edinger, W.; Daniel, P.; Monk, P. Selective effects of sulfur dioxide and yeast starter culture addition on indigenous yeast populations and sensory characteristics of wine. *J. Appl. Microbiol.* **1998**, *84*, 865–876. [CrossRef]

55. Domizio, P.; Romani, C.; Lencioni, L.; Comitini, F.; Gobbi, M.; Mannazzu, I.; Ciani, M. Outlining a future for non-*Saccharomyces* yeasts: Selection of putative spoilage wine strains to be used in association with *Saccharomyces cerevisiae* for grape juice fermentation. *Int. J. Food Microbiol.* **2011**, *147*, 170–180. [CrossRef] [PubMed]

56. Shekhawat, K.; Porter, T.J.; Bauer, F.F.; Setati, M.E. Employing oxygen pulses to modulate *Lachancea thermotolerans–Saccharomyces cerevisiae* Chardonnay fermentations. *Ann. Microbiol.* **2018**, *68*, 93–102. [CrossRef]

57. Ciani, M.; Maccarelli, F. Oenological properties of non-*Saccharomyces* yeasts associated with wine-making. *World J. Microbiol. Biotechnol.* **1998**, *14*, 199–203. [CrossRef]

58. Nieuwoudt, H.H.; Prior, B.A.; Pretorius, I.S.; Bauer, F.F. Glycerol in South African table wines: An assessment of its relationship to wine quality. *S. Afr. J. Enol. Vitic.* **2002**, *23*, 22–30. [CrossRef]

59. Capozzi, V.; Garofalo, C.; Chiriatti, M.A.; Grieco, F.; Spano, G. Microbial terroir and food innovation: The case of yeast biodiversity in wine. *Microbiol. Res.* **2015**, *181*, 75–83. [CrossRef] [PubMed]

60. Pinto, C.; Pinho, D.; Cardoso, R.; Custódio, V.; Fernandes, J.; Sousa, S.; Pinheiro, M.; Egas, C.; Gomes, A.C. Wine fermentation microbiome: A landscape from different Portuguese wine appellations. *Front. Microbiol.* **2015**, *6*, 905. [CrossRef] [PubMed]

61. Bokulich, N.A.; Collins, T.S.; Masarweh, C.; Allen, G.; Heymann, H.; Ebeler, S.E.; Mills, D.A. Associations among wine grape microbiome, metabolome, and fermentation behavior suggest microbial contribution to regional wine characteristics. *mBio* **2016**, *7*, e00631-16. [CrossRef] [PubMed]

62. Morata, A.; Loira, I.; Suárez Lepe, J.A. Influence of Yeasts in Wine Colour. In *Grape and Wine Biotechnology*; Morata, A., Loira, I., Eds.; IntechOpen: London, UK, 2016; pp. 285–305. Available online: https://www.intechopen.com/books/grape-and-wine-biotechnology/influence-of-yeasts-in-wine-colour (accessed on 30 June 2018).

63. Morata, A.; Gómez-Cordovés, M.C.; Colomo, B.; Suárez, J.A. Pyruvic acid and acetaldehyde production by different strains of *Saccharomyces cerevisiae*: Relationship with vitisin A and B formation in red wines. *J. Agric. Food Chem.* **2003**, *51*, 6475–6481. [CrossRef] [PubMed]

64. Morata, A.; Gómez-Cordovés, M.C.; Calderón, F.; Suárez, J.A. Effects of pH, temperature and SO$_2$ on the formation of pyranoanthocyanins during red wine fermentation with two species of *Saccharomyces*. *Int. J. Food Microbiol.* **2006**, *106*, 123–129. [CrossRef] [PubMed]

65. Morata, A.; González, C.; Suárez-Lepe, J.A. Formation of vinylphenolic pyranoanthocyanins by selected yeasts fermenting red grape musts supplemented with hydroxycinnamic acids. *Int. J. Food Microbiol.* **2007**, *116*, 144–152. [CrossRef] [PubMed]

66. Suárez-Lepe, J.A.; Morata, A. New trends in yeast selection for winemaking. *Trends Food Sci. Technol.* **2012**, *23*, 39–50. [CrossRef]

67. Escott, C.; Morata, A.; Loira, I.; Tesfaye, W.; Suarez-Lepe, J.A. Characterization of polymeric pigments and pyranoanthocyanins formed in microfermentations of non-*Saccharomyces* yeasts. *J. Appl. Microbiol.* **2016**, *121*, 1346–1356. [CrossRef] [PubMed]

68. Morata, A.; Gómez-Cordovés, M.C.; Suberviola, J.; Bartolomé, B.; Colomo, B.; Suárez, J.A. Adsorption of anthocyanins by yeast cell walls during the fermentation of red wines. *J. Agric. Food Chem.* **2003**, *51*, 4084–4088. [CrossRef] [PubMed]

69. Morata, A.; Gómez-Cordovés, M.C.; Colomo, B.; Suárez, J.A. Cell wall anthocyanin adsorption by different *Saccharomyces* strains during the fermentation of *Vitis vinifera* L. cv Graciano grapes. *Eur. Food Res. Technol.* **2005**, *220*, 341–346. [CrossRef]
70. Magyar, I.; Tóth, T. Comparative evaluation of some oenological properties in wine strains of *Candida stellata*, *Candida zemplinina*, *Saccharomyces uvarum* and *Saccharomyces cerevisiae*. *Food Microbiol.* **2011**, *28*, 94–100. [CrossRef] [PubMed]

fermentation

MDPI

Review

Schizosaccharomyces pombe: A Promising Biotechnology for Modulating Wine Composition

Iris Loira *, Antonio Morata, Felipe Palomero, Carmen González and José Antonio Suárez-Lepe

Departamento de Química y Tecnología de Alimentos, Universidad Politécnica de Madrid, Av. Puerta de Hierro, nº 2, 28040 Madrid, Spain; antonio.morata@upm.es (A.M.), felipe.palomero@upm.es (F.P.); carmen.gchamorro@upm.es (C.G.); joseantonio.suarez.lepe@upm.es (J.A.S.-L.)
* Correspondence: iris.loira@upm.es

Received: 26 July 2018; Accepted: 21 August 2018; Published: 23 August 2018

Abstract: There are numerous yeast species related to wine making, particularly non-*Saccharomyces*, that deserve special attention due to the great potential they have when it comes to making certain changes in the composition of the wine. Among them, *Schizosaccharomyces pombe* stands out for its particular metabolism that gives it certain abilities such as regulating the acidity of wine through maloalcoholic fermentation. In addition, this species is characterized by favouring the formation of stable pigments in wine and releasing large quantities of polysaccharides during ageing on lees. Moreover, its urease activity and its competition for malic acid with lactic acid bacteria make it a safety tool by limiting the formation of ethyl carbamate and biogenic amines in wine. However, it also has certain disadvantages such as its low fermentation speed or the development of undesirable flavours and aromas. In this chapter, the main oenological uses of *Schizosaccharomyces pombe* that have been proposed in recent years will be reviewed and discussed.

Keywords: *Schizosaccharomyces pombe*; oenological uses; maloalcoholic fermentation; stable pigments; wine safety

1. Origin and Features of *Schizosaccharomyces pombe*

Schizosaccharomyces pombe, also known as fission yeast, was discovered by Lindner in 1983 [1]. The cells of this species have a characteristic rod shape with sizes varying between 3–5 × 5–24 μm (Figure 1). However, immediately after cell division, new cells formed have a more rounded shape due to the turgor pressure [2]. It has a peculiar mode of vegetative reproduction by fission (cross-wall formation) instead of budding, which is more common among yeasts [3]. Cells are separated by the formation of a transverse septum. The spores are formed as a result of sexual reproduction by conjugation of the cells when adverse conditions occur, such as nutrient starvation, and, in the case of *S. pombe*, between two and four (most often) haploid spores originate in the ascus [4].

Its growth rate is very slow, with a long lag phase and high vitamin requirement. However, it has a low nitrogen requirement [5]. In normal minimal or complex media, the generation time is between 2 and 4 h [6]. Usually, *S. pombe* does not develop properly in most culture media due to its aforementioned low growing rate, thus making its isolation from the environment more difficult. A selective-differential medium based on the resistance of *S. pombe* to actidione (antibiotic) and to benzoic acid (inhibitory agent) has been recently proposed to isolate strains of this genus from media with high sugar content [7]. *S. pombe* strains have been isolated from grape juice, molasses, and kombucha tea [1,8]. In addition to glucose, *S. pombe* can also use glycerol, sucrose, raffinose, and maltose as carbon sources [9].

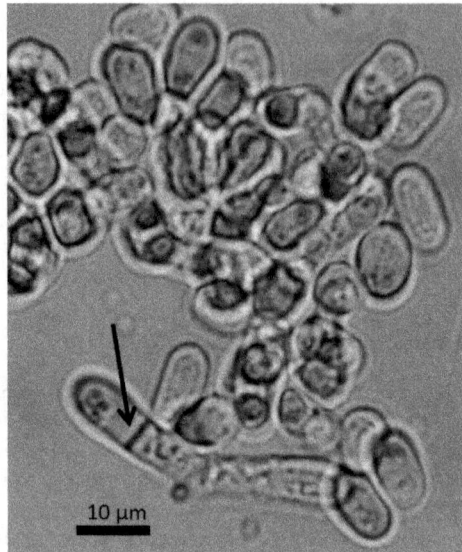

Figure 1. Optical microscope image of *Schizosaccharomyces pombe* (*S. pombe*). The transverse septum formed during the asexual reproduction is indicated by an arrow.

Another peculiarity of *S. pombe* is that it can grow in environments with low water activity, that is, it is an osmophilic yeast, and therefore can be found in media with high sugar content [1]. It can also develop in very low pH environments and in a wide range of temperatures [10]. Moreover, it is somewhat resistant to food preservatives, such as sulphur dioxide, actidione, benzoic acid, and dimethyl dicarbonate [10,11].

Regarding its fermentative performance, it is able to ferment glucose to an alcoholic degree of around 10–15% *v/v* ethanol, depending on the yeast strain and the aeration conditions [10]. As already mentioned, the genus *Schizosaccharomyces* is known for its slow growth rate and excessive production of hydrogen sulphide during fermentation [12]. These two features, together with high volatile acidity, are the main limitations for its use in winemaking. The production of acetic acid is strain-dependent, usually ranging between 0.8 and 1.4 g/L [13]. Nevertheless, through the selection of strains and their use in combination with yeasts of the genus *Saccharomyces*, wines of quality can be obtained from unbalanced musts with high total acidity.

Currently, thanks to recent research that presents new possibilities for their use, non-*Saccharomyces* yeasts are shedding their bad reputation, and it is possible to find *S. pombe* yeasts encapsulated in alginate beads being marketed as an alternative to malolactic fermentation or chemical deacidification [14]. An advantage of using these encapsulated yeasts is that they can be removed from the medium at a desired time and, in addition, the same capsules can be reused in several cycles (up to 5 times), although with a slight loss of degrading activity [15,16]. Regarding sensory properties, the wines obtained by sequential fermentation of *S. pombe* and *S. cerevisiae* were full-bodied, with better structure, balance, and length than the controls made without using this deacidification technique [16].

2. Wine Acidity Modulation

Wine acidity is mainly responsible for freshness. After L-tartaric acid, L-malic acid is the second organic acid in wine that contributes significantly to its total acidity. Its average content in wine highly depends on the grape variety and the climate, varying widely between 1 and 10 g/L [17]. Reaching an appropriate balance between the sugar content and the total acidity of the wine is fundamental to ensure its optimum quality. In addition, excessive amounts of malic acid may cause microbiological instability in wine. These are the two main reasons to modulate wine pH.

The biological deacidification of wine through the use of *Schizosaccharomyces pombe* has been studied thoroughly [18,19], since its ability to transform malic acid into ethanol and carbon dioxide was discovered in the early 20th century [20]. Thanks to this ability of *S. pombe* to develop maloalcoholic fermentation (MAF) (Figure 2), it is possible to modulate the pH of the wine by the consumption of practically all the malic acid present in the must with the corresponding stoichiometric production of ethanol. Unlike *S. cerevisiae*, in which the malic enzyme is located in the mitochondria (organelle in low numbers and dysfunctional under winemaking conditions), *S. pombe* has an active transport system for the uptake of extracellular malic acid in addition to a malic enzyme located in the cytosol with a very high substrate affinity [21]. The degree of degradation of malic acid is strain-dependent, normally varying between 75% and 100% [5,10]. *Issatchenkia orientalis* has been also proved to have this strong malic acid degradative metabolism [22,23]. However, this yeast species is only present in small quantities at the beginning of fermentation due to its sensitivity to ethanol [24,25]. Kim, Hong, & Park (2008) [26] have also reported the effectiveness of using a mixed culture of *Issatchenkia orientalis* and *Saccharomyces cerevisiae* to reduce the malic acid content during fermentation. When trying to improve the quality of the wine through the combination in mixed or sequential fermentation of different non-*Saccharomyces* and *Saccharomyces* yeast species, it is not only important to know the contribution of each species or strain but also to select the adequate inoculum ratio [27]. Other yeast species, including some *Saccharomyces* spp. (commercially available strains are generally unable to degrade L-malic acid effectively during fermentation), are able to consume malic acid, but to a lesser extent (usually <25%). Another possibility is to use genetic engineering to improve the ability of *S. cerevisiae* to degrade malic acid, for example, through the incorporation of genes responsible for the transport of malic acid in *S. pombe* or the malolactic enzyme from *Oenococcus oeni* [28].

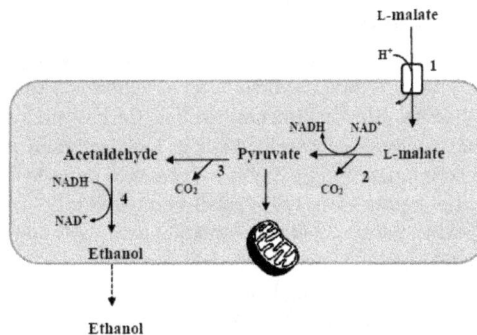

Figure 2. Schematic representation of malic acid degradation by *Schizosaccharomyces pombe*: maloalcoholic fermentation (MAF) and alternative use of pyruvate in mitochondria for cellular biosynthesis. The enzymes involved in the biochemical transformations of the MAF are the following: 1: malate permease (active transport); 2: malic enzyme (malate decarboxylase); 3: pyruvate decarboxylase; 4: alcohol dehydrogenase. The arrows indicate the direction of the metabolic pathway involved in the transport and degradation of malic acid, identifying the main substrates, products and intermediaries.

Although it has been mentioned on numerous occasions that the aroma produced by the fermentative metabolism of *S. pombe* is not suitable for a quality wine, no particular off-flavour has been described yet. In general, only atypical aroma [19] and a loss in fruity character [29] have been identified by some authors. This could be directly related to the fact that it is a slow fermenting yeast [12]. Sometimes, more than 30 days are necessary to finish the fermentation [29]. Moreover, immobilization techniques have been developed (e.g., alginate beads) in order to avoid negative side effects such as high levels of acetic acid and other off-flavours [30]. Following this technique, once the desired malic acid content is reached in the wine thanks to the demalication activity of *S. pombe*, this yeast is removed, and the fermentation is finished by *S. cerevisiae*. Snow & Gallander (1979) [31] suggested that one or two days of *S. pombe* fermentation is adequate to obtain a quality wine like that obtained by *S. cerevisiae* in pure fermentation, avoiding at the same time an excessive deacidification.

In addition to biological deacidification, either by yeast (MAF) or lactic acid bacteria (MLF), there are also other methods to regulate wine acidity such as blending, carbonic maceration, or chemical deacidification using carbonate salts (usually calcium carbonate, $CaCO_3$; potassium carbonate, K_2CO_3; and potassium bicarbonate, $KHCO_3$) [17].

3. Influence on Wine Colour

The colour of the wine is one of the main sensory properties indicative of quality, especially in red wines. It is highly dependent on winemaking technology, especially influenced by the maceration time and the mechanical processes performed during the vinification process (e.g., punch downs, pump overs, délestage...) [32], but the microorganisms used, both yeast and bacteria, also play an important role in the development of the final colour of the wine and its stability.

First, fermentative yeast may influence the colour by modifying the pH. In this sense, oenological yeasts usually produce wines with pH between 3.2 and 3.8, and under these conditions the anthocyanins tend to have a deep red colour, while the higher the pH, the greater bathochromic shift from red to purple and later to blue [33]. In the case of pure fermentations with *S. pombe*, pH control would be essential to avoid changes in colour to some extent, because this yeast, as mentioned above, makes a high consumption of malic acid, and therefore the pH of the wine increases slightly.

Second, through the release of secondary metabolites of fermentation that can react with the grape anthocyanins and thus create more stable forms of colour such as pyranoanthocyanins. The formation of pyranoanthocyanins is of great interest to preserve intense colour in the wine during aging processes, since these pigments that originated during fermentation are much more long-lasting due to their greater stability against pH changes and discoloration by SO_2. The strain effect is quite significant when several *S. cerevisiae* are used, so yeast selection is a useful tool to obtain higher amounts of pyranoanthocyanins during fermentation [34]. Certain strains of *S. pombe* can lead to improvements in the colour stability of red wine thanks to its high formation of vitisin A and derivatives thereof [35]. It is a direct consequence of its high production of pyruvic acid during fermentation (ranging 150–350 mg/L). The higher values in *S. pombe* compared to *S. cerevisiae* (<100 mg/L; [35]) are probably due to the specific maloalcoholic fermentation pathway in which pyruvate is involved as metabolic intermediate (Figure 2). The amount of pyruvate released by *S. pombe* can favour the formation of 2–4-fold more vitisin A and acetyl vitisin A than *S. cerevisiae* during the fermentation of red wines (Figure 3). Concerning the production of Vitisin B derivatives, a lower formation of vitisin B is observed with respect to *S. cerevisiae* fermentations [35]. However, the synthesis of vitisin A is more interesting in wines than vitisin B, because the latter is a red-brown pigment (495 nm compared with 515 nm vitisin A [34], and the hue is less suitable for wine-colour quality).

The enzymatic activity of the yeast is also fundamental, especially regarding the β-glucosidase and hydroxycinnamate decarboxylase activities. The first enzyme leads to a colour loss, because it catalyses the breakage of the bond between the glucose and the anthocyanidin moieties, also known as anthocyanase activity [36]. These same authors reported that no β-glucosidase activity was detected for the four strains of *S. pombe* that were evaluated. Some strains of *S. pombe* have also shown the ability to transform the hydroxycinnamic acids present in the must (e.g., *p*-coumaric, caffeic and ferulic acids) into vinyl phenols that can condense with monomeric anthocyanins to form vinylphenolic pyranoanthocyanins [37,38]. These pyranoanthocyanins display same enological properties as vitisins concerning to their stability. Vinyl phenols are formed during fermentation by yeasts with hydroxycinnamate decarboxylase activity (HCDA). It is possible to find this activity in many non-*Saccharomyces* strains [39]. It has been observed that some strains of *S. pombe* exhibit a strong HCDA that can increase the formation of malvidin-3-*O*-glucoside-4-vinylphenol and malvidin-3-*O*-(6'-*p*-coumaroylglucoside)-4-vinylphenol by 10–30% compared with *S. cerevisiae* strains selected for their performance in this activity [35]. Therefore, the selection and use of *S. pombe* in fermentation is a powerful tool for increasing the formation of stable pyranoanthocyanins, including vitisin A derivatives and vinylphenolic pyranoanthocyanins.

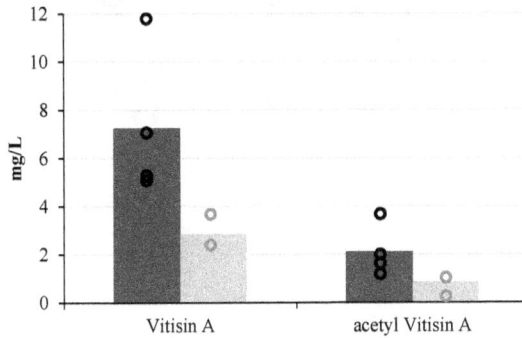

Figure 3. Formation of vitisin A and acetyl vitisin A during the fermentation with *S. pombe* (4 strains; dark grey bars) and *Saccharomyces cerevisiae* (*S. cerevisiae*) (2 strains; light grey bars). Bars are average concentrations, dots are single values for each strain (Adapted from [35]).

Polymeric pigments are formed by chemical condensation between grape anthocyanins and other flavanols in a slow process during ageing that is affected by precursor contents, pH, and temperature and oxygen levels. Some of these pigments have red-orange colour, but others can absorb at 540 nm or higher wavelength showing red-bluish colours. It has been observed that some yeasts are able to promote the formation of these pigments more quickly during fermentation [40,41]. *S. pombe* has shown that high performance increases the formation of polymeric pigments derived from malvidin-3-*O*-glucoside and catechin or procyanidin B2 [40]. On average, increments in the polymeric pigments can vary widely between 35.9 and 88.0% with respect to their counterparts with *S. cerevisiae*, depending on the par of yeast species used in the sequential fermentation [40]. Yeast influence can be produced by the fermentative release of acetaldehyde, which can favour the formation of ethyl linked dimers [42]. Furthermore, some synergic effects were seen when *S. pombe* was used in sequential fermentations with *Lachancea thermotolerans* [40].

Finally, yeasts are able to trap significant amounts of pigments in their cell wall [43,44]. During fermentation, yeast cell walls can reach a specific surface of 10 m^2/L of must when population is 10^8–10^9 CFU/mL [43], and cell wall adsorbed anthocyanins can represent 1.6–6% of wine content reaching 28% for some derivatives in some yeast strains [44]. The number of anthocyanins adsorbed into yeast cell walls can be evaluated using plating media dosed with grape anthocyanins [45]. Alternatively, and with higher precision, it is possible to recover adsorbed anthocyanins from yeast cell walls and analyse the extracts by LC-DAD or LC-DAD-MS [43]. When adsorption of anthocyanins was studied in non-*Saccharomyces* cell walls, a differential adsorption was observed according to yeast species but also according to strain. In a comparative view of adsorption in the plating system with anthocyanins added to the agar formulation, it is possible to see a medium adsorption capacity for *S. pombe* (strain 938) compared with other species with high adsorption (*L. thermotolerans*, formerly *Kluyveromyces thermotolerans*) or the low adsorption of *Metschnikowia pulcherrima* (Figure 4).

S. cerevisiae (7VA) S. ludwigii (979)

T. delbrueckii (291) S. pombe (938)

K. thermotolerans (KT) M. pulcherrima (MP)

Figure 4. Adsorption of grape anthocyanins in yeast cell walls (*Saccharomyces* and non-*Saccharomyces*) during growth in a specific plating media containing pigments.

The protective effect of ageing on lees (AOL) on wine oxidation by the release of reductive compounds from yeast structures such as glutathione can also produce some preservation on wine anthocyanins. The high release of cell wall compounds from *S. pombe* during ageing on lees [46] can favour and enhance this protective effect, thus better preserving wine colour.

4. Large Release of Polysaccharides during Ageing on Lees

As previously stated, *Schizosaccharomyces* yeast genus physiology and metabolism present some peculiarities. *S. pombe* cell wall carbohydrate composition and distribution of polysaccharides and other wall constituents is quite particular, which is something which should not be surprising bearing in mind we are dealing with an osmophilic yeast that reproduces itself/asexually by binary fission thanks to the formation of a wall, from centre to centre of the cell (Figure 1; [47,48]). Early in the seventies, some authors studied the structure and composition of their cell walls [49]. Indeed, the qualitative composition and formation of cell wall polysaccharides can vary greatly among yeasts, and these differences are useful for taxonomic classification purposes [50,51]. Later electron microscopy studies after enzyme treatments, aiming to deeply understand the molecular organization of *S. pombe* cell walls, highlighted the presence of galacto-mannoproteins in the outer layer of the cell wall as its main qualitative discriminating feature [52]. A comparative illustration of the cell walls of *Schizosaccharomyces pombe* and *Saccharomyces cerevisiae* based on these studies permits the observation of these qualitative differences among these species [46]. Besides these special features of *S. pombe*, a simple optical microscopy lets us appreciate its thickened cell walls. That significant thickness (average: 10–200 nm [2,53]) and the previously mentioned particular molecular organization of biopolymers in *S. pombe* cell walls give them structural strength enough to resist high osmotic pressures.

Ageing over lees has been traditionally used to produce white wines [54,55], but red wines can undergo this process too, and its repercussion in the sensory profile of the wine is nowadays better understood. The autolytic release of yeast cell wall polysaccharides in wine making and their contribution to aspects such as mouthfeel and tactile properties, wine aroma, body, and physicochemical stability has been widely studied, discussed, and demonstrated [56]. In particular, mannoproteins act as colloidal stabilizers, having a positive effect on tartrate stability [57], decreasing tannin aggregation and precipitation [58], and protecting wine form protein haze in white wines [59,60]. The production and release of these macromolecules depends on the yeast strain, and according to Vidal et al. in 2003 [61], they can reach up to 35% of total polysaccharides in wine.

Ageing over lees of red wines has been usually described as a technique that helps to stabilize colour and modulate astringency. Several interactions between yeast lees and phenols could take place simultaneously, and factors such as yeast lees reactivity towards oxygen [62,63], the strain capacity to absorb anthocyanins [43,64,65], or the presence of β-glucosidase activity explain why contradictory results and hypotheses can be easily found in the literature.

The autolytic release of polysaccharides by *Schizosaccharomyces pombe* was studied for the first time in synthetic media by Palomero et al., in 2009 [46]. From the first moment, results showed a high quantity of cell wall polysaccharides. At 28 days, both osmophilic yeasts *Schizosaccharomyces pombe* and *Saccharomycodes ludwigii* released cell-wall fragments in concentrations more than ten times greater than those produced by the *Saccharomyces* and *Pichia* strains studied. Two months into the over lees ageing process, the polysaccharide concentrations of the osmophilic yeast autolysates were those that would be reached at six or seven months by *Saccharomyces*. This rapid release could entail an important competitive advantage for wineries due to the reduction of ageing periods. Besides this, an early elution peak was observed in the HPLC-RI chromatograms corresponding to biopolymers of over 788 kDa. These fragments are therefore larger than most of those observed from *Saccharomyces* and should be studied in order to understand its potential oenological interest to modulate astringency, improve palatability, and preserve colour [46]. Domizio et al. in 2017 [66] have recently found larger polysaccharide molecules in earlier peaks by HPLC-RI when studying and characterizing *S. pombe* polysaccharides but were not detected in the chromatograms of the corresponding *S. japonicus*. These authors obtained similar results when studying the release of polysaccharides during alcoholic fermentation. They conclude that all the *Schizosaccharomyces* strains studied released a quantity of polysaccharides approximately 3 to 7 times higher than that released by a commercial *Saccharomyces cerevisiae*. Subsequent studies have obtained similar results, underlining the polysaccharide overproduction of yeast belonging to the *Schizosaccharomyces* genus [67].

5. Bio-Tool for Ensuring Wine Safety

The metabolic characteristics of *S. pombe* make it the ideal bio-tool to reduce the content of certain unwanted compounds in musts and wines, either because they are responsible for long-term microbiological instability, because they adversely affect the sensory quality of the wine, or because they can lead to toxic compounds [10]. Some examples of compounds that can be controlled by *S. pombe* are malic acid, biogenic amines, gluconic acid, and ethyl carbamate.

The ability of *S. pombe* to metabolize malic acid was previously discussed (see section "Wine pH modulation" above). This organic acid is not only a source of microbiological instability in the wine but also contributes a hard and green acidity.

Biogenic amines may pose health problems. An advantage of using *S. pombe* instead of lactic acid bacteria (LAB) for the elimination of malic acid is that this yeast does not promote the formation of biogenic amines, a health issue associated with the traditional malolactic fermentation performed by LAB [68,69]. It is worth mentioning that, unlike wild malolactic bacteria, the available commercial LABs have been selected according to the criterion of low production of biogenic amines and ethyl carbamate. The combination of *Lachancea thermotolerans* and *S. pombe* for the sequential fermentation of a red must with high pH has been shown to be effective in the control of the biogenic amines formation, with a reduction in the concentration of histamine of up to four times with respect to the same grape must that underwent a malolactic fermentation [68]. In addition, the organoleptic quality of the wine was not compromised by the employment of two non-*Saccharomyces* yeasts for the fermentation.

Gluconic acid decreases the microbiological stability of the wine, since LAB can metabolize it, increasing the acetic acid levels and, therefore, damaging the quality of the wine [70]. High contents of this acid are related to the development of the fungal disease called gray rot. It has been shown that *S. pombe* can metabolize gluconic acid and in this way favour the biological aging of wines by preventing the development of LAB [71].

Despite belonging to the ascomycetes group, *S. pombe* exhibits a strong urease activity [72]. Urease catalyses the hydrolysis of urea into ammonia and carbamate; the latter product is spontaneously hydrolysed to carbonic acid and ammonia under oenological conditions. Ethyl carbamate is a well-known carcinogen produced during fermentation and ageing, and urea is one of its precursors [73]. The safety limits established for ethyl carbamate in wines widely vary between 15 and 60 μg/L depending on the country and the type of wine, with dessert wines constituting the upper limit. Therefore, fermenting with *S. pombe* may prevent ethyl carbamate production.

6. Sparkling Wines and Other Fermented Beverages (Ice Wines, Beers)

In the winemaking of sparkling wines, there is a second fermentation in bottle in which it is necessary to use a yeast that can ferment in the presence of around 10% *v/v* ethanol (base wine). In addition to complying with this requirement, *S. pombe* can be an interesting species for this process due to its high release rate of polysaccharides during ageing on lees [74]. The higher the polysaccharide content in the wine, the better the mouth-feel sensations (reducing astringency, enhancing sweetness and roundness) and the aromatic persistence and quality [75]. It also has a protective effect on wine colour.

Moreover, the employment of *S. pombe* for the second fermentation and ageing on lees of sparkling wines production was suggested to obtain differentiation [76]. With the use of *S. pombe*, red sparkling wines with higher pyranoanthocyanin concentrations and higher colour intensity were obtained [76]. Concerning the sensory evaluation, both white and red wines were rated as high-quality without notable differences in relation to the control of *S. cerevisiae* in taste characters but with some differences at the aromatic and visual level. *S. pombe* seems more suitable for winemaking in red than in white, partially losing the fruity and floral character in the latter. The red sparkling wines made with *S. pombe* were the ones that obtained the highest score for the colour intensity and aromatic intensity. Although the herbal, buttery, and yeasty notes stood out significantly, the wines had a good balance on the nose, and no aromatic defects were perceived. However, these results come from a single strain of *S. pombe* (selected for its good fermentative behaviour); it would certainly be interesting to try new strains.

S. pombe was also tested as yeast responsible for the bottle fermentation in brewing [13]. The amount of sugar metabolized in beer production is markedly lower than in wine and, therefore, the amount of acetic acid synthesized by *S. pombe* should also be lower. *S. pombe* also has potential application in the production of ice wines, especially when it comes to obtaining wines with a better balance of acidity [77].

7. Conclusions

Schizosaccharomyces pombe is a useful tool for total or partial deacidification of grape musts and achieves high quality in the final wine when immobilization techniques are used. This species has shown promising results regarding the production of stable colour forms, thereby ensuring colour preservation in aged wines. Its use in winemaking can also prevent potential health risks associated with the metabolism of the lactic acid bacteria responsible for malolactic fermentation. As a general conclusion, due to its great resistance to pH, temperature, and preservatives, and its ability to ferment in media with high sugar content, *S. pombe* is a strain of great versatility with potential utility not only in oenology but also in other sectors of the food industry.

Author Contributions: I.L.: literature review, writing, and editing; A.M.: literature review, writing, and images design; F.P.: literature review and writing; C.G.: critical reading; and J.A.S.-L.: critical reading.

Conflicts of Interest: The authors declare no conflicts of interest.

References

1.	Kurtzman, C.P.; Fell, J.W.; Boekhout, T. *The Yeasts: A Taxonomic Study*; Elsevier: London, UK, 2010.
2.	Atilgan, E.; Magidson, V.; Khodjakov, A.; Chang, F. Morphogenesis of the fission yeast cell through cell wall expansion. *Curr. Biol.* **2015**, *25*, 2150–2157. [CrossRef] [PubMed]
3.	Vaughan Martini, A. Evaluation of phylogenetic relationships among fission yeast by nDNA/nDNA reassociation and conventional taxonomic criteria. *Yeast* **1991**, *7*, 73–78. [CrossRef] [PubMed]

4. Krapp, A.; Del Rosario, E.C.; Simanis, V. The role of *Schizosaccharomyces pombe* dma1 in spore formation during meiosis. *J. Cell Sci.* **2010**, *123*, 3284–3293. [CrossRef] [PubMed]

5. Benito, S.; Palomero, F.; Morata, A.; Calderón, F.; Suárez-Lepe, J.A. New applications for *Schizosaccharomyces pombe* in the alcoholic fermentation of red wines. *Int. J. Food Sci. Technol.* **2012**, *47*, 2101–2108. [CrossRef]

6. Moreno, S.; Klar, A.; Nurse, P. Molecular genetic analysis of fission yeast *Schizosaccharomyces pombe*. *Methods Enzymol.* **1991**, *194*, 795–823. [CrossRef] [PubMed]

7. Benito, S.; Galvez, L.; Palomero, F.; Calderón, F.; Morata, A.; Palmero, D.; Suárez-Lepe, J.A. Schizosaccharomyces selective differential media. *Afr. J. Microbiol. Res.* **2013**, *7*, 3026–3036. [CrossRef]

8. Teoh, A.L.; Heard, G.; Cox, J. Yeast ecology of Kombucha fermentation. *Int. J. Food Microbiol.* **2004**, *95*, 119–126. [CrossRef] [PubMed]

9. Petersen, J.; Russell, P. Growth and the environment of *Schizosaccharomyces pombe*. *Cold Spring Harb. Protoc.* **2016**, *3*, 210–226. [CrossRef]

10. Suárez-Lepe, J.A.; Palomero, F.; Benito, S.; Calderón, F.; Morata, A. Oenological versatility of *Schizosaccharomyces* spp. *Eur. Food Res. Technol.* **2012**, *235*, 375–383. [CrossRef]

11. Escott, C.; Loira, I.; Morata, A.; Bañuelos, M.A.; Suárez-Lepe, J.A. Wine spoilage yeasts: Control strategy. In *Yeast-Industrial Applications*; InTech: London, UK, 2017; pp. 89–116.

12. Rankine, B.C. The importance of yeasts in determining the composition and quality of wines. *Vitis* **1968**, *7*, 22–49.

13. Callejo, M.J.; González, C.; Morata, A. Use of non-Saccharomyces yeasts in bottle fermentation of aged beers. In *Brewing Technology*; InTech: London, UK, 2017; pp. 101–119.

14. Proenol ProMalic. Biological Deacidification of Musts and Wines. Available online: https://www.proenol.com/web/produtos/leveduras-encapsuladas/promalic-detail (accessed on 22 August 2018).

15. Ramon-Portugal, F.; Silva, S.; Taillandier, P.; Strehaiano, P. Inmovilización de Levaduras. Usos Enológicos Actuales. Vinidea.net., 2003; pp. 1–8. Available online: https://www.infowine.com/intranet/libretti/libretto922-01-1.pdf (accessed on 22 August 2018).

16. Silva, S.; Ramón-Portugal, F.; Andrade, P.; Abreu, S.; de Fatima Texeira, M.; Strehaiano, P. Malic acid consumption by dry immobilized cells of *Schizosaccharomyces pombe*. *Am. J. Enol. Vitic.* **2003**, *54*, 50–55.

17. Su, J.; Wang, T.; Wang, Y.; Li, Y.-Y.; Li, H. The use of lactic acid-producing, malic acid-producing, or malic acid-degrading yeast strains for acidity adjustment in the wine industry. *Appl. Microbiol. Biotechnol.* **2014**, *98*, 2395–2413. [CrossRef] [PubMed]

18. Rankine, B.C. Decomposition of L-malic acid by wine yeasts. *J. Sci. Food Agric.* **1966**, *17*, 312–316. [CrossRef] [PubMed]

19. Gallander, J.F. Deacidification of eastern table wines with *Schizosaccharomyces pombe*. *Am. J. Enol. Vitic.* **1977**, *28*, 65–68.

20. Kluyver, A.J. Biochemische Suikerbepalingen. Ph.D. Thesis, Delft University of Technology, Delft, The Netherlands, January 1914.

21. Volschenk, H.; van Vuuren, H.J.J.; Viljoen–Bloom, M. Malo-ethanolic fermentation in *Saccharomyces* and *Schizosaccharomyces*. *Curr. Genet.* **2003**, *43*, 379–391. [CrossRef] [PubMed]

22. Seo, S.-H.; Rhee, C.-H.; Park, H.-D. Degradation of malic acid by *Issatchenkia orientalis* KMBL 5774, an acidophilic yeast strain isolated from Korean grape wine pomace. *J. Microbiol.* **2007**, *45*, 521–527. [PubMed]

23. Hong, S.K.; Lee, H.J.; Park, H.J.; Hong, Y.A.; Rhee, I.K.; Lee, W.H.; Choi, S.W.; Lee, O.S.; Park, H.D. Degradation of malic acid in wine by immobilized *Issatchenkia orientalis* cells with oriental oak charcoal and alginate. *Lett. Appl. Microbiol.* **2010**, *50*, 522–529. [CrossRef] [PubMed]

24. Clemente-Jimenez, J.M.; Mingorance-Cazorla, L.; Martínez-Rodríguez, S.; Las Heras-Vázquez, F.J.; Rodríguez-Vico, F. Molecular characterization and oenological properties of wine yeasts isolated during spontaneous fermentation of six varieties of grape must. *Food Microbiol.* **2004**, *21*, 149–155. [CrossRef]

25. Zott, K.; Claisse, O.; Lucas, P.; Coulon, J.; Lonvaud-Funel, A.; Masneuf-Pomarede, I. Characterization of the yeast ecosystem in grape must and wine using real-time PCR. *Food Microbiol.* **2010**, *27*, 559–567. [CrossRef] [PubMed]

26. Kim, D.-H.; Hong, Y.-A.; Park, H.-D. Co-fermentation of grape must by *Issatchenkia orientalis* and *Saccharomyces cerevisiae* reduces the malic acid content in wine. *Biotechnol. Lett.* **2008**, *30*, 1633–1638. [CrossRef] [PubMed]

27. Comitini, F.; Gobbi, M.; Domizio, P.; Romani, C.; Lencioni, L.; Mannazzu, I.; Ciani, M. Selected non-Saccharomyces wine yeasts in controlled multistarter fermentations with *Saccharomyces cerevisiae*. *Food Microbiol.* **2011**, *28*, 873–882. [CrossRef] [PubMed]

28. Volschenk, H.; van Vuuren, H.J.J.; Viljoen-Bloom, M. Malic acid in wine: Origin, function and metabolism during vinification. *S. Afr. J. Enol. Vitic.* **2006**, *27*, 123–136. [CrossRef]

29. Redzepovic, S.; Orlic, S.; Majdak, A.; Kozina, B.; Volschenk, H.; Viljoen-Bloom, M. Differential malic acid degradation by selected strains of *Saccharomyces* during alcoholic fermentation. *Int. J. Food Microbiol.* **2003**, *83*, 49–61. [CrossRef]

30. Yokotsuka, K.; Otaki, A.; Naitoh, A.; Tanaka, H. Controlled simultaneous deacidification and alcohol fermentation of a high-acid grape must using two immobilized yeasts, *Schizosaccharomyces pombe* and *Saccharomyces cerevisiae*. *Am. J. Enol. Vitic.* **1993**, *44*, 371–377.

31. Snow, P.G.; Gallander, J.F. Deacidification of white table wines through partial fermentation with *Schizosaccharomyces pombe*. *Am. J. Enol. Vitic.* **1979**, *30*, 45–48.

32. Morata, A.; González, C.; Tesfaye, W.; Loira, I.; Suárez-Lepe, J.A. Maceration and fermentation. New technologies to increase extraction. In *Red Wine Technology*; Morata, A., Ed.; Elsevier-Academic Press: Cambridge, MA, USA, 2018.

33. Khoo, H.E.; Azlan, A.; Tang, S.T.; Lim, S.M. Anthocyanidins and anthocyanins: Colored pigments as food, pharmaceutical ingredients, and the potential health benefits. *Food Nutr. Res.* **2017**, *61*, 1361779. [CrossRef] [PubMed]

34. Morata, A.; Loira, I.; Suárez-Lepe, J.A. Influence of yeasts in wine colour. In *Grape and Wine Biotechnology*; Morata, A., Loira, I., Eds.; InTech: London, UK, 2016; pp. 288–289.

35. Morata, A.; Benito, S.; Loira, I.; Palomero, F.; González, M.C.; Suárez-Lepe, J.A. Formation of pyranoanthocyanins by *Schizosaccharomyces pombe* during the fermentation of red must. *Int. J. Food Microbiol.* **2012**, *159*, 47–53. [CrossRef] [PubMed]

36. Manzanares, P.; Rojas, V.; Genovés, S.; Vallés, S. A preliminary search for anthocyanin-β-D-glucosidase activity in non-*Saccharomyces* wine yeasts. *Int. J. Food Sci. Technol.* **2000**, *35*, 95–103. [CrossRef]

37. Morata, A.; Gómez-Cordovés, M.C.; Calderón, F.; Suárez, J.A. Effects of pH, temperature and SO$_2$ on the formation of pyranoanthocyanins during red wine fermentation with two species of *Saccharomyces*. *Int. J. Food Microbiol.* **2006**, *106*, 123–129. [CrossRef] [PubMed]

38. Morata, A.; González, C.; Suárez-Lepe, J.A. Formation of vinylphenolic pyranoanthocyanins by selected yeasts fermenting red grape musts supplemented with hydroxycinnamic acids. *Int. J. Food Microbiol.* **2007**, *116*, 144–152. [CrossRef] [PubMed]

39. Suárez-Lepe, J.A.; Morata, A. Capítulo IV.-El atractivo visual del color del vino tinto: Implicaciones microbiológicas en nuevas formas estables y pérdidas de antocianos durante la fermentación. In *Levaduras Para Vinificación en tinto [Yeasts for Red Winemaking]*; Tecnovino: Bilbao, Spain, 2015.

40. Escott, C.; Del Fresno, J.M.; Loira, I.; Morata, A.; Tesfaye, W.; González, M.D.C.; Suárez-Lepe, J.A. Formation of polymeric pigments in red wines through sequential fermentation of flavanol-enriched musts with non-Saccharomyces yeasts. *Food Chem.* **2018**, *239*. [CrossRef] [PubMed]

41. Escott, C.; Morata, A.; Loira, I.; Tesfaye, W.; Suarez-Lepe, J.A. Characterization of polymeric pigments and pyranoanthocyanins formed in microfermentations of non-Saccharomyces yeasts. *J. Appl. Microbiol.* **2016**, *121*, 1346–1356. [CrossRef] [PubMed]

42. Dallas, C.; Ricardo-da-Silva, J.M.; Laureano, O. Products formed in model wine solutions involving anthocyanins, procyanidin B2, and acetaldehyde. *J. Agric. Food Chem.* **1996**, *44*, 2402–2407. [CrossRef]

43. Morata, A.; Gómez-Cordovés, M.C.; Suberviola, J.; Bartolomé, B.; Colomo, B.; Suárez, J.A. Adsorption of anthocyanins by yeast cell walls during the fermentation of red wines. *J. Agric. Food Chem.* **2003**, *51*, 4084–4088. [CrossRef] [PubMed]

44. Morata, A.; Gómez-Cordovés, M.C.; Colomo, B.; Suárez, J.A. Cell wall anthocyanin adsorption by different *Saccharomyces* strains during the fermentation of *Vitis vinifera* L. cv Graciano grapes. *Eur. Food Res. Technol.* **2005**, *220*, 341–346. [CrossRef]

45. Caridi, A.; Sidari, R.; Kraková, L.; Kuchta, T.; Pangallo, D. Assessment of color adsorption by yeast using grape skin agar and impact on red wine color. *OENO One* **2015**, *49*, 195–203. [CrossRef]

46. Palomero, F.; Morata, A.; Benito, S.; Calderón, F.; Suárez-Lepe, J.A. New genera of yeasts for over-lees aging of red wine. *Food Chem.* **2009**, *112*, 432–441. [CrossRef]

47. Kreger-Van Rij, N.J.W. *The Yeasts: A Taxonomic Study*; Elsevier: New York, NY, USA, 1984.

48. Lodder, J. *The Yeasts; a Taxonomic Study*; North-Holland Pub. Co.: Amsterdam, The Netherlands, 1970.

49. Manners, D.J.; Meyer, M.T. The molecular structures of some glucans from the cell walls of *Schizosaccharomyces pombe*. *Carbohydr. Res.* **1977**, *57*, 189–203. [CrossRef]

50. Bartnicki-Garcia, S. Cell Wall Chemistry, Morphogenesis, and Taxonomy of Fungi. *Annu. Rev. Microbiol.* **1968**, *22*, 87–108. [CrossRef] [PubMed]

51. Weijman, A.C.M.; Golubev, W.I. Carbohydrate patterns and taxonomy of yeast and yeast-like fungi. In *The Expanding Realm of Yeast-Like Fungi*; Hoong, G.S., Smith, M.T., Weijman, A.C.M., Eds.; Elsevier: Amsterdam, The Netherlands, 1987; pp. 361–371.

52. Kopecká, M.; Fleet, G.H.; Phaff, H.J. Ultrastructure of the Cell Wall of *Schizosaccharomyces pombe* Following Treatment with Various Glucanases. *J. Struct. Biol.* **1995**, *114*, 140–152. [CrossRef] [PubMed]

53. Maclean, N. Electron microscopy of a fission yeast, *Schizosaccharomyces pombe*. *J. Bacteriol.* **1964**, *88*, 1459–1466. [PubMed]

54. Vivas, N.; Vivas de Gaulejac, N.; Nonier, M.F.; Nedjma, M. Les phénomènes colloïdaux et l'interêt des lies dans l'élevage des vins rouges: Une nouvelle approche technologique et méthodologique. 2° partie-Méthodes destinés aux élevages en cuves de grande capacité. *Revue Française D'oenologie* **2001**, *190*, 32–35.

55. Vivas, N.; Vivas de Gaulejac, N.; Nonier, M.F.; Nedjma, M. Les phénomènes colloïdaux et l'interêt des lies dans l'élevage des vins rouges: Une nouvelle approche technologique et méthodologique. 1° partie-Methodes traditionnelles d'élevage sur lie destinés aux vins en fûts. *Revue Française D'oenologie* **2001**, *189*, 33–38.

56. Pérez-Serradilla, J.A.; de Castro, M.D.L. Role of lees in wine production: A review. *Food Chem.* **2008**, *111*, 447–456. [CrossRef] [PubMed]

57. Lubbers, S.; Leger, B.; Charpentier, C.; Feuillat, M. Effet colloide-protecteur d'extraits de parois de levures sur la stabilité tartrique d'une solution hydro-alcoolique modele. *J. Int. des Sci. la Vigne du Vin* **1993**, *27*, 13–22. [CrossRef]

58. Riou, V.; Vernhet, A.; Doco, T.; Moutounet, M. Aggregation of grape seed tannins in model wine-Effect of wine polysaccharides. *Food Hydrocoll.* **2002**, *16*, 17–23. [CrossRef]

59. Moine-Ledoux, V.; Dubourdieu, D. An invertase fragment responsible for improving the protein stability of dry white wines. *J. Sci. Food Agric.* **1999**, *79*, 537–543. [CrossRef]

60. Waters, E.J.; Pellerin, P.; Brillouet, J.-M. A Saccharomyces mannoprotein that protects wine from protein haze. *Carbohydr. Polym.* **1994**, *23*, 185–191. [CrossRef]

61. Vidal, S.; Williams, P.; Doco, T.; Moutounet, M.; Pellerin, P. The polysaccharides of red wine: Total fractionation and characterization. *Carbohydr. Polym.* **2003**, *54*, 439–447. [CrossRef]

62. Frankel, E.N.; German, J.B.; Kinsella, J.; Parks, E.; Kanner, J.E. Inhibition of oxidation of human low-density lipoprotein by phenolic substances in red wine. *Lancet* **1993**, *341*, 454–457. [CrossRef]

63. Salmon, J.M.; Fornairon-Bonnefond, C.; Mazauric, J.P. Interactions between Wine Lees and Polyphenols: Influence on Oxygen Consumption Capacity during Simulation of Wine Aging. *J. Food Sci.* **2002**, *67*, 1604–1609. [CrossRef]

64. Mazauric, J.-P.; Salmon, J.-M. Interactions between Yeast Lees and Wine Polyphenols during Simulation of Wine Aging: I. Analysis of Remnant Polyphenolic Compounds in the Resulting Wines. *J. Agric. Food Chem.* **2005**, *53*, 5647–5653. [CrossRef] [PubMed]

65. Mazauric, J.-P.; Salmon, J.-M. Interactions between Yeast Lees and Wine Polyphenols during Simulation of Wine Aging: II. Analysis of Desorbed Polyphenol Compounds from Yeast Lees. *J. Agric. Food Chem.* **2006**, *54*, 3876–3881. [CrossRef] [PubMed]

66. Domizio, P.; Liu, Y.; Bisson, L.F.; Barile, D. Cell wall polysaccharides released during the alcoholic fermentation by *Schizosaccharomyces pombe* and *S. japonicas*: Quantification and characterization. *Food Microbiol.* **2017**, *61*, 136–149. [CrossRef] [PubMed]

67. Romani, C.; Lencioni, L.; Gobbi, M.; Mannazzu, I.; Ciani, M.; Domizio, P. *Schizosaccharomyces japonicus*: A Polysaccharide-Overproducing Yeast to Be Used in Winemaking. *Fermentation* **2018**, *4*, 14. [CrossRef]

68. Benito, Á.; Calderón, F.; Palomero, F.; Benito, S. Combine use of selected *Schizosaccharomyces pombe* and *Lachancea thermotolerans* yeast strains as an alternative to the traditional malolactic fermentation in red wine production. *Molecules* **2015**, *20*, 9510–9523. [CrossRef] [PubMed]

69. De Fatima, M.; Centeno, F.; Palacios, A. Desacidificación biológica de mosto a través de la inoculación de levadura *Schizosaccharomyces pombe* encapsulada como alternativa a la no producción de aminas biógenas. In Proceedings of the International Symposium of Microbiology and Food Safety in Wine "Microsafetywine", Vilafranca del Penedès, Spain, 20–21 November 2007.

70. Peinado, R.A.; Moreno, J.J.; Maestre, O.; Ortega, J.M.; Medina, M.; Mauricio, J.C. Gluconic acid consumption in wines by *Schizosaccharomyces pombe* and its effect on the concentrations of major volatile compounds and polyols. *J. Agric. Food Chem.* **2004**, *52*, 493–497. [CrossRef] [PubMed]

71. Peinado, R.A.; Mauricio, J.C.; Medina, M.; Moreno, J.J. Effect of *Schizosaccharomyces pombe* on aromatic compounds in dry sherry wines containing high levels of gluconic acid. *J. Agric. Food Chem.* **2004**, *52*, 4529–4534. [CrossRef] [PubMed]

72. Lubbers, M.W.; Rodriguez, S.B.; Honey, N.K.; Thornton, R.J. Purification and characterization of urease from *Schizosaccharomyces pombe*. *Can. J. Microbiol.* **1996**, *42*, 132–140. [CrossRef] [PubMed]

73. Uthurry, C.A.; Suárez-Lepe, J.A.; Lombardero, J.; García Del Hierro, J.R. Ethyl carbamate production by selected yeasts and lactic acid bacteria in red wine. *Food Chem.* **2006**, *94*, 262–270. [CrossRef]

74. Kulkarni, P.; Loira, I.; Morata, A.; Tesfaye, W.; González, M.C.; Suárez-Lepe, J.A. Use of non-Saccharomyces yeast strains coupled with ultrasound treatment as a novel technique to accelerate ageing on lees of red wines and its repercussion in sensorial parameters. *LWT-Food Sci. Technol.* **2015**, *64*, 1255–1262. [CrossRef]

75. Loira, I.; Vejarano, R.; Morata, A.; Ricardo-da-Silva, J.M.; Laureano, O.; González, M.C.; Suárez-Lepe, J.A. Effect of Saccharomyces strains on the quality of red wines aged on lees. *Food Chem.* **2013**, *139*, 1044–1051. [CrossRef] [PubMed]

76. Ivit, N.N.; Loira, I.; Morata, A.; Benito, S.; Palomero, F.; Suárez-Lepe, J.A. Making natural sparkling wines with non-Saccharomyces yeasts. *Eur. Food Res. Technol.* **2018**, *244*, 925–935. [CrossRef]

77. Wang, J.; Li, M.; Li, J.; Ma, T.; Han, S.; Morata, A.; Suárez Lepe, J.A. Biotechnology of ice wine production. In *Advances in Biotechnology for Food Industry*; Elsevier: New York, NY, USA, 2018; pp. 267–300.

fermentation

MDPI

Review

The Yeast *Torulaspora delbrueckii*: An Interesting But Difficult-To-Use Tool for Winemaking

Manuel Ramírez * and Rocío Velázquez

Departamento de Ciencias Biomédicas (Área de Microbiología), Facultad de Ciencias, Universidad de Extremadura, 06006 Badajoz, Spain; rociovelazquez1981@gmail.com
* Correspondence: mramirez@unex.es; Tel.: +34-924-289426

Received: 27 September 2018; Accepted: 9 November 2018; Published: 12 November 2018

Abstract: *Torulaspora delbrueckii* is probably the non-*Saccharomyces* yeast that is currently most used for winemaking. Multiple advantages have been claimed for it relative to conventional *S. cerevisiae* strains. However, many of these claimed advantages are based on results in different research studies that are contradictory or non-reproducible. The easiest way to explain these discrepancies is to attribute them to the possible differences in the behaviour of the different strains of this yeast that have been used in different investigations. There is much less knowledge of the physiology, genetics, and biotechnological properties of this yeast than of the conventional yeast *S. cerevisiae*. Therefore, it is possible that the different results that have been found in the literature are due to the variable or unpredictable behaviour of *T. delbrueckii*, which may depend on the environmental conditions during wine fermentation. The present review focusses on the analysis of this variable behaviour of *T. delbrueckii* in the elaboration of different wine types, with special emphasis on the latest proposals for industrial uses of this yeast.

Keywords: *Torulaspora delbrueckii*; winemaking; yeast inoculation; yeast dominance; wine quality; genetic improvement

---

1. Introduction

Of the non-*Saccharomyces* yeasts, *Torulaspora delbrueckii* is probably the most suitable for use in winemaking. This is because it has a good fermentation performance compared to other non-*Saccharomyces* yeasts that might be considered for winemaking, such as *Hanseniaspora uvarum*, *H. vineae*, *Candida zemplinina*, *C. pulcherrima*, *C. stellata*, *Schizosaccharomyces pombe*, *Hansenula anomala*, *Metschnikowia pulcherrima*, *Lachancea thermotolerans*, *Pichia fermentans*, *P. kluyveri*, and *Kazachstania aerobia*. Additionally, it has been claimed that *T. delbrueckii* can be used to optimise some wine parameters with respect to usual *S. cerevisiae* wines such as a low amount of acetic acid, lower ethanol concentration, increased amount of glycerol, greater mannoprotein and polysaccharide release, promoted malolactic fermentation, increased amounts of interesting aromatic compounds (fruity esters, lactones, thiols, and terpenes), and decreased amounts of unwanted aromatic compounds (such as higher alcohols); these may improve wine quality or complexity (reviewed by Benito in [1]). These additional features constitute the main reason why modern œnologists are conducting trials with this yeast, as well as with other non-*Saccharomyces* yeasts, for industrial winemaking. However, this option entails additional complications and economic costs in controlling must fermentation with respect to the use of conventional selected strains of *S. cerevisiae*, which is undoubtedly the most reliable yeast species for this purpose. Some of the complications of using *T. delbrueckii* in commercial wineries come from its physiological properties under the stressing conditions that are usual in the winemaking process. In this work, we review these specific properties, with a focus on their relevance for some applied aspects of winemaking.

2. The *Torulaspora* Yeasts

Torulaspora are fermentative yeasts that can be found in wild and anthropic habitats, where they may coincide with other fermentative yeasts such as *Saccharomyces* and *Zygosaccharomyces*. In the past, there has been some misclassification of some species in these three genera, because they share various morphological and physiological features. The genus *Torulaspora* includes at least six species: *T. delbrueckii* (anamorph *Candida colliculosa*), *T. franciscae*, *T. pretoriensis*, *T. microellipsoides*, *T. globosa*, and *T. maleeae* [2]. Other species have also been proposed for inclusion in this genus following their characterisation by molecular tools used to discriminate new isolates, examples being *T. indica* [3] and *T. quercuum* [4]. The taxonomy of *Torulaspora* is changing rapidly, and one can expect species reassignments and new species in the near future. In this sense, based on the nucleotide sequence of the internal transcribed spacer (ITS) region located between the 18S and 28S rRNA genes, it has already been proposed that *Zygosaccharomyces mrakii* and *Z. microellipsoides* might be reclassified in the genus *Torulaspora* [5]. Similarly, four strains presumed to be *T. delbrueckii* were found to be considerably different from the type strain, and were reclassified into the genera *Debaryomyces* and *Saccharomyces* [6]. In particular, one must bear in mind that some error in the presumptive assignment of a new isolate to *Torulaspora* may explain some of the controversial results found for the biotechnological properties that have been recently claimed for these yeasts (see Section 5 below).

Torulaspora cells mostly have a spherical shape (*torulu*), although ovoid and ellipsoidal shapes are also frequent. Its cellular size, 2–6 × 3–7 μm, uses to be smaller than that of *S. cerevisiae* (Figure 1A,E). *Torulaspora* may rarely produce pseudohyphae, but never true hyphae. Its life cycle has yet to be elucidated. In contrast with wild *S. cerevisiae* strains, which are diploids or polyploids during vegetative propagation, *T. delbrueckii* was long believed to be a haploid yeast, mainly because of its small cell size [7], and because tetrads are absent or rarely observed in sporulation media. However, it has recently been suggested that *T. delbrueckii* may actually be mostly diploid and homothallic [8]. If this is the case, the small cell size of *Torulaspora* may still be explained by its having only 16 chromosomes in the diploid stage, instead of the 32 chromosomes that are present in *S. cerevisiae* diploid yeasts. It is believed that *T. delbrueckii* diverged from the *S. cerevisiae* lineage before the latter underwent its genome polyploidisation [9].

All species of *Torulaspora* reproduce asexually by multilateral budding (mitosis). Sexual reproduction may occur in sporulation media through asci that contain one to four spherical ascospores (Figure 1B). The ascus can be originated by the heterogamic conjugation between a cell and its bud, or between two independent cells. Three types of conjugation have been proposed in the *Torulaspora* genus [10]: (i) between mother and daughter cells (the daughter cell may be one small bud still attached to the mother cell, but separated from it by a cross-wall); (ii) between two cells with their respective conjugation tubes; and (iii) between a cell with conjugation tube and another cell without it. The ascospores (one to four per ascus, frequently two, 2–4 μm diameter) are spherical with a smooth or warty cell wall. It should be noted that starvation is not required for *T. delbrueckii* to sporulate [8,11]; i.e., spores can be formed on rich media such as yeast extract peptone dextrose (YEPD) agar (Figure 1D), as also may occur for some wine strains of *S. cerevisiae*.

Cells with conjugation tubes are frequent in *T. delbrueckii*, mainly in non-growing cells in a stationary growth stage or in sporulation media (Figure 1B), but most of them seem uninvolved in the yeast conjugation process [12]. Probably, most of the several buds that have been frequently observed in single non-growing yeasts are actually incipient cell wall protuberances destined to potentially become a conjugation tube for yeast mating.

Given this lack of precise knowledge about the life cycle of *Torulaspora*, it is complicated to design strategies for the biotechnological improvement of *T. delbrueckii* by using classical genetic techniques such as those proposed for *S. cerevisiae* wine yeasts [13–16].

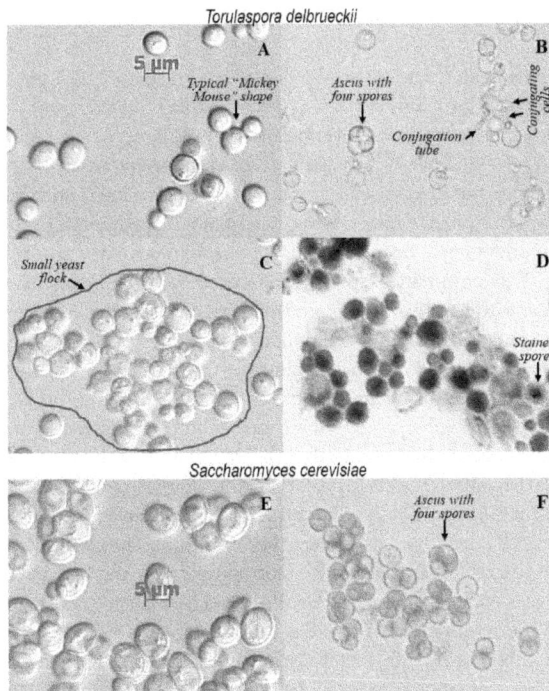

Figure 1. Vegetative cells (**A,C,E**) and spores (**B,D,F**) of *T. delbrueckii* and *S. cerevisiae* for comparison. Photomicrographs were taken using Nomarski (vegetative cells) or bright-field illumination (spores) with a Nikon Eclipse 600 microscope equipped with a 60× objective. (**B,F**) Spores from Minimal Sporulation Medium. (**D**) Spores from YEPD agar stained with malachite green.

3. Growth and Fermentation Capability

With respect to the features required to perform industrial alcoholic fermentation, of the non-*Saccharomyces* yeasts, *T. delbrueckii* is the most similar to the referent, best-considered, *S. cerevisiae*. This is the main reason why *T. delbrueckii* was the first non-*Saccharomyces* yeast proposed for industrial use in wine fermentation [17]. However, slight but relevant differences in the physiological characteristics of the two yeasts affect their choice as options for different industrial applications. In particular, the specific rates of CO_2 production and O_2 consumption are higher in *T. delbrueckii* than in *S. cerevisiae*, which results in lower biomass yields from batch cultures, and may represent a handicap for the large-scale production of *T. delbrueckii* with respect to *S. cerevisiae* [18]. The two yeasts' sugar utilisation and regulation patterns are very similar, with similar biomass yields in glucose-limited oxygen-sufficient chemostat cultures, as is consistent with fully respiratory growth. Interestingly, *S. cerevisiae* is the first to switch to a respirofermentative metabolism as the oxygen feed rate decreases, showing a lower biomass yield at low oxygen tensions with respect to *T. delbrueckii*, which is still able to sustain full respiration under these conditions [19]. Unfortunately, however, *T. delbrueckii* shows less growth than *S. cerevisiae* under strict anaerobic conditions [20,21], and has less fermentation vigour and slower growth rate than *S. cerevisiae* under usual wine fermentation conditions, being quickly overcome by wild or inoculated *S. cerevisiae* strains [18,22]. Obviously, this is an important issue for winemaking, which is usually done under strict anaerobic conditions for white and sparkling wines, or in the presence of very low amounts of oxygen for red wines, where the cap of grape skins is punched down into the must several times a day. A poorer performance of *T. delbrueckii* with respect to *S. cerevisiae* should always be expected under these working conditions, especially for white and sparkling wines.

When inoculated in fresh grape must, the dominance ratio of *T. delbrueckii* depends on many factors: the size of the inoculum, amount and types of viable wild microbes initially present in the must, fermentation stage, sugar and ethanol concentrations, killer phenotype and killer sensitivity of the inoculated yeast and the wild yeasts, SO_2 (sulfur dioxide) concentration, Cu (copper) concentration, pesticides coming in with the grapes, etc. Some of these have been analysed in several related works. The *T. delbrueckii* dominance ratio during the must fermentation reported in the literature of specialised research works ranges from 0 to 100%, being in most cases around 50% in tumultuous fermentation [22–27]. This dominance ratio can be expected to be lower in commercial wineries, where the working conditions are not as controlled as in research laboratories or experimental wineries. This ratio is much lower than that which is expected for the reference yeast *S. cerevisiae*, which is assumed to be close to 100%, even considering that this yeast is usually inoculated at a lower concentration (around 10^6 CFU/mL) than *T. delbrueckii* (around 10^7 CFU/mL).

A strong dominance of *T. delbrueckii* is usually achieved by using sterile or much clarified must, and a killer strain for inoculation. In this way, the relative initial growth of the *T. delbrueckii* population is high, and favours its becoming the clear protagonist during fermentation. Nonetheless, since this yeast is less resistant than *S. cerevisiae* to high ethanol concentrations (Figure 2), the fermentation rate slows down, and cell death increases after the tumultuous fermentation of sugar-rich musts (Figure 4). As a consequence, fermentation may stop, slow down and become sluggish, or continue mainly because of the presence of some contaminating wild *Saccharomyces* yeasts, which may continue must fermentation if the required nutrients are still available. Obviously, any such sequence of events will reduce the domination ratio of *T. delbrueckii* by the end of fermentation, even down to its disappearance [22,24]. This situation is more evident and occurs more rapidly when mixed-inoculating with the two yeasts (co-inoculation or sequential inoculation) in grape must be supplemented with yeast nutrients. In this case, the dominance of *T. delbrueckii*, if it occurs at all, is usually ephemeral, lasting no more than two days [18,22,24].

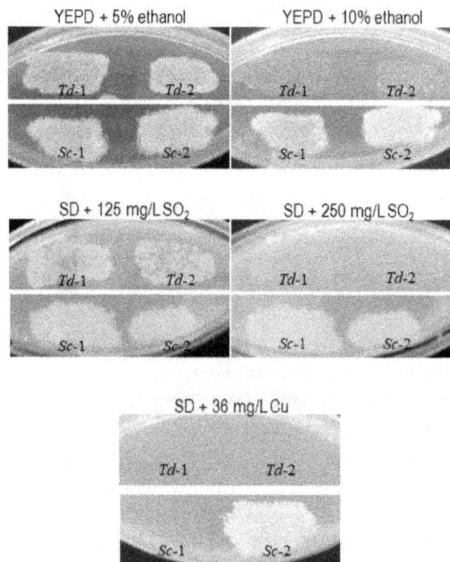

Figure 2. Resistance of *T. delbrueckii* (Td) and *S. cerevisiae* (Sc) to ethanol, SO_2, and Cu. Two strains of each yeast species are shown (Td-1 and Td-2, and Sc-1 and Sc-2, respectively). YEPD: rich YEPD medium. SD: minimal SD medium.

This situation makes it difficult to reliably assign a specific feature to *T. delbrueckii*, and may explain why some disagreement is found in the literature about the effect of this yeast on winemaking and wine quality. For example, the effects related to the production of some aroma compounds remain confusing. Mixed *T. delbrueckii/S. cerevisiae* inoculation has been described as increasing the total ester concentration (mainly isoamyl acetate and ethyl hexanoate, octanoate, and 3-hydroxybutanoate) relative to inoculation with *S. cerevisiae* [28], while the contrary has also been reported [29,30]. Such contradictory results may occur just because the production and degradation of these compounds depend on the proportion of each species of yeast during must fermentation. Likewise, even when a putative technological improvement of *T. delbrueckii* affecting the quality of wine has been accepted, it is still very complicated to achieve the desired result in commercial wineries if the dominance of this yeast cannot be ensured, and this may dissuade œnologists from spending money purchasing new non-*Saccharomyces* yeasts for must inoculation. Consequently, it is necessary to thoroughly control the level of dominance of *T. delbrueckii* throughout the wine fermentation process, analysing adequate numbers of samples and colonies of viable microorganisms sampled at different times during that process. Frequently, one finds in the literature that few samples and yeast colonies were analysed (10 colonies, only twice per fermentation [31]), or simply that the inoculated *T. delbrueckii* strain was not monitored, but was simply assumed to dominate the wine fermentation [32]. Moreover, in some cases, the authors themselves recognise the presence of contaminating yeasts in the pasteurised must; these are yeasts that, in parallel experiments in which there was no dominance of the inoculated yeast, were able to complete the fermentation. Even more so, in some cases, the fermentations were carried out by continuous mixing of the inoculated must in an orbital shaker to increase the oxygen supply [31], contrary to what is usually done in commercial cellars where the must is not continuously agitated. Therefore, in sum, the different environmental conditions used in the experimental fermentations of *T. delbrueckii*, and the uncertainty about the desired dominance of this yeast during fermentation, suggest that the observed effect of a given strain on the quality of wine may be different, or even contradictory, depending on the working conditions. Consequently, it would be necessary to establish criteria that are more precise regarding the experimental conditions and the minimum level of dominance of *T. delbrueckii* in order to obtain more homogeneous and reliable conclusions about any of the advantages of this yeast in winemaking.

In the absence of general agreement on these criteria, one possibility that might explain some of the conflicting results found is that different *T. delbrueckii* strains might have different fermentation capabilities. This would not be too surprising, given the possible lack of a reliable and definitive taxonomic classification. However, to confirm this argument, it would be necessary to test in parallel the different yeast strains under identical experimental conditions, and achieve a clear dominance of each strain tested. Moreover, in addition to the idea that the production of esters by *T. delbrueckii* might be strain-dependent, it has been proposed that this production may even vary when this yeast is associated with *S. cerevisiae* in mixed cultures [25,33,34]. This situation may become even more complicated, although clearly very interesting, when considering the use of *T. delbrueckii* for red winemaking, since this differs from white winemaking in certain aspects that affect this yeast's growth during must fermentation, and hence its effect on wine quality. Namely, oxygen availability, alcohol content, the amount of initial wild microorganisms, and nutrient availability are usually greater in red than in white wine fermentation. Given these differences, and since fermentations dominated by *T. delbrueckii* are slower than those dominated by *S. cerevisiae*, the eventual development of lactic acid bacteria may promote malolactic fermentation occurring simultaneously with alcoholic fermentation. This promotion is enhanced in situations where these bacteria may be more frequent, as when using poorly clarified musts or crushed grapes as starting material for winemaking. Indeed, we found that in such situations all of the *T. delbrueckii* wines, but none of the *S. cerevisiae* wines, underwent malolactic fermentation, and putative lactic acid bacteria were always found in the *T. delbrueckii* wines, but none or very few were found in the *S. cerevisiae* wines. The highest malic acid degradation was found in those wines where *T. delbrueckii* reached the greatest dominance ratios, and at the same time, had the slowest fermentation kinetics [23,24]. Likewise, some companies, such as Chr. Hansen (Hoersholm, Denmark), recommend inoculation with *T. delbrueckii* to promote malolactic fermentation, although in this case, the effect is claimed to be due to a decreased production of toxic medium-chain fatty acids that favour the growth of lactic acid bacteria.

I.E., the confluence of so many environmental factors in winemaking that may change the behaviour of *T. delbrueckii* with respect to the production of aroma compounds makes it difficult to design a reliable strategy for wine quality improvement in commercial wineries. In any case, until the handicap of low dominance of *T. delbrueckii* is resolved, one cannot determine with precision the real usefulness of these yeasts in the cellar. To address this problem, further knowledge is required of how *T. delbrueckii* interacts with other wine microorganisms during wine fermentation so as to be able to ensure the dominance of this yeast species.

4. Interaction with Other Fermentative Yeasts and Ability to Dominate Must Fermentation

As noted above, the ability of *S. cerevisiae* to overcome other non-*Saccharomyces* yeasts and dominate wine fermentation can be explained mainly by its excellent biological fitness and its tolerance to the stress factors that converge in this process. This circumstance makes it possible that an almost undetectable contamination of *S. cerevisiae* in a must that has been previously inoculated with *T. delbrueckii* can be responsible for carrying out most of the fermentation process. This may occur because *T. delbrueckii* dies prematurely, or because its metabolic activity declines markedly as a result of the environmental stress. However, the opposite case is clearly unlikely, so that non-*Saccharomyces* yeasts usually play an irrelevant part in the fermentations of musts inoculated with *S. cerevisiae*. Biotechnologically, this situation will not change until *T. delbrueckii* strains whose biological efficacy is similar to that of *S. cerevisiae* are available, which is an aspect that we shall return to later in this review. Apart from this, other types of direct interactions between the two yeast species have been described that can influence, or even determine, the preponderance of one yeast over the other.

The yeast interactions in pure and mixed fermentations have been the subject of a thorough review [35] which clearly stated that the competitiveness of any given strain can be influenced by many factors, which are both abiotic (pH, temperature, ethanol, osmotic pressure, nitrogen availability, molecular sulfur dioxide concentration, etc.) and biotic (type of microorganisms, presence of antimicrobial molecules, cell-to-cell contact, grape variety, etc.). This competitiveness will change depending on the specific environmental conditions of a given food fermentation, which will in turn influence the ability of a given strain to out-compete other yeasts. Moreover, in mixed fermentation, the choice of the specific strains of *T. delbrueckii* and *S. cerevisiae* can be another biotic factor to consider as an origin of variability in the competitiveness of each yeast strain. However, the use of a fixed couple of yeast strains could produce predictable results related to the dynamic of each yeast population and, thereby to the wine quality.

With respect to the biotic factors, a new wine *T. delbrueckii* killer strain has recently been characterised. It secretes a killer toxin (Kbarr-1) that is encoded in a double-stranded RNA virus (ScV-Mbarr-1) with broad antifungal activity against *S. cerevisiae* (killer and non-killer strains) and other non-*Saccharomyces* yeasts [36,37]. This negative interaction has already been applied to promote the dominance of *T. delbrueckii* in white [24] and red [23] table wines. This killer strain had the advantage of dominating must fermentation in the presence of *S. cerevisiae* relative to the non-killer isogenic strains. However, full dominance is not ensured if a fairly large population of *S. cerevisiae* yeasts (about 10^6 CFU/mL) is co-inoculated or was already present as wild yeasts in the grape must. Therefore, a large inoculum (about 10^7 CFU/mL) of *T. delbrueckii* should be considered to achieve satisfactory dominance in these situations.

Apart from this, yeast growth inhibition by cell-to-cell contact is another biotic factor that may influence the competitiveness of each strain in multi-starter fermentations. Physical contact between *S. cerevisiae* and *T. delbrueckii* seems to induce the rapid death of the latter yeast. However, when the two yeasts were separated from each other in a double-compartment fermenter, *T. delbrueckii* maintained its viability and metabolic activity [38]. The details of this mechanism of cell-to-cell contact inhibition are still unknown. It has recently been proposed that in the case of the early death of *L. thermotolerans* during anaerobic fermentation mixed-inoculated with *S. cerevisiae*, this phenomenon is caused by a combined effect of cell-to-cell contact plus antimicrobial peptides [39]. Similarly, it has been found that cell-free *S. cerevisiae* supernatants induce fast culturability loss in other non-*Saccharomyces* yeasts, meaning that this inhibitory effect is mainly caused by some metabolites secreted by *S. cerevisiae*. To further complicate the issue, the culturability loss of non-*Saccharomyces* yeasts induced by *S. cerevisiae* is species-dependent and strain-dependent [40]. The effect of these biotic factors on the competitiveness of each strain can be influenced by an abiotic factor, an example being the addition of assimilable nitrogen, which was found to partially mitigate the inhibitory interaction between different yeasts [41].

Given this situation in which many different factors may influence a given yeast's competitiveness with respect to others, and the lack of precise knowledge about the mechanisms of this influence, the degree of compatibility of co-inoculated strains of *T. delbrueckii* and *S. cerevisiae* must be known with precision in order to program and control the industrial use of that combination. This is especially important if a specific proportion of the fermentation must be performed by *T. delbrueckii* in order to obtain a determined effect on the final wine's organoleptic quality. Mixed inoculation results may not be fully satisfactory if the production of aromas by *T. delbrueckii* varies when it is associated in a variable and difficult-to-control proportion with *S. cerevisiae*, or when the type and amount of the aromas produced in mixed cultures depend on a synergistic interaction between these two yeasts [33]. If this is the case, the use of *T. delbrueckii* in mixed-inoculated fermentation could be a situation that is far too complicated to control satisfactorily at an industrial level. However, if total or major dominance of *T. delbrueckii* is desired, this can be achieved using killer strains, a large yeast inoculum (approximately 10^7 CFU/mL), and grape must containing only a discrete amount of wild *Saccharomyces* yeasts (less than 10^5 CFU/mL) [23,24,42]. In this situation, if the effect of *T. delbrueckii* on the quality of the wine is more intense than desired, this could be resolved by mixing with conventional *S. cerevisiae* wine to obtain the required wine quality.

5. Influence on the Chemical Composition and Aroma of Table Wine

As mentioned in the Introduction section, several wine parameters can be improved by inoculating the must with *T. delbrueckii*. However, some of these claimed improvements can be achieved only in certain specific circumstances. For example, the production of reduced amounts of acetic acid is usually found only in high-sugar fermentations, which are conditions where *S. cerevisiae* produces high amounts of this compound [43]. It is possible that no such effect will be achieved with usual sugar concentrations, because *T. delbrueckii*-dominated fermentation takes a long time to complete, and this may lead to the additional oxidation of the wine [23,24]. Sometimes, the claimed improvement may not be very relevant; an example is the reduction in ethanol concentration, which is usually low under anaerobic conditions, and correlates with increased acetic acid and residual sugar contents [44,45], both of which may be undesirable. Moreover, the supposed improvement may not be generally reproducible at commercial wineries, given that contradictory results have been described in experimental vinifications, as is the case for the claimed increased amount of important aromatic compounds: fruity esters, lactones, thiols, and terpenes [23–26,28–30,33,34,46,47].

These circumstances may result in eventual disappointment for œnologists attempting to improve some wine types, such as young white or rosé wines, by using *T. delbrueckii*. The *T. delbrueckii* wines may contain slightly greater amounts of volatile acidity and residual sugars, have roughly the same amount of ethanol, have decreased fresh fruity aromas, or be slightly oxidised with respect to *S. cerevisiae* wines. Considering that using *T. delbrueckii* for winemaking is more expensive and time consuming that using conventional *S. cerevisiae* yeasts, and that most of the commercially available wine strains of the latter yeast are of proven virtuosity for winemaking, most œnologists may refuse to further try *T. delbrueckii* to improve wine quality. Therefore, wine yeast researchers and producer firms should be very careful in advising very precisely those winemaking conditions that may really be appropriate for wine quality improvement by using *T. delbrueckii*, perhaps even more careful than when the original recommendations were given for inoculating with *S. cerevisiae* yeasts (about one century ago) to improve winemaking with respect to traditional spontaneous must fermentation. Otherwise, the use of *T. delbrueckii* for commercial winemaking may become just anecdotal and circumscribed to experimental research, as will probably end up being the case for many other non-conventional wine yeast species.

Given that these important prior considerations must be kept in mind, we shall now look at some specific approaches using *T. delbrueckii* in terms of its most common effects on wine quality and aroma composition, and which have so far been described as being independent of the winemaking conditions. These are decreased acetic acid production in grape must with a high sugar concentration, decreased concentrations of the main esters, and increased concentrations of lactones and some minor ethyl esters, which are all relative to *S. cerevisiae*-fermented wines. Overall, these effects have led to decreased fresh-fruit aroma intensity, but increased aroma complexity with a dried-fruit/pastry flavour, which is achieved mainly when the *T. delbrueckii* strains reached a high dominance ratio (above 70% of viable cells) at the tumultuous fermentation stage [23,24,26,29,30]. It should be noted that part of this effect on the aroma complexity of wine may be due to the growth of lactic acid bacteria [23]. According to these common effects, the use of *T. delbrueckii* for young (white or rosé) table wines should not be recommended mostly because of the loss of fresh-fruit aromas, which are usually wanted in these types of wine. However, an exception should be considered for those specific situations in which a low dominance ratio of *T. delbrueckii* does not reduce the fresh-fruit aroma intensity of the wine, or it interacts with *S. cerevisiae* to increase the amounts of specific aromas such as volatile thiols (3-sulfanylhexan-1-ol and 3-sulfanylhexyl acetate) that clearly improve the quality of certain wines, with Sauvignon Blanc being an example [33,48]. Besides this, we think that inoculation with *T. delbrueckii* can mostly be recommended for full-bodied red wines made from grapes with undesirably high malic acid content, natural sweet wines made from grapes with high sugar content, or any aged wine in which the dried-fruit aromas and oxidised character that are usually produced by this yeast may be greatly appreciated by the consumers.

6. Traditional Sparkling Wine Making

There have also recently been trials of the use of *T. delbrueckii* for sparkling wine making. The sequential inoculation of this yeast with *S. cerevisiae* has been shown to be an interesting tool with which to obtain base wines (first fermentation of the grape must) with different characteristics [22]. However, those authors did not analyse the usefulness of *T. delbrueckii* to finally complete the production of sparkling wine (second fermentation of the base wine), probably because they suspected that this yeast could not complete this task. We confirmed these results in a recent work, finding that *T. delbrueckii* could indeed be used to improve base wine quality, but was unable to complete second fermentation, because it cannot survive above 3.5 atm of CO_2 pressure [49]. However, we should point out that base wine that was single-inoculated with *T. delbrueckii* was able to complete second fermentation when there was a very small contamination (4.3×10^{-5} CFU/mL) of *Saccharomyces* yeasts, which became dominant by the time CO_2 pressure increased above 3 atm, and *T. delbrueckii* quickly became inviable (Figure 3). This clearly indicates that CO_2 pressure is another abiotic factor that may decrease the competitiveness of *T. delbrueckii* relative to other potential competitor microorganisms such as *S. cerevisiae*, and that its resistance to high pressure should be genetically improved to get a dominance of this yeast during the second fermentation of sparkling wine.

It should also be mentioned that two *T. delbrueckii* strains (DiSVA 130 and DiSVA 313) single-inoculated in a base wine with high ethanol concentration (11.65% vol) were able to complete the secondary fermentation of sparkling wine even more efficiently than single-inoculated *S. cerevisiae* [32]. This was an unexpected result, mainly because there had been no previous data reported related to the resistance of these DiSVA strains at high ethanol concentrations. Whichever the case, these authors unfortunately did not confirm the dominance of the inoculated strains during second fermentation.

Therefore, some contamination with *S. cerevisiae* yeasts may have occurred, with these being the yeast that was actually responsible for second fermentation completion, as was the case mentioned above (Figure 3).

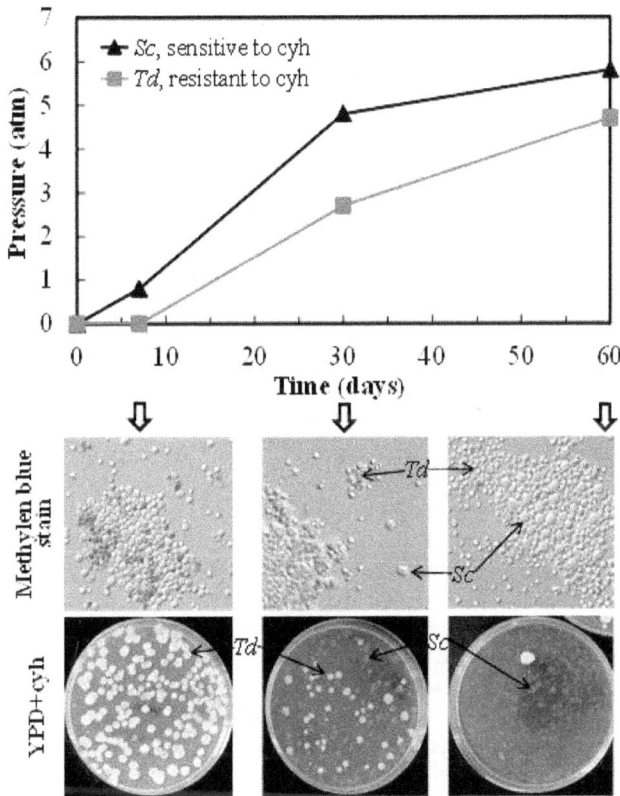

Figure 3. (**Top**) Evolution of CO_2 pressure during the second fermentation of cava wine inoculated with *S. cerevisiae* (*Sc*) sensitive to cycloheximide (cyh) and *T. delbrueckii* (*Td*) resistant to cyh. (**Bottom**) Methylene blue stain to show cell death, and cyh-resistance of viable cells (YEPD+cyh) to monitor the inoculated yeast, for the *T. delbrueckii* inoculated fermentation (7 days, 30 days, and 60 days of fermentation).

7. Proposals for Genetic Improvement of *T. delbrueckii*

The main disadvantages of *T. delbrueckii* that hinder it from becoming as good as *S. cerevisiae* for industrial use are probably its small cell size [50], slow growth rate, and low fermentation vigour under the usual alcoholic fermentation conditions [18,22,24]. These peculiarities result in small harvest quantities in industrial biomass production, and extra difficulties in recovering, washing, and dehydrating the yeast biomass using conventional microfiltration, which may take a long time because the filters become clogged by the small yeast cells, and have to be replaced frequently.

The small cell size of *T. delbrueckii* cells seems to be because it propagates vegetatively as a haploid yeast [7], while the industrial strains of *S. cerevisiae* propagate as diploid or polyploid yeasts. This cell size can be enlarged by increasing the number of chromosomes (ploidy) inducing diploids through perturbed protoplast regeneration or the UV irradiation of intact cells [51–53]. Stable diploid strains have already been constructed from baker's haploid *T. delbrueckii* strains. These diploid strains were about three times larger than the original parental haploid, while both types had similar biomass yield, stress resistance, gassing power, and sweet dough-leavening ability. The propagation and manipulation of this artificially constructed diploid seemed to be enhanced under industrial conditions, and its stress tolerance and CO_2 production after direct freezing make it possible to reduce the yeast dose in the formulation of frozen sweet bakery products [51,53]. These interesting results for baker's *T. delbrueckii* strains should serve to encourage attempts to perform similar genetic improvements with wine strains, which is a task that remains to be completed.

With regard to the properties that are desired of yeasts for them to perform industrial alcoholic fermentations efficaciously, the tolerance to high glucose concentrations (up to 700 g/L) of *S. cerevisiae* (a poor osmotolerant) was increased by fusing it with the heat-treated protoplasts of an osmotolerant *T. delbrueckii* strain. The interspecific hybrids were able to grow in high glucose concentration media, producing enhanced amounts of ethanol. Further investigation of the practical application of these hybrids is pending [54]. Other than this one-off achievement, there has been little published research related to this issue. In particular, there remains the problem of improving *T. delbrueckii* with respect to the properties of *S. cerevisiae* that make this latter yeast the best choice for most industrial alcoholic fermentations. It is accepted that *T. delbrueckii* may be more resistant than *S. cerevisiae* to various stressors such as freeze–thaw or growth in high osmolality media, which can be very important properties for a baker's yeast used in frozen and frozen sweet dough technology [52,53]. While these properties may be interesting for certain specific wine fermentations, there are other stress resistance properties that are generally considered to be of major importance in winemaking technology, such as the resistance to ethanol, SO_2, or Cu, each of which affects the grape must fermentation vigour of *T. delbrueckii*.

Based on our experience, some strains of *T. delbrueckii* can fully dominate and complete crushed grape fermentation to reach above 14° GL, which is a very high alcohol content for non-*Saccharomyces* yeast fermentation (Figure 4). However, this achievement occurs only rarely, under certain favourable environmental conditions: the presence of only very low amounts of competitor *Saccharomyces* yeasts; a very large inoculum of healthy *T. delbrueckii* cells to start must fermentation quickly and overcome the growth of any other yeasts; frequent agitation to provide extra oxygen; the addition of some extra amounts of nutrients that are usually contained in crushed grapes but which may be scarce in clarified grape must; low amounts of toxic compounds such as SO_2, Cu, pesticides, etc. Even under these favourable conditions, *T. delbrueckii*-dominated fermentation took much longer to complete than *S. cerevisiae*-dominated fermentation, because fermentation vigour and cell viability declined more quickly in *T. delbrueckii* than in *S. cerevisiae* after the wine reached about 5% alcohol (Figure 4B). The differences in ethanol resistance between the two yeast species can be seen on YEPD agar plates supplemented with different ethanol concentration (Figure 2). Similarly, while *T. delbrueckii* is able to complete fermentation in the presence of 50 mg/L SO_2 (Figure 5), although a partial lethal effect cannot be ruled out, it is clearly less resistant to this compound than *S. cerevisiae*. A concentration of 125 mg/L SO_2, which is frequently used in winemaking, is clearly lethal for *T. delbrueckii* (Figure 2). Moreover, there may be an undesirable amount of Cu in grape juice, because Cu compounds are frequently used as fungicides in the vineyard. Again, *S. cerevisiae* is more resistant to Cu than *T. delbrueckii* (Figure 2), meaning that the possible presence of high concentrations (20–30 mg/L) of Cu would reduce fermentation vigour more drastically in the latter yeast than in the former. Experiments to obtain new *T. delbrueckii* strains resistant to ethanol, SO_2, and Cu (as well as to high CO_2 pressures) are currently underway to improve the overall fermentation performance of this species of yeast to bring it as close as possible to that usually shown by selected *S. cerevisiae* wine yeasts. *A priori*, new strains with improved fermentation capacities could be isolated and selected, since there could be intraspecific variations in *T. delbrueckii* as they exist in other yeast species such as *S. cerevisiae*.

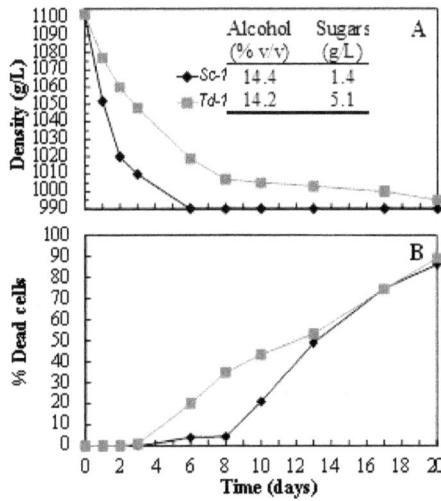

Figure 4. Must fermentation kinetics (**A**) and proportion of dead cells (**B**) of Garnacha grape fermentation inoculated with *S. cerevisiae* (*Sc*) and *T. delbrueckii* (*Td*). The concentrations of ethanol and residual sugars in the wines are given.

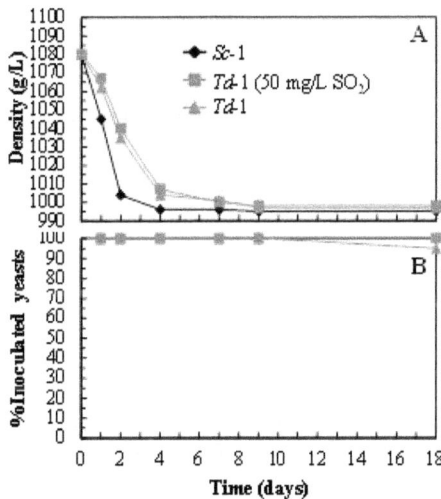

Figure 5. Must fermentation kinetics (**A**) and yeast population dynamics (**B**) of Pinot Noir grape fermentation inoculated with *S. cerevisiae* (*Sc*) and *T. delbrueckii* (*Td*) in the absence and presence of 50 mg/L SO₂.

8. Concluding Remarks

Given that *T. delbrueckii* is not as good a grape must fermenter as the selected *S. cerevisiae* wine yeasts that are currently available in the market at reasonable prices, further research effort will be required to select new *T. delbrueckii* strains or genetically improve the strains that have already been described before this yeast can become generally useful for commercial winemaking. The currently available strains of this yeast present certain handicaps that both limit their performance as good wine fermenters and increase their production costs (and hence their market selling price).

Perhaps most of the considerations that we have made here regarding the possible erroneous assignment of interesting properties to non-*Saccharomyces* yeasts such as *T. delbrueckii* should also have been made for strains of *S. cerevisiae* when they were originally selected for winemaking. However, in this latter case, any possible misassignment would not have been so critical as to dissuade the œnologist from using the selected yeast industrially, because the fermentation vigour of most *S. cerevisiae* strains is fairly good, and one could have expected certain success with the inoculation of these yeasts in the winery compared with the traditional spontaneous fermentation. In particular, it would be rare for any relevant growth of a contaminating non-*Saccharomyces* yeast in must fermentations inoculated with *S. cerevisiae* to occur, and such contaminations should not have any negative impact on wine quality. However, on the contrary, one may expect the massive growth of *Saccharomyces* yeasts in fermentations inoculated with non-*Saccharomyces* such as *T. delbrueckii* to occur frequently, with the consequence of failing to obtain the desired effect on wine quality. Given this current situation, in order to obtain the required improvement of wine quality, we would propose for the time being to use single inoculation with *T. delbrueckii* under appropriate conditions to achieve its full dominance, and then to mix the resulting wines with other *S. cerevisiae* wines.

Author Contributions: Conceptualization, M.R. and R.V.; Methodology, M.R. and R.V.; Formal Analysis, M.R. and R.V.; Investigation, M.R. and R.V.; Resources, M.R.; Writing-Original Draft Preparation, M.R.; Writing-Review & Editing, M.R.; Supervision, M.R. and R.V.; Project Administration, M.R.; Funding Acquisition, M.R.

Funding: This research was funded by grants GR18117 of the Extremadura Regional Government and AGL2017-87635-R of the Spanish Ministry of Economy, Industry and Competitiveness (co-financed with FEDER funds).

Conflicts of Interest: The authors declare no conflict of interest.

References

1. Benito, S. The impact of *Torulaspora delbrueckii* yeast in winemaking. *Appl. Microbiol. Biotechnol.* **2018**, *102*, 3081–3094. [CrossRef] [PubMed]
2. Kurtzman, C.P. Torulaspora lindner (1904). In *The Yeasts: A Taxonomic Study*, 5th ed.; Kurtzman, C.P., Fell, J.W., Boekhout, T., Eds.; Elsevier: London, UK, 2011; Volume 2, pp. 867–874.
3. Saluja, P.; Yelchuri, R.K.; Sohal, S.K.; Bhagat, G.; Paramjit; Prasad, G.S. *Torulaspora indica* a novel yeast species isolated from coal mine soils. *Antonie van Leeuwenhoek* **2012**, *101*, 733–742. [CrossRef] [PubMed]
4. Wang, Q.M.; Xu, J.; Wang, H.; Li, J.; Bai, F.Y. *Torulaspora quercuum* sp. Nov. and *Candida pseudohumilis* sp. Nov., novel yeasts from human and forest habitats. *FEMS Yeast Res.* **2009**, *9*, 1322–1326. [CrossRef] [PubMed]
5. James, S.A.; Collins, M.D.; Roberts, I.N. Use of an rRNA internal transcribed spacer region to distinguish phylogenetically closely related species of the genera *Zygosaccharomyces* and *Torulaspora*. *Int. J. Syst. Bacteriol.* **1996**, *46*, 189–194. [CrossRef] [PubMed]
6. Oda, Y.; Yabuki, M.; Tonomura, K.; Fukunaga, M. Reexamination of yeast strains classified as *Torulaspora delbrueckii* (lindner). *Int. J. Syst. Bacteriol.* **1997**, *47*, 1102–1106. [CrossRef] [PubMed]
7. Yarrow, D. Genus 29. *Torulaspora* lindner. In *The Yeasts, A Taxonomic Study*, 3rd ed.; Kreger-van Rij, N.J.W., Ed.; Elsevier Science: Amsterdam, The Netherlands, 1984; pp. 434–439.
8. Albertin, W.; Chasseriaud, L.; Comte, G.; Panfili, A.; Delcamp, A.; Salin, F.; Marullo, P.; Bely, M. Winemaking and bioprocesses strongly shaped the genetic diversity of the ubiquitous yeast *Torulaspora delbrueckii*. *PLoS ONE* **2014**, *9*, e94246. [CrossRef] [PubMed]
9. Wolfe, K.H.; Shields, D.C. Molecular evidence for an ancient duplication of the entire yeast genome. *Nature* **1997**, *387*, 708–713. [CrossRef] [PubMed]
10. Kreger-van Rij, N.J.; Veenhuis, M. Ultrastructure of the ascospores of some species of the *Torulaspora* group. *Antonie Leeuwenhoek* **1976**, *42*, 445–455. [CrossRef] [PubMed]
11. Codón, A.C.; Gasent-Ramírez, J.M.; Benítez, T. Factors which affect the frequency of sporulation and tetrad formation in *Saccharomyces cerevisiae* baker's yeasts. *Appl. Environ. Microbiol.* **1995**, *61*, 630–638. [PubMed]
12. Limtong, S.; Imanishi, Y.; Jindamorakot, S.; Ninomiya, S.; Yongmanitchai, W.; Nakase, T. *Torulaspora maleeae* sp. Nov., a novel ascomycetous yeast species from Japan and Thailand. *FEMS Yeast. Res.* **2008**, *8*, 337–343. [CrossRef] [PubMed]

13. Ramírez, M.; Ambrona, J. Construction of sterile *ime1Δ* transgenic *Saccharomyces cerevisiae* wine yeasts unable to disseminate in nature. *Appl. Environ. Microbiol.* **2008**, *74*, 2129–2134. [CrossRef] [PubMed]

14. Ramírez, M.; Pérez, F.; Regodón, J.A. A simple and reliable method for hybridization of homothallic wine strains of *Saccharomyces cerevisiae*. *Appl. Environ. Microbiol.* **1998**, *64*, 5039–5041. [PubMed]

15. Ramírez, M.; Rebollo, J.E. Technological wine yeast improvement by classical genetic techniques. In *Recent Research Developments in Microbiology*; Pandalai, S.G., Ed.; Research Signpost: Kerala, India, 2003; pp. 87–101.

16. Ramírez, M.; Regodón, J.A.; Pérez, F.; Rebollo, J.E. Wine yeast fermentation vigor may be improved by elimination of recessive growth-retarding alleles. *Biotechnol. Bioeng.* **1999**, *65*, 212–218. [CrossRef]

17. Castelli, T. Yeasts of wine fermentations from various regions of Italy. *Am. J. Enol. Vitic.* **1955**, *6*, 18–20.

18. Mauricio, J.C.; Millán, C.; Ortega, J.M. Influence of oxygen on the biosynthesis of cellular fatty acids, sterols and phospholipids during alcoholic fermentation by *Saccharomyces cerevisiae* and *Torulaspora delbrueckii*. *World J. Microbiol. Biotechnol.* **1998**, *14*, 405–410. [CrossRef]

19. Alves-Araujo, C.; Pacheco, A.; Almeida, M.J.; Spencer-Martins, I.; Leao, C.; Sousa, M.J. Sugar utilization patterns and respiro-fermentative metabolism in the baker's yeast *Torulaspora delbrueckii*. *Microbiology* **2007**, *153*, 898–904. [CrossRef] [PubMed]

20. Visser, W.; Scheffers, W.A.; Batenburg-van der Vegte, W.H.; van Dijken, J.P. Oxygen requirements of yeasts. *Appl. Environ. Microbiol.* **1990**, *56*, 3785–3792. [PubMed]

21. Hanl, L.; Sommer, P.; Arneborg, N. The effect of decreasing oxygen feed rates on growth and metabolism of *Torulaspora delbrueckii*. *Appl. Microbiol. Biotechnol.* **2005**, *67*, 113–118. [CrossRef] [PubMed]

22. González-Royo, E.; Pascual, O.; Kontoudakis, N.; Esteruelas, M.; Esteve-Zarzoso, B.; Mas, A.; Canals, J.M.; Zamora, F. Oenological consequences of sequential inoculation with non-*Saccharomyces* yeasts (*Torulaspora delbrueckii* or *Metschnikowia pulcherrima*) and *Saccharomyces cerevisiae* in base wine for sparkling wine production. *Eur. Food Res. Technol.* **2014**, *240*, 999–1012. [CrossRef]

23. Ramírez, M.; Velázquez, R.; Maqueda, M.; Zamora, E.; López-Piñeiro, A.; Hernández, L.M. Influence of the dominance of must fermentation by *Torulaspora delbrueckii* on the malolactic fermentation and organoleptic quality of red table wine. *Int. J. Food Microbiol.* **2016**, *238*, 311–319. [CrossRef] [PubMed]

24. Velázquez, R.; Zamora, E.; Alvarez, M.L.; Hernández, L.M.; Ramírez, M. Effects of new *Torulaspora delbrueckii* killer yeasts on the must fermentation kinetics and aroma compounds of white table wine. *Front. Microbiol.* **2015**, *6*, 1222. [CrossRef] [PubMed]

25. Renault, P.; Coulon, J.; de Revel, G.; Barbe, J.-C.; Bely, M. Increase of fruity aroma during mixed *T. delbrueckii/S. cerevisiae* wine fermentation is linked to specific esters enhancement. *Int. J. Food Microbiol.* **2015**, *207*, 40–48. [CrossRef] [PubMed]

26. Azzolini, M.; Fedrizzi, B.; Tosi, E.; Finato, F.; Vagnoli, P.; Scrinzi, C.; Zapparoli, G. Effects of *Torulaspora delbrueckii* and *Saccharomyces cerevisiae* mixed cultures on fermentation and aroma of Amarone wine. *Eur. Food Res. Technol.* **2012**, *235*, 303–313. [CrossRef]

27. Azzolini, M.; Tosi, E.; Lorenzini, M.; Finato, F.; Zapparoli, G. Contribution to the aroma of white wines by controlled *Torulaspora delbrueckii* cultures in association with *Saccharomyces cerevisiae*. *World J. Microbiol. Biotechnol.* **2015**, *31*, 277–293. [CrossRef] [PubMed]

28. Herraiz, G.; Reglero, M.; Herraiz, P.; Alvarez, M.; Cabezudo, M. The influence of the yeast and type of culture on the volatile composition of wine fermented without sulphur dioxide. *Am. J. Enol. Vitic.* **1990**, *41*, 313–318.

29. Comitini, F.; Gobbi, M.; Domizio, P.; Romani, C.; Lencioni, L.; Mannazzu, I.; Ciani, M. Selected non-*Saccharomyces* wine yeasts in controlled multistarter fermentations with *Saccharomyces cerevisiae*. *Food Microbiol.* **2011**, *28*, 873–882. [CrossRef] [PubMed]

30. Sadoudi, M.; Tourdot-Maréchal, R.; Rousseaux, S.; Steyer, D.; Gallardo-Chacón, J.-J.; Ballester, J.; Vichi, S.; Guérin-Schneider, R.; Caixach, J.; Alexandre, H. Yeast–yeast interactions revealed by aromatic profile analysis of Sauvignon Blanc wine fermented by single or co-culture of non-*Saccharomyces* and *Saccharomyces* yeasts. *Food Microbiol.* **2012**, *32*, 243–253. [CrossRef] [PubMed]

31. Escribano, R.; Gonzalez-Arenzana, L.; Portu, J.; Garijo, P.; Lopez-Alfaro, I.; Lopez, R.; Santamaria, P.; Gutierrez, A.R. Wine aromatic compound production and fermentative behaviour within different non-*Saccharomyces* species and clones. *J. Appl. Microbiol.* **2018**, *124*, 1521–1531. [CrossRef] [PubMed]

32. Canonico, L.; Comitini, F.; Ciani, M. *Torulaspora delbrueckii* for secondary fermentation in sparkling wine production. *Food Microbiol.* **2018**, *74*, 100–106. [CrossRef] [PubMed]

33. Renault, P.; Coulon, J.; Moine, V.; Thibon, C.; Bely, M. Enhanced 3-sulfanylhexan-1-ol production in sequential mixed fermentation with *Torulaspora delbrueckii/Saccharomyces cerevisiae* reveals a situation of synergistic interaction between two industrial strains. *Front. Microbiol.* **2016**, *7*, 293. [CrossRef] [PubMed]

34. Renault, P.; Miot-Sertier, C.; Marullo, P.; Hernández-Orte, P.; Lagarrigue, L.; Lonvaud-Funel, A.; Bely, M. Genetic characterization and phenotypic variability in *Torulaspora delbrueckii* species: Potential applications in the wine industry. *Int. J. Food Microbiol.* **2009**, *134*, 201–210. [CrossRef] [PubMed]

35. Ciani, M.; Capece, A.; Comitini, F.; Canonico, L.; Siesto, G.; Romano, P. Yeast interactions in inoculated wine fermentation. *Front. Microbiol.* **2016**, *7*, 555. [CrossRef] [PubMed]

36. Ramírez, M.; Velázquez, R.; Maqueda, M.; López-Piñeiro, A.; Ribas, J.C. A new wine *Torulaspora delbrueckii* killer strain with broad antifungal activity and its toxin-encoding double-stranded RNA virus. *Front. Microbiol.* **2015**, *6*, 983. [CrossRef] [PubMed]

37. Ramirez, M.; Velazquez, R.; Lopez-Pineiro, A.; Naranjo, B.; Roig, F.; Llorens, C. New insights into the genome organization of yeast killer viruses based on "atypical" killer strains characterized by high-throughput sequencing. *Toxins* **2017**, *9*, 292. [CrossRef] [PubMed]

38. Renault, P.E.; Albertin, W.; Bely, M. An innovative tool reveals interaction mechanisms among yeast populations under oenological conditions. *Appl. Microbiol. Biotechnol.* **2013**, *97*, 4105–4119. [CrossRef] [PubMed]

39. Kemsawasd, V.; Branco, P.; Almeida, M.G.; Caldeira, J.; Albergaria, H.; Arneborg, N. Cell-to-cell contact and antimicrobial peptides play a combined role in the death of *Lachanchea thermotolerans* during mixed-culture alcoholic fermentation with *Saccharomyces cerevisiae*. *FEMS Microbiol. Lett.* **2015**, *362*, fnv103. [CrossRef] [PubMed]

40. Wang, C.; Mas, A.; Esteve-Zarzoso, B. The interaction between *Saccharomyces cerevisiae* and non-*Saccharomyces* yeast during alcoholic fermentation is species and strain specific. *Front. Microbiol.* **2016**, *7*, 502. [CrossRef] [PubMed]

41. Taillandier, P.; Lai, Q.P.; Julien-Ortiz, A.; Brandam, C. Interactions between *Torulaspora delbrueckii* and *Saccharomyces cerevisiae* in wine fermentation: Influence of inoculation and nitrogen content. *World J. Microbiol. Biotechnol.* **2014**, *30*, 1959–1967. [CrossRef] [PubMed]

42. Velázquez, R.; Zamora, E.; Álvarez, M.L.; Álvarez, M.L.; Ramírez, M. Using mixed inocula of new killer strains of *Saccharomyces cerevisiae* to improve the quality of traditional sparkling-wine. *Food Microbiol.* **2016**, *59*, 150–160. [CrossRef] [PubMed]

43. Bely, M.; Stoeckle, P.; Masneuf-Pomarède, I.; Dubourdieu, D. Impact of mixed *Torulaspora delbrueckii–Saccharomyces cerevisiae* culture on high-sugar fermentation. *Int. J. Food Microbiol.* **2008**, *122*, 312–320. [CrossRef] [PubMed]

44. Contreras, A.; Hidalgo, C.; Schmidt, S.; Henschke, P.A.; Curtin, C.; Varela, C. The application of non-*Saccharomyces* yeast in fermentations with limited aeration as a strategy for the production of wine with reduced alcohol content. *Int. J. Food Microbiol.* **2015**, *205*, 7–15. [CrossRef] [PubMed]

45. Arslan, E.; Celik, Z.D.; Cabaroglu, T. Effects of pure and mixed autochthonous *Torulaspora delbrueckii* and *Saccharomyces cerevisiae* on fermentation and volatile compounds of Narince wines. *Foods* **2018**, *7*, 147. [CrossRef] [PubMed]

46. Zhang, B.-Q.; Luan, Y.; Duan, C.-Q.; Yan, G.-L. Use of *Torulaspora delbrueckii* co-fermentation with two *Saccharomyces cerevisiae* strains with different aromatic characteristic to improve the diversity of red wine aroma profile. *Front. Microbiol.* **2018**, *9*, 606. [CrossRef] [PubMed]

47. Belda, I.; Navascues, E.; Marquina, D.; Santos, A.; Calderon, F.; Benito, S. Dynamic analysis of physiological properties of *Torulaspora delbrueckii* in wine fermentations and its incidence on wine quality. *Appl. Microbiol. Biotechnol.* **2015**, *99*, 1911–1922. [CrossRef] [PubMed]

48. Belda, I.; Ruiz, J.; Beisert, B.; Navascues, E.; Marquina, D.; Calderon, F.; Rauhut, D.; Benito, S.; Santos, A. Influence of *Torulaspora delbrueckii* in varietal thiol (3-sh and 4-msp) release in wine sequential fermentations. *Int. J. Food Microbiol.* **2017**, *257*, 183–191. [CrossRef] [PubMed]

49. Velázquez, R.; Zamora, E.; Álvarez, M.L.; Ramírez, M. Using *Torulaspora delbrueckii* killer yeasts in the elaboration of base wine and traditional sparkling. *Int. J. Food Microbiol.* **2018**, in press.

50. Spencer, J.F.T.; Spencer, D.M. Taxonomy: The names of the yeasts. In *Yeasts in Natural and Artificial Habitats*; Spencer, J.F.T., Spencer, D.M., Eds.; Springer: Berlin/Heidelberg, Germany, 1997; pp. 11–32.

51. Ohshima, Y.; Sugaura, T.; Horita, M.; Sasaki, T. Industrial application of artificially induced diploid strains of *Torulaspora delbrueckii*. *Appl. Environ. Microbiol.* **1987**, *53*, 1512–1514. [PubMed]

52. Sasaki, T.; Ohshima, Y. Induction and characterization of artificial diploids from the haploid yeast *Torulaspora delbrueckii*. *Appl. Environ. Microbiol.* **1987**, *53*, 1504–1511. [PubMed]

53. Hernandez-Lopez, M.J.; Pallotti, C.; Andreu, P.; Aguilera, J.; Prieto, J.A.; Randez-Gil, F. Characterization of a *Torulaspora delbrueckii* diploid strain with optimized performance in sweet and frozen sweet dough. *Int. J. Food Microbiol.* **2007**, *116*, 103–110. [CrossRef] [PubMed]

54. Lucca, M.E.; Loray, M.A.; de Figueroa, L.I.C.; Callieri, D.A. Characterisation of osmotolerant hybrids obtained by fusion between protoplasts of *Saccharomyces cerevisiae* and heat treated protoplasts of *Torulaspora delbrueckii*. *Biotechnol. Lett.* **1999**, *21*, 343–348. [CrossRef]

fermentation

MDPI

Review

The Impact of Non-*Saccharomyces* Yeast on Traditional Method Sparkling Wine

Nedret Neslihan Ivit [1,2] and Belinda Kemp [3,4,*]

[1]   Perennia Food and Agriculture Inc., 32 Main Street, Kentville, NS B4N 1J5, Canada; nivit@perennia.ca
[2]   Office of Industry and Community Engagement, Acadia University, 210 Horton Hall, 18 University Ave,
      Wolfville, NS B4P 2R6, Canada
[3]   Cool Climate Oenology and Viticulture Institute (CCOVI), Brock University, 1812 Sir Isaac Brock Way,
      St. Catharines, ON L2S 3A1, Canada
[4]   Department of Biological Science, Faculty of Maths and Science, Brock University, 1812 Sir Isaac Brock Way,
      St. Catharines, ON L2S 3A1, Canada
*   Correspondence: bkemp@brocku.ca

Received: 1 August 2018; Accepted: 23 August 2018; Published: 1 September 2018

Abstract: The interest in non-*Saccharomyces* yeast for use in sparkling wine production has increased in recent years. Studies have reported differences in amino acids and ammonia, volatile aroma compounds (VOCs), glycerol, organic acids, proteins and polysaccharides. The aim of this review is to report on our current knowledge concerning the influence of non-*Saccharomyces* yeast on sparkling wine chemical composition and sensory profiles. Further information regarding the nutritional requirements of each of these yeasts and nutrient supplementation products specifically for non-*Saccharomyces* yeasts are likely to be produced in the future. Further studies that focus on the long-term aging ability of sparkling wines made from non-*Saccharomyces* yeast and mixed inoculations including their foam ability and persistence, organic acid levels and mouthfeel properties are recommended as future research topics.

Keywords: non-*Saccharomyces*; yeast; sparkling wine; nitrogen; aroma

---

1. Introduction

One of the main ways to make sparkling wine is the traditional method, also known as *Methodé champenoise* in Champagne [1,2], and other parts of the world as the classic method, *Methode traditionale* or bottle-fermented [3]. The traditional method of sparkling wine production differs from other winemaking processes in so much as the second alcoholic fermentation and the aging on yeast lees, both take place in sealed bottles (Figure 1). These are the same bottles that customers purchase from shops and restaurants [4–7]. Sparkling wines produced by the traditional method include Champagne wines in France, Cava in Spain, Brazil, Italy, USA, Australia, New Zealand, England, South Africa and Canada [3,4,6–10].

The first alcoholic fermentation to produce base wine is typically started by yeast inoculation. It occurs at controlled temperatures usually below 20 °C [1,11]. The malolactic fermentation (MLF) of the base wine is optional, depends on the decision of the producer, the malic acid level, intended wine style and the desired flavor profile [1,3]. Each producer aims to bring together the characteristics of different grape varieties, different base wines and different years by blending base wines [12]. At this stage, the wine becomes a "*cuvée*," which refers to blended base wines, which go on to be fermented in bottles [1]. Wines also undergo tartaric stabilization and filtration to remove prior to bottling [8,11].

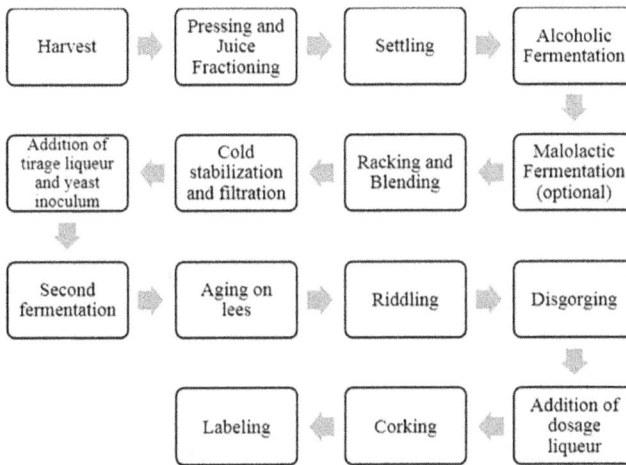

Figure 1. Traditional method sparkling wine flowchart that depicts the steps of production, which described in Section 1 of this review.

Liqueur de tirage is a concentrated sucrose solution that may contain necessary yeast nutrients or adjuvants. The composition of the *liqueur de tirage* may vary depending on the producer [3]. Generally, 24 g/L of sucrose is added to each bottle [1], since 4–4.30 g of sugar results in one atmosphere (atm) of carbon dioxide (CO_2). The aim is to reach between 5–6 atm of CO_2 by the end of the bottle fermentation [7]. Each bottle is inoculated with an initial yeast population of 1.5×10^6 cells/mL [13]. The second fermentation in the bottle, also referred to as *"prise de mousse,"* is followed by aging in contact with the yeast lees at low temperatures, approximately 12–16 °C [5]. Once the wines are deemed ready, the bottles are riddled to move the yeast cells to the neck of the bottles. The next stage is disgorging, which involves freezing the neck of the bottle in a brine at −25 °C. The frozen yeast deposit is then eliminated under pressure when the bottle is opened and the *dosage*, is immediately added then the bottle is the closed with a cork/closure and labelled [14].

1.1. Non-Saccharomyces Yeasts

Alcoholic fermentation for winemaking is usually carried out by inoculating the juice *Saccharomyces cerevisiae* yeast, the most widely used yeast, often referred to as "conventional" wine yeast. Throughout the past decade, "non-conventional" yeasts have become popular [15] and the effects of non-*Saccharomyces* yeasts on wine quality has been extensively studied [16]. Traditionally, only negative impacts of non-*Saccharomyces* were investigated, since they were believed to be reason for the microbial-related problems during winemaking. These included a high production of undesirable compounds i.e., acetic acid, ethyl acetate and acetaldehyde [17–19]. It has been suggested that widespread use of commercialized *S. cerevisiae* strains in winemaking creates uniformity in wines [2]. However, the exclusion of non-*Saccharomyces* yeasts from the fermentation process may result in a loss of complexity and result in wines lacking distinctive character [20,21]. Non-*Saccharomyces* yeasts, which are naturally found on grapes may positively affect wine quality [22]. Spontaneous wine fermentations initially start with indigenous yeasts, the native microbiota of the grape juice but *Saccharomyces cerevisiae* will, in most cases, take over and complete the fermentation [23]. Researchers have highlighted the positive role of non-*Saccharomyces* yeast, as well as their negative role on the chemical composition and sensory profile of resultant wines [19,24,25]. The increased knowledge of yeast biochemistry and physiology has made the selection of yeast strains with specific traits possible [21]. The potential benefits for using non-*Saccharomyces* yeast in winemaking has been well documented for still wines but far less so for sparkling wines [18]. The current demand for innovative wines in the competitive international market creates new opportunities to make wines with unique characteristics [26,27]. This increased knowledge combined with the interest from the wine market has produced commercially available non-*Saccharomyces* yeasts in the form of both dry active and frozen yeasts, i.e., *Torulaspora delbrueckii*, *Metschnikowia pulcherrima* and *Pichia kluyveri* [18,22]. The main oenological properties of non-*Saccharomyces* yeasts have been reviewed [28], as well as the technological and safety issues that these yeasts can solve, such as volatile acidity, alcohol reduction, high glycerol content, enhanced varietal aromas and the reduction of contaminants [28]. The current purpose of using non-*Saccharomyces* yeasts in winemaking varies. It is possible to use non-*Saccharomyces* as a sole yeast, although most studies demonstrate that sequential inoculation, or co-inoculation with mixed starter cultures of a non-*Saccharomyces* with a *S. cerevisiae* are preferred [29]. Recently, Comitini et al. [27] reviewed the specific purposes and benefits of non-*Saccharomyces* species in winemaking, which included the improvement of the complexity of the aromatic profile of wines, control of undesirable microflora, alcohol reduction and low sulfur dioxide (SO_2), hydrogen sulfide (H_2S) and acetaldehyde concentrations, along with the reduction in copper. The main non-*Saccharomyces* species found in vineyards and wineries include *Aureobasidium*, *Brettanomyces*, *Debaryomyces*, *Hanseniaspora*, *Metschnikowia*, *Lachancea*, *Torulaspora*, *Pichia*, *Rhodotorula*, *Starmerella* and *Zygosaccharomyces*, which were recently reviewed by Varela and Borneman [30]. Further studies have investigated specific effects of non-*Saccharomyces* yeast on wine quality. The yeasts that were studied include *Hanseniaspora uvarum* [31,32], *Hanseniaspora vineae* [26,33], *Torulaspora delbrueckii* [34–38], *M. pulcherrima* [29,39–41], *Starmerella bacillaris* [22,42], *Schizosaccharomyces pombe* [43,44], *Schizosaccharomyces japonicus* [16], *Lachancea thermotolerans* [29,40,45], *Zygotorulaspora florentina* [46], *P. kluyveri* [22,47], and *Zygosaccharomyces bailii* [48,49].

T. delbrueckii, was one of the first commercial non-*Saccharomyces* yeast on the market [18]. *T. delbrueckii* produces low concentrations of glycerol, acetaldehyde, acetic acid and ethyl acetate. In the mixed or sequential fermentation with *S. cerevisiae*, it has improved faults particularly volatile acidity [50]. Studies that used it for first fermentations found that it increased glycerol concentration compared to other yeasts in the trial, decreased volatile acidity and had positive effects on foaming properties [35]. It has also been utilized for second fermentation in bottle both in solitary and mixed fermentation, and found suitable for sparkling wines. However, the wines produced had a different aroma composition and sensory profile compared to those of the *S. cerevisiae* strains [51].

M. pulcherrima is a high producer of β-glucosidase and its presence in mixed cultures can decrease volatile acidity, yet increase the production of medium-chain fatty acids (MCFA), alcohol, esters, terpenols and glycerol. Its contribution to base wine aroma profiles and improved foaming characteristics has been established [35].

Schizosaccharomyces pombe species is extremely useful in cold viticultural regions due its ability to transform malic acid into ethanol [50]. This ability makes it possible to significantly reduce of the levels of biogenic amines and ethyl carbamate precursors without the need for any MLF [44]. It is able to ferment sparkling base wine to dryness without producing any aromatic defects [52].

L. thermotolerans enhances wine acidity and increases the overall perceived wine quality [53]. Its' ability to produce L-lactic acid from glucose and fructose could be used to increase acidity in sparkling wines produced in warm viticultural regions [54].

As the characteristics of yeast species differ, the importance of strain/isolate differences must be considered. The effect of yeast strain differences on the chemical composition of sparkling wines, as well as on still wines, has been studied. Martínez-Rodríguez et al. [55] showed the different influence of five yeast strains of the species *S. cerevisiae* and *S. bayanus* on the content of free amino acids and peptides during ageing in contact with yeast lees, for traditional method sparkling wines [55]. Perpetuini et al. [56] studied 28 strains of *S. cerevisiae* that had different flocculation abilities. These authors showed that using new starting strains to improve sparkling wine was possible. Another study investigated six strains of *S. cerevisiae* and reported phenotypic differences in the concentration of aroma compounds of the finished sparkling wines [57]. These studies illustrate the importance of yeast species and strain selection, on the final attributes of sparkling wines.

1.2. Non-Saccharomyces Yeasts and Sparkling Wine

The fermentation media for the second fermentation for traditional method wines creates a hostile and unfavorable environment for yeast [57,58]. These conditions include high ethanol content (10–12% v/v), low pH (2.9–3.2), low temperatures (10–15 °C), SO_2 concentrations (50–80 mg/L) high total acidity (5–7 g/L H_2SO_4), nitrogen starvation and CO_2 pressure [58,59]. Their ability to re-ferment base wines to produce sparkling wines is crucial but also critical for post fermentation, since the second alcoholic fermentation is followed by a period of aging on yeast lees [2,60]. Several characteristics are taken into account when choosing yeast for the second fermentation, amongst them are their resistance to high pressure (5–6 atmosphere), resistance to ethanol and the ability to ferment at low temperatures [11]. It can also increase the quality of the product due to its autolytic ability, good flocculation ability (to facilitate lees movement during riddling) and organoleptic properties [61].

The interest in non-*Saccharomyces* yeast for use in sparkling wine production has increased in recent years, and hence generated a new research area for wine scientists (Table 1). To the best of our knowledge, the focus of the studies on non-*Saccharomyces* yeasts, for traditional method sparkling wine, include: *T. delbrueckii*, *M. pulcherrima*, *S. pombe* and *Saccharomycodes ludwigii*. Differences in amino acids, ammonia, volatile aroma compounds (VOCs), glycerol and proteins, which impact sparkling wine flavor and foaming ability, have been reported (Table 1). Due to a lack of research specific to sparkling wines, studies concerning the influence of non-*Saccharomyces* yeast on still wines have been used in the following section to predict the possible effects on sparkling wines. Nitrogenous compounds with a focus on amino acids, ammonia, biogenic amines and VOCs, proteins, organic acids and sensory qualities have been considered. Despite the substantial potential of non-*Saccharomyces* for sparkling wine production, further investigation is required to understand the possible effects of non-*Saccharomyces* for second fermentation and aging [60]. Therefore, the following section of this review discusses our current knowledge concerning the impact of non-*Saccharomyces* yeasts on traditional method sparkling wine.

Table 1. Effects of non-*Saccharomyces* yeasts studied in base wines and sparkling wines.

Yeast Name	Production Stage	Isolation Method	Grape Variety Used	Studied Parameters	Effect on Sparkling Wine	Reference
T. delbrueckii	Second fermentation	Originally isolated from natural matrices from different environments.	White base wine made of Verdicchio from Ancona, Italy.	Ethanol, volatile acidity, sensory analysis and VOCs.	Positive, distinctive effects on overall aroma and sensory characteristics of wines were reported.	[51]
S. pombe *S. ludwigii*	Second fermentation	Previously isolated yeasts from the archive of Chemistry and Food Technology Department of Universidad Politecnica de Madrid.	White base wine made of Airen grapes and red base wine made of Tempranillo grapes, Spain.	Alcohol, total acidity, pH, sugars, organic acids, glycerol, anthocyanins, VOCs, amino acids, biogenic amines, sensory analysis.	Changes on color, acidity, volatile compounds, biogenic amines of the final products as well as on the sensorial evaluation was observed.	[52]
T. delbrueckii (sequential inoculation with *S. cerevisiae*)	First fermentation followed by a second fermentation	Base wine from the study done by González-Royo et al. [35] fermented with *S. cerevisiae bayanus* (EC1118 Lallemand Inc., Montreal, QC, Canada)	*V. vinifera* cv. Macabeo.	Proteins, polysaccharides and foaming properties.	Better foaming properties were observed.	[36]
T. delbrueckii (sequential inoculation with *S. cerevisiae*) *M. pulcherrima* (sequential inoculation of *S. cerevisiae*)	First fermentation	Commercial yeasts of *T. delbrueckii* (Biodiva™) and *M. pulcherrima* (Flavia®) from Lallemand Inc., Montreal, QC, Canada.	*V. vinifera* cv. Macabeo.	Ethanol, titratable acidity, pH, volatile acidity, glycerol, proteins, foaming properties, polysaccharides, VOCs, sensory analysis.	Base wines with different characteristics were obtained. Positive effects on foaming properties were observed.	[35]

2. Nitrogenous Compounds and Non-*Saccharomyces* Yeast

Grapes and wine contain many nitrogenous compounds in both inorganic forms (ammonia and nitrate) and organic forms (amines, amides, amino acids, pyrazines, nitrogen bases, pyrimidins, proteins and nucleic acids) [62]. For sparkling wines, the nitrogen fraction consists mostly of peptides, free amino acids and proteins. Different aspects of sparkling wines are influenced by juice and base wine nitrogen composition including foam quality, aroma profile and organoleptic characteristics [6,63]. Nitrogen compounds, mainly peptides and amino acids are considered the major compounds that are released into wine during yeast autolysis [55,64,65].

The interactions between base wine, temperature and yeast strain have the strongest effect on fermentation kinetics [66,67]. During second fermentation in bottle, the wines are kept at a relatively stable temperature, preferably between 10 °C and 15 °C [1]. The beginning of yeast autolysis may differ by up to several months depending on the storage conditions. Although there is no agreement about the time needed for yeast autolysis [68], it has been reported that it commences after 2–4 months after the second fermentation finishes [5,64,69].

2.1. Yeast Acclimation and Nitrogen Requirements

The stressful environment for yeast growth the acclimatization of yeast to the base wine is crucial. For this reason, yeasts are cultured in media that contains increasing ethanol concentrations [66,70]. Total assimilable nitrogen (YAN mg N/L) includes all nitrogen sources (amino acids, ammonia and peptides up to five amino acids in length) that can be assimilated and metabolized by yeasts [62]. The YAN content of grape must and its' usage by yeast during the first fermentation will affect the YAN concentration of the base wine. The nitrogen content of the base wines is highly variable (17–75 mg/L). Nevertheless, nitrogen requirements for the second fermentation are very low [66]. During the second fermentation in bottle, the YAN content decreases halfway through due to the yeast consumption in the early stages of the fermentation but increases at the end of the fermentation. This is due to the physiological responses of the yeast to the lack of nutrition, which they restore by means of passive release of nitrogen compounds [64,71].

Liqueur de Tirage ingredients vary depending on the producer and there are different adjuvants listed in literature. Thiamine and nitrogen, usually diammonium phosphate (DAP), at rates of 0.5–100 mg/L respectively were to base wine. Martí-Raga et al. [66] reported that the addition of nitrogen before fermentation effects fermentation kinetics. However, this is only when the levels of nitrogen are below 30 mg N/L. Nitrogen addition in the yeast acclimation media had a strong impact on yeast growth and significantly affected second fermentation kinetics [72]. The effects of non-*Saccharomyces* yeasts on amino acids and ammonia, biogenic amines and VOCs in sparkling wines are discussed in the following sections.

2.2. Amino Acids and Ammonia

Amino acids improve the aroma potential of sparkling wines since they are the precursors of several aroma compounds produced from deamination or decarboxylation reactions [73]. Ammonia is the nitrogen form most directly assimilable by yeasts [74] and NH_4^+ ammonium cation is the form most directly assimilable by yeasts. Amino acids such as α-alanine, serine, arginine, proline, glutamic acid and its amide form, glutamine, known to be an ammonia transporter are some of the amino acids that are predominant in must. Concentration of arginine and proline depends on the grape variety, although during fermentation most *S. cerevisiae* yeasts and lactic bacteria use arginine [74]. Moreno-Arribas et al. [68] studied the amino acid composition of peptides present in sparkling wines. They reported that threonine and serine are a major presence but these were not found in the base wines. Glutamic acid, glutamine and arginine are among the assimilable nitrogen that are the specific nutrients for alcoholic fermentation and microbial metabolism. Amino acids such as leucine, isoleucine, threonine, valine and phenylalanine are directly involved in the production of higher alcohols, which have an effect on organoleptic properties [75].

The yeast species and strain influences the content of free amino acids and peptides during ageing in contact with lees [55,71,76,77]. When yeasts perform the second fermentation, amino acids and proteins decrease while the peptides are liberated. Next, while nitrogen compounds are used as nutrients for viable cells, proteins are degraded to peptides and converted to amino acids. Finally, when there are no viable cells left, the release of proteins and peptides prevail. After 270 days of aging with yeast lees, the amino acid content of some wines decreases [55]. The changes observed in the amino acid content of wines may be the result of the assimilation and excretion process by yeast during fermentation [78], as well as the adsorption of amino acids by bentonite (if used as an adjuvant). Although, Martínez-Rodríguez and Polo [63] reported that the use of bentonite as a co-adjuvant in the concentrations used (3 g/hL) did not affect the amino acid concentration in the sparkling wines.

Gobert et al. [67] investigated the ability of non-*Saccharomyces* yeasts to consume nitrogen. Sequential fermentation using non-*Saccharomyces* yeasts (*Starmerella bacillaris*, *M. pulcherrima* and *Pichia membranifaciens*) in grape juice was undertaken and specific amino-acid consumption profiles of non-*Saccharomyces* yeasts were revealed. Cysteine was found to be the preferred nitrogen source for all non-*Saccharomyces* yeasts. Histidine, methionine, threonine and tyrosine were not consumed by *S. bacillaris*, aspartic acid was consumed very slowly by *M. pulcherrima* and glutamine was not utilized by *P. membranifaciens*.

These results suggest that a specific addition of amino acids in must should be considered for non-*Saccharomyces* yeast. Hence specific nitrogen products for non-*Saccharomyces* fermentations is likely in the future. In the context of indigenous fermentations, the study indicated that non-*Saccharomyces* yeasts compete with *Saccharomyces* for nitrogen sources. In contrast, Llexià et al. [79] found that nitrogen limitation increased the time of the fermentation as well as the proportion of non-*Saccharomyces* yeast at the mid and final stages of fermentation. The authors suggested that under conditions where nitrogen limitation occurs, *S. cerevisiae* should be co-inoculated to ensure nitrogen availability for this yeast.

Ivit et al. [52] used *S. ludwigii* (979) and *S. pombe* (938) for second fermentation of sparkling wines and compared them to *S. cerevisiae* (7VA) fermentations. Both base wines were produced from *Vitis vinifera* cv. Airen and *V. vinifera* cv. Tempranillo grapes, fermented in bottle and aged on lees for four months. The amino acid content of the base wines changed during the second fermentation and some amino acids decreased and others increased. Although no significant change was seen between the base wines and sparkling wines produced with *S. cerevisiae*, the total amino acids increased for the sparkling wines produced by non-*Saccharomyces* yeast. This difference in amino acid content between the yeasts could be due to their different release mechanisms, their different amino acid consumption during fermentation and/or the structural composition of the yeasts [67,80].

2.3. Biogenic Amines

Biogenic amines (BAs) are a class of nitrogenous compounds in wine that have oenological importance due to their adverse effect on human health and negative impact on wine quality [81–85]. Raw material and fermentation processes are the two sources of BAs in wine [86]. The main BAs found in wine are histamine, tyramine and cadaverine and those reported in grape must include putrescine, ethylamine, 2-phenylethylamine, spermine and spermidine [87]. The three main BAs associated with MLF are histamine, tyramine and putrescine, which are formed mainly by lactic acid bacteria by the decarboxylation of free amino acids [88]. Some authors suggest that certain yeast strains (*S. cerevisiae*, *Brettanomyces bruxellens*, *Kloeckera apiculata*, *Candida stellate* and *M. pulcherrima*) can produce BAs in wine [89,90]. BAs are also important in wine from an economical point of view, since they may cause problems for import and export processes due to official legal limits in some countries [81]. In many countries, no official maximum limits exist for histamine content in wines, though many wine importers require wines to be analyzed for histamine levels. The current limits are 2 mg/L in Germany, 4 mg/L in Holland, 6 mg/L in Belgium, 8 mg/L in France and 10 mg/L in

Switzerland [91,92]. In sparkling wines, histamine levels of twenty-six Austrian wines were found to be in the range of 0.001–1.9 mg/L (mean 0.30 ± 0.55, median 0.02) without significant differences between grape varieties [93]. Konakovsky et al. [94] reported the same range when twenty-nine German, Spanish and Italian sparkling wines were analyzed. Additionally, Caruso et al. [90] studied the effect of non-*Saccharomyces* yeasts on BAs and their precursors. These authors compared the formation of BAs from different yeast species and strains. *B. bruxellensis* formed the highest amount of total BAs (average of 15 mg/L), followed by *S. cerevisiae* (average of 12.1 mg/L). The other yeast species included *K. apiculata* and *M. pulcherrima* formed less than 10 mg/L of BAs. *S. pombe* is of particular interest as it reduced the risk of BA formation. Malic acid consumption by *S. pombe* yeast enables a non-bacterial biological de-acidification, which reduces the possibility of lactic acid bacteria growing; thereby reducing the risk of biogenic amine formation [44]. Wines made from Chardonnay grapes, produced with *H. vineae* and *S. cerevisiae* have both shown reduced BA content [33].

While some of the processes during traditional method sparkling wine production can increase the BA content, others can decrease it. The base wines that are subjected to MLF have more risk of high BA content, since BA concentration increases due to lactic acid bacteria [95]. The amount of BAs increases during contact with yeast lees [81]. The addition of clarification substances and oenological adjuvants such as bentonite or polyvinylpolypirrolidone (PVPP), have been found to reduce BA content, due to their ability to absorb them [96]. Ivit et al. [52] compared two non-*Saccharomyces* yeasts to one another. In white sparkling wines, total BA concentrations were significantly lower in comparison to the base wine. However, the red sparkling wines produced from *S. pombe* showed significantly lower total BAs in comparison to wines fermented with *S. cerevisiae*. This was likely due to different adsorption characteristics from different type of yeast lees during the aging process, or during the fermentation [89].

2.4. Volatile Aroma Compounds

Volatile aroma compounds (VOCs) produced by non-*Saccharomyces* yeasts can be grouped into higher alcohols, esters, aldehydes, volatile fatty acids, volatile phenols and sulfur compounds [24]. Higher alcohols, mainly 3-methyl butanol, 2-methyl propanol and 1-propanol among many others, can contribute aromatic complexity to wine at concentrations below 300 mg/L. However, it can cause a negative effect when the concentration is higher than 400 mg/L [97]. 2-phenylethyl alcohol has been attributed to floral and rose aromas and isoamyl alcohol with marzipan aromas [75]. Esters are produced by yeasts during fermentation and contribute positively to wine aroma by bestowing fruit characteristics to wine. The main ester in wine, ethyl acetate, causes spoilage at levels of 150–200 mg/L. There are esters that produce pleasant aromas including isoamyl acetate with banana and pear odors, 2-phenethyl acetate with rose, honey, fruity and flowery odors, ethyl hexanoate with apple and violet, and ethyl octanoate with pineapple and pear [75]. Aldehydes, mainly acetaldehyde contributes apple-like odors to wine, and at high levels can also cause spoilage. With regards to carbonyl compounds, diacetyl, is produced by yeast metabolism and the resultant buttery aromas are perceptible at concentrations between at 1–4 mg/L [75].

Non-*Saccharomyces* yeasts can influence the primary and the secondary aromas by the production of enzymes and metabolites, respectively. They have been described in literature as producers of enzymes involved in the release of primary aroma compounds from grape precursors [24]. Secondary aromas formed during fermentation by non-*Saccharomyces* yeasts include higher alcohols, esters, aldehydes (acetaldehyde), volatile phenols and sulfur compounds [24]. However, non-*Saccharomyces* yeasts also effect VOCs during aging in contact with yeast lees since yeast autolysis leads to significant changes in wine aroma composition [64]. During their autolysis, yeasts lead to the formation, or degradation of VOCs, which modifies the aroma profile of sparkling wines.

The possibility of modifying wine sensory profile by using combinations of non-*Saccharomyces* yeasts with *Saccharomyces* yeast strains has been studied extensively [37,98,99]. Englezos et al. [42] studied the effect of mixed fermentations of *Starmerella bacillaris* (synonym *Candida zemplinina*) with *S. cerevisiae* on the aroma profile and composition of Barbera wines. The wines produced from mixed cultures contained higher amounts of pleasant esters compared to the wine fermented with *S. cerevisiae* alone. Lencioni et al. [46] compared mixed fermentation of *Z. florentina* with *S. cerevisiae* to a fermentation with only *S. cerevisiae* and showed that the mixed fermentation produced a higher concentration of 2-phenylethanol. Furthermore, Belda et al. [37] reported that *T. delbrueckii*, used for white wine fermentation, increased the concentration of volatile thiols, with 4-methyl-4-sulfanylpentan-2-one (4-MSP) found over its sensory threshold level. Reduced alcohol wines fermented with *M. pulcherrima* and *S. uvarum* were described as having red fruit and berry aromas [39]. *M. pulcherrima* wines had higher concentrations of ethyl acetate, total esters and total higher alcohols [39]. The effect of *H. uvarum* in mixed fermentations with *S. cerevisiae* found that simultaneous fermentations and extracellular extract of *H. uvarum*, improved the overall quality of wine aromas, by increasing fruity and floral traits and enhancing terpenes, C13-norisoprenoids, acetate esters, ethyl esters and fatty acids [32]. Romani et al. [16] studied the effect of *S. japonicus* and found that the sequential fermentation with *S. cerevisiae* enhanced wine complexity and aroma by increasing higher alcohols (isobutanol, amylic and isoamylic alcohols), acetate esters (isoamyl acetate, phenyl ethyl acetate) and alcohols (β-phenyl ethanol) above their threshold levels. The effect of co-inoculation of *T. delbrueckii* and *S. cerevisiae* on Moscato Branco sparkling wines found that wines produced from co-fermentation had higher concentrations of 2-phenyl ethanol, acetate and ethyl esters with fruity and flowery descriptors while a decrease in concentration of undesirable compounds such as volatile fatty acids occurred [38].

Escribano et al. [49] determined the fermentation behavior and aroma formation from several non-*Saccharomyces* yeasts. Twenty-five yeasts from nine species were studied (*Candida zeylanoides, Cryptococcus uzbekistanensis, Debaryomyces hansenii, L. thermotolerans, M. pulcherrima, T. delbrueckii, Williopsis pratensis, Zygosaccharomyces bailii* and *S. cerevisiae*) in pasteurized grape juice. These authors suggested that *M. pulcherrima* was a good candidate for wine fermentation due to the formation of high concentrations of 2-phenyl ethyl alcohol and 2-phenyl acetate in the resultant wine. They further demonstrated the possibility of using *M. pulcherrima* and *L. thermotolerans* as an inoculum. First and second fermentations significantly alter the volatile composition of sparkling wines and aroma profile [5,6,100]. Some VOCs (i.e., esters, aldehydes and terpenes) can be adsorbed onto yeast lees, reducing their concentration in aged sparkling wines, although this depends on the structure of the yeast cell walls [101].

Yeast undergo important changes during second fermentation in sealed bottles under CO_2 pressure, before aging with yeast lees [5]. Different native yeast strains have been used for in-bottle fermentation to overcome uniformity in sparkling wine [2]. The study stated that an increase in the choice of available yeast strains for second fermentation in bottle would be useful for the differentiation of sparkling wines. Even so, few studies exist on the effect of non-*Saccharomyces* yeasts on VOCs in base wines and sparkling wines. Chardonnay and Pinot noir musts were fermented with *Pichia membranaefaciens, Kloeckera apiculata, Candida valida* and *S. cerevisiae* [102]. This study demonstrated the differences in the production of VOCs including ethyl acetate, isoamyl acetate, acetaldehyde, 2-methyl propanol and 3-methyl butanol. *P. membranaefaciens* was deemed to be the most suitable yeast to produce sparkling wines due to the sensory panels' preference, in comparison to *S. cerevisiae* wines. González-Royo et al. [35] studied sequential fermentations of *T. delbrueckii* and *M. pulcherrima* with *S. cerevisiae* to determine their effect on the chemical composition on base wine of *V. vinifera* cv. Macabeo grapes. Some significant differences were observed in comparison to solitary fermentations with *S. cerevisiae*: higher alcohols (3-methyl-1-butanol), acetaldehyde and major esters, minor esters (butyl acetate) and lactones (g-decalactone). Nevertheless, no major differences were reported in the aromatic profile of the wines. Canonico et al. [51] studied the effect of *T. delbrueckii* in second fermentation of Verdicchio base wine. Two *T. delbrueckii* strains and a strain of *S. cerevisiae* were used in both pure and mixed cultures. Significant differences were detected in hexanol, ethyl hexanoate, isoamyl acetate, ethyl octanoate and ethyl butyrate production. VOCs of white and red sparkling wines produced with *S. pombe* and *S. ludwigii* were compared to *S. cerevisiae* [52] and no significant differences were reported in the total concentration of VOCs, although specific VOC families (esters, acetoin metabolites and higher alcohols) were found to be significantly different (Table 2). Differences in VOCs reported in studies could be a result of the nitrogen preferences of the yeasts due to their specific amino acid consumption profiles [67].

Table 2. Volatile aroma compounds (VOCs) studied in base wines and sparkling wines produced with non-*Saccharomyces* yeasts.

Yeast	Production Stage	Higher Production in Comparison to *S. cerevisiae*	Lower Production in Comparison to *S. cerevisiae*	Reference
T. delbrueckii + *S. cerevisiae*	First fermentation for base wine production	Total higher alcohols Total lactones 3-methyl-1-butanol 2-methylpropanol 1-butanol 2-phenyl ethyl acetate	β-phenylethanol 1-hexanol Ethyl lactate Ethyl decanoate Ethyl octanoate	[35]
M. pulcherrima + *S. cerevisiae*		Total higher alcohols Total major esters Total minor esters Total lactones 3-methyl-1-butanol 2-methylpropanol Diethyl succinate 2-phenyl ethyl acetate Ethyl isovalerate Methyl vanillate	Methionol Acetaldehyde Ethyl lactate Ethyl decanoate Ethyl acetate Ethyl octanoate Butyl acetate Linalool acetate	
S. ludwigii 979	Second fermentation in bottle + 4 months of aging on lees	Diacetyl Acetoin 2-methyl-1-butanol Ethyl acetate	Acetaldehyde 2.3-butan-ediol Isoamyl acetate	[52]
S. pombe 938		Acetoin	Isoamyl acetate 2.3-butan-ediol	
T. delbrueckii 130	Second fermentation in bottle + 12 months of aging on lees	Ethyl hexanoate Ethyl octanoate Isoamyl acetate Hexanol	Acetaldehyde n-propanol Isobutanol Isoamyl alcohol	[51]
T. delbrueckii 313		Ethyl hexanoate Ethyl octanoate Isoamyl acetate Hexanol	Acetaldehyde Ethyl butyrate n-propanol Isobutanol Isoamyl alcohol	
S. cerevisiae + *T. delbrueckii* 130		Ethyl hexanoate Ethyl octanoate	Acetaldehyde Ethyl butyrate Ethyl acetate n-propanol Isobutanol Isoamyl alcohol	
S. cerevisiae + *T. delbrueckii* 313		Ethyl hexanoate Ethyl octanoate Isoamyl acetate Hexanol	Acetaldehyde Ethyl butyrate Ethyl acetate n-propanol Isobutanol Isoamyl alcohol	

3. Yeast-Derived Proteins

It is important to note that the definition of yeast mannoproteins is contentious in literature because they are referred to as both mannoproteins and polysaccharides [103]. Yeast mannoproteins are glycoproteins, which belong to the proteoglycan family and contain 10% protein and 90%

mannose [104,105]. Proteins are important compounds in foaming ability and foaming stability even though they are present at low concentrations in sparkling wines (4–16 mg/L) [103].

The yeast mannoproteins/polysaccharides are located in the outermost layer of the yeast cell wall linked by β-1,3 glycan chains to the innermost fibrous layer and formed from β-1,3 glycan and chitin [106–108]. They are released from the yeast cell wall during alcoholic fermentation and aging in contact with yeast lees. They represent one of the major polysaccharides found in wine [109]. *S. pombe* has only been considered for use in sparkling wine production as a base wine deacidification tool. However, its' ability to reduce lees aging has already been demonstrated in red wines through the rapid release of its' cell wall mannoproteins, due to its fast autolytic activity [110].

The unique mannoprotein profiles of non-*Saccharomyces* yeast (*Hansensiaspora osmophila*, *L. thermotolerans*, *M. pulcherrima*, *Pichia fermentans*, *S. ludwigii*, *Starmerella bacillaris*, *T. delbrueckii* and *Zygosaccharomyces florentinus*) have been isolated and characterized by Domizio et al. [109]. A synthetic, polysaccharide-free grape juice was used to characterize their release during alcoholic fermentation of still table wines. All strains showed a different intensity for each glycan and a core of N-glycans with a mass ranging in size from 8–15 mannoses was determined. We know that mannoproteins increase the mouthfeel properties of wine [111] and proteins contribute to sweet and bitter tastes as well as playing an important role in foam stability [112]. However, it appears vague from current literature, which non-Saccharomyces yeast, species, proteins and/or strains, contribute to the sweet, and/or bitter tastes. It is also unclear how their unique protein profiles contribute to sparkling wine composition during aging, particularly with relation to haze-related proteins.

T. delbrueckii had a positive effect on the foaming properties of cava wines, when used for the first fermentation, while *M. pulcherrima* increased foam stability [35]. Sequential inoculation (*T. delbrueckii* and *S. cerevisiae*) produced base wines with higher foaming potential than *S. cerevisiae* alone [36]. This was undoubtedly due to the greater release of proteins from *T. delbrueckii* cells compared to *S. cerevisiae*, particularly the low molecular weight (LMW) fraction.

Further consideration for future sparkling wine studies could be the enzyme production by non-*Saccharomyces* yeasts and their effect on sparkling wine, although it is difficult to distinguish between compounds synthesized from enzyme activity, and the compounds released from cells through yeast autolysis. Most noticeably missing from our current knowledge is the effect of non-*Saccharomyces* yeast on the protein evolution and foam qualities of sparkling wines following a long period of cellar aging when wines are in contact with yeast lees. Current studies have only reported on results of non-*Saccharomyces* influence on wine after short aging periods but many traditional method sparkling wines can spend long periods of time in the cellar (i.e., 2–10 years). Importantly, the practical implications during wine production, of high protein concentrations in bottles concerning haze formation, riddling, disgorging and gushing have so far, not been investigated.

4. Organic Acids

Some non-*Saccharomyces* yeasts have the ability to reduce alcohol [113], tolerate low temperature fermentations [114], influence aroma compounds and the wines' sensory profile, increase glycerol yield and specific strains can be used to target particular organic acids [115,116]. MLF can be carried out in sparkling base wines to alter the organic acid ratio by converting malic acid to lactic acid. Because of its malic dehydrogenase activity, the non-*Saccharomyces* yeast most exploited for this is *S. pombe*. Additionally, *S. pombe* produces less urea and more pyruvic acid than *Saccharomyces* species. Three of four *Schizosaccharomyces* strains completed the breakdown of malic acid by day four of a red wine fermentation [110]. The main negative effect of *S. pombe* is strong acetic acid production, which is most likely the reason for its lack of use in second fermentation. Nevertheless, a recent study by Ivit et al. [52] reported that 78% of malic acid was metabolized by *S. pombe* during the fermentation in bottle. In contrast to *S. pombe*, *L. thermotolerans* is a low producer of acetic acid though the level of production tends to be strain dependent and it does not always complete sugar consumption, which could be problematic for in-bottle fermentation [19,117]. Mixed-fermentations of non-*Saccharomyces* yeast produce less MCFA, known inhibitors of MLF [118]. *T. delbrueckii* has also been found to slightly reduce malic acid levels in wine by approximately 20% [119] and 25% [40] but Loira et al. [34] reported no effect, demonstrating the influence of yeast strain on malic acid consumption. *T. delbrueckii* increased succinic acid in sequential inoculations but this acid contributes to undesirable bitter/salty flavors to wine [120,121]. On the other hand, pyruvic acid is produced at high concentrations by *T. delbrueckii* [119] and this acid can improve MLF performance by *Oenococcus oeni* by acting as an external electron acceptor facilitating the production of NAD+ [118,122]. Benito et al. [123] and Ivit et al. [52] both reported differences in pyruvic acid production amongst non-*Saccharomyces* yeast. Pyruvic acid is involved in the formation of stable pigments i.e., pyranoanthocyanins [124], which could have implications for the color stability of rosé sparkling wines. Significant differences in total acidity were found (reported as tartaric acid) between yeasts in the study by Ivit et al. [52], due to malic acid changes during fermentation.

5. Effect of Non-*Saccharomyces* Yeasts on Sparkling Wine Sensory Profiles

Sensory analysis continues to be an efficient tool for assessing the sensory properties of sparkling wines [125]. However, there is lack of an internationally accepted or recognized sensory analysis method specifically for sparkling wines, as well as published criteria to evaluate effervescence and foam properties of sparkling wines [126]. Sensory evaluation of sparkling wines has been carried out using tasting cards, proposed by the OIV (Office International de la Vigne et du Vin, 1994) for international wine competitions, then partially modified by the Instituto Nacional de Denominaciones de Calidad of the Spanish Ministry of Agriculture, Fisheries and Food [63,127,128]. The method evaluates attributes by scoring them according to the following method: visual aspects carried a weight of 1, intensity and quality of aroma and intensity of taste 2 and finally quality of taste and harmony carried a weight of 3. Visual aspect of the wines, along with the color and foam characteristics were also evaluated [127].

McMahon et al. [129] evaluated sparkling wines, which were sweetened to different sugar levels, using a trained panel and consumer panel. Aroma attributes (nasal pungency, fruity, floral, green, yeasty and toasted); flavor attributes (fruity, floral, green, yeasty and toasted); taste attributes (sweet, sour and bitter); and mouthfeel attributes (bubble pain, foamy and creamy) were used. López de Lerma et al. [130] used descriptive analysis to evaluate color, odor and taste descriptors of sparkling wines that were produced with different yeast strains. The sensory attributes used in their study included: color quality, aroma quality, aroma intensity, fruity, yeasty and mold aroma and in terms of the taste, intensity and quality, acidity, body and bitterness. The authors classified into nine aroma groups; chemistry, fruity, toasty, green fruit, citrus, floral, fatty, creamy, herbaceous [130]. Few studies take into consideration how the CO_2 in sparkling wine may affect odor detection when compared to the same VOC in an aqueous, still wine or ethanolic solution [3]. The concentration of dissolved CO_2 effects the sensory properties including the frequency of bubble formation in the glass, the growth rate of rising bubbles, mouthfeel and the aromatic perception [131]. The attributes used for describing foam quality and effervescence include; the initial quantity of foam formed upon pouring, the appearance of the foam across the surface of the wine, the presence of foam collar, bubble size and duration of the bubble formation and foam stability [132,133].

5.1. Sensory Effects of Non-Saccharomyces Yeasts on Sparkling Wines

The influence of non-*Saccharomyces* yeasts on the sensory profile of wines is of great interest in current research [99]. Non-*Saccharomyces* yeasts possess special metabolic characteristics that affect the organoleptic profile of wines [24]. The sensory effects of non-*Saccharomyces* yeasts on base wines, second fermentation and aging on lees have been investigated [35,36,51,52].

González-Royo et al. [35] conducted two triangle tests and a preference test to evaluate the effect of sequential inoculations with *T. delbrueckii* or *M. pulcherrima* and *S. cerevisiae* versus *S. cerevisiae* fermentation only. A group of nine people from the Rovira i Virgili University conducted the sensory tests. Six out of nine tasters were able to distinguish between the base wines produced with sequential inoculation by *T. delbrueckii* and the *S. cerevisiae* wines. Five of the six tasters successfully differentiated the wines and preferred wines produced from the sequential inoculation of *T. delbrueckii*. In the case of *M. pulcherrima*, eight out of nine tasters were able to distinguish between the wines, the preference was equal; four preferred the *S. cerevisiae* wines, while the other four preferred the wine fermented using a sequential inoculation. Five of the eight tasters who successfully differentiated the wines, associated smoky and flowery *aromas* with the wine fermented by sequential inoculation with *M. pulcherrima*. The smoky perception was associated with the higher production of 2,6-dimethoxyphenol, flowery notes could not be associated with any of the measured VOCs [35].

The base wines produced in the study by González-Royo et al. [35] and were used to produce sparkling wines by Medina-Trujillo et al. [36]. A triangle test and a preference test with a group of twelve oenologists from the Rovira i Virgili University were conducted, to compare sparkling wines produced from the base wine of the sequential fermentation of *T. delbrueckii* and *S. cerevisiae* with those fermented only with *S. cerevisiae*. Nine out of twelve tasters were able to identify the wines produced by sequential inoculation with *T. delbrueckii*. Additionally, six of the nine tasters who successfully identified the wines, preferred the sparkling wine made from the sequential inoculation. This was because they found that the effervescence was more integrated and the wines had a less aggressive mouthfeel. The results concerning effervescence was related to improved foam properties specifically higher maximum foam height and higher amounts of proteins, especially LMW fraction [36].

The two studies of González-Royo et al. [35] and Medina-Trujillo et al. [36] confirmed that the base wine characteristics carry through to the finished sparkling wines. Unfortunately, though, Medina-Trujillo et al. [36] did not include base wines produced by a sequential inoculation of *M. pulcherrima* in the study from González-Royo et al. [35]. Similar to González-Royo et al. [35], the triangle test followed a preference test in the study of Medina-Trujillo et al. [36]. However, there was a higher number of panelists in one [36] than in the other [35]. In both studies of the base and finished sparkling wines, the majority of the panelists were able to distinguish wine produced from the sequential inoculation by *T. delbrueckii*. Sensory evaluation of white sparkling wines made from *V. vinifera* cv. Airén grapes, as well as red sparkling wines of *V. vinifera* cv. Tempranillo grapes, both made with either non-*Saccharomyces* yeasts or *S. cerevisiae* for the second fermentation [52]. To assess the final wines a descriptive sensory analysis was conducted using pre-determined scorecards. The scorecards consisted of fifteen attributes, including visual, olfactory and mouthfeel attributes as well as the overall perceived quality. The panel consisted of 11 experienced people from Polytechnic University of Madrid (age range from 27 to 57 years, four women and seven men). In both white and red wines, those produced from non-*Saccharomyces* yeasts showed significantly more limpidity compared to those produced from *S. cerevisiae*. In the red wines, highest effervescence was found in the sparkling wines produced with *S. ludwigii*, while the highest color intensity was reported for those produced with *S. pombe*. The white sparkling wines produced using *S. cerevisiae* were perceived as having significantly higher aroma quality compared to those produced from non-*Saccharomyces* yeasts. Higher aroma intensity scores were also reported for white sparkling wines produced from *S. ludwigii*, while in red samples those with *S. pombe*. White sparkling wines produced with *S. pombe* had higher scores for buttery and yeasty aromas and lower scores for flowery and fruity aromas. The red wines made from *S. pombe* had the highest scores for herbal, buttery, yeasty, acetic acid and oxidation aromas. Higher scores for buttery aromas were related to higher diacetyl production (characterized by buttery aromas with a threshold value of 0.1–5 mg/L) [75]. Wines made from non-*Saccharomyces* yeasts scored lower for fruity aromas in white sparkling wines, purportedly due to lower ester production by the yeasts. Crucially, the length of time aging on lees was only 4 months, while traditional method sparkling wines are subjected to longer aging periods. Further studies over longer periods are necessary to be able to evaluate the effect of non-*Saccharomyces* yeasts on the organoleptic characteristics of the wines.

Verdicchio base wine fermented in bottle with fermentations of two different *T. delbrueckii* strains, a mixed fermentation of *T. delbrueckii* strains with *S. cerevisiae* versus wines fermented only with

S. cerevisiae was carried out. The sensory analysis was carried out using a pre-determined list of descriptors and a scale of 1 to 10. The aromatic attributes (e.g., floral, fruity, toasty) and the main structural features (e.g., sweet, acidity, flavor, astringency, bitterness, olfactory persistence) were evaluated. The panel consisted of 10-trained tasters. For the main sensorial descriptors, significant differences were reported for mixed fermentations and pure fermentation of *T. delbrueckii* strains in comparison to *S. cerevisiae* wines. The sparkling wines produced with pure fermentation of *T. delbrueckii* 130 strain was characterized by the sensorial attributes of white flowers, bread crust, sapidity and acidity and were significantly different from the other wines, except for the attribute "sapidity" (the savory flavor associated with wine). Sparkling wines produced with both *T. delbrueckii* strains in pure and mixed fermentations obtained higher scores for the aromatic descriptors of white flowers, citrus, honey, odor intensity and softness in comparison with the control sparkling wines. These results were in agreement with the respective volatile compounds measured, since samples showed higher amounts of ethyl butyrate, ethyl hexanoate, ethyl octanoate and isoamyl acetate. Wines produced from the pure fermentation of *S. cerevisiae* obtained significantly higher scores for astringency [51]. However, components that contribute to astringency in wine such as phenolic compounds were not measured so it is unclear which compounds were responsible for these results.

The following paragraph discusses the sensory effects of non-*Saccharomyces* yeasts on both base wines and sparkling wines beginning with sequential fermentation (Table 3). González-Royo et al. [35] showed the effect of sequential fermentations of *T. delbrueckii* and *M. pulcherrima* on base wines. The majority of the panelists preferred base wines produced from sequential fermentations of *T. delbrueckii* over the wines made only with *S. cerevisiae*. The effect of sequential fermentations using *T. delbrueckii* for first fermentation on the corresponding traditional method sparkling wine was investigated by Medina-Trujillo et al. [36]. In this case, the majority of panelists preferred sparkling wines produced from sequential fermentations of *T. delbrueckii*.

Table 3. Summary of the impact of non-*Saccharomyces* yeasts on sensory profiles of base wines and sparkling wines.

Yeast	Production Stage	Sensory Evaluation	Effect on the Sensory Profile	Reference
T. delbrueckii + *S. cerevisiae*	First fermentation for base wine production	Sensory triangle test, panel with 9 tasters	It was distinguishable by 6 of the 9 tasters and 5 of them preferred them over control wine.	[35]
M. pulcherrima + *S. cerevisiae*			It was distinguishable by 8 of the 9 tasters and 4 of them preferred them over control wine. Smoky and flowery aromas.	
T. delbrueckii (sequential inoculation with *S. cerevisiae*)	First fermentation followed by a second fermentation	Sensory triangle test, panel with 12 tasters	It was distinguishable by 9 of the 12 tasters and 8 of them preferred them over control wine. Better integrated effervescence and less aggressiveness in the mouth.	[36]
S. ludwigii 979	Second fermentation in bottle + 4 months of aging on lees	Prepared evaluation sheet, panel with 11 tasters	In the red sparkling wines, higher limpidity and effervescence, in white sparkling wines higher limpidity but lower aroma intensity and quality in comparison to control.	[52]
S. pombe 7VA			In red sparkling wines, higher aroma intensity and higher scores for herbal, buttery, yeasty, acetic acid and oxidation aromas, in white sparkling wines higher limpidity; lower aroma quality, higher buttery, yeasty and reduction; lower flowery and fruity aromas in comparison to control.	
T. delbrueckii 130	Second fermentation in bottle + 12 months of aging on lees	Prepared evaluation sheet, panel with 11 tasters	It was characterized for the sensorial attributes of white flowers, bread crust, sapidity and acidity, with significant differences from other sparkling wines, except the attribute of sapidity.	[51]
T. delbrueckii 313			Significant differences were detected in the main sensory attributes in comparison to control wine. Higher scores for the aromatic descriptors (white flowers, citrus, honey, odor intensity, softness). Control wine showed significantly higher astringency in comparison to all other studied fermentations.	
S. cerevisiae + *T. delbrueckii* 130				
S. cerevisiae + *T. delbrueckii* 313				

5.2. Sensorial Influence from Lees Aging

During aging on lees, the organoleptic properties of sparkling wines evolve due to yeast autolysis, wine chemical composition, enzyme activity and the subsequent range of compounds that are released during storage [64]. Vannier et al. [125] and Torrens et al. [134] both evaluated the olfactory descriptors of panelists, who evaluated champagne and cava wines respectively. Champagne and cava sparkling wines both age on yeast lees but for different lengths of time, depending on their styles. The grape varieties in the two differ because, champagne wines are mainly made from Chardonnay, Pinot noir and Pinot meunier [135], while Macabeu, Xarel·lo and Parellada are the main varieties used in cava production [8]. According to Vannier et al. [125], the herbaceous and exotic fruit aromas decreased in champagne wines, while chemical, yeasty, butter and toasty notes increased during aging. Descriptors of base wine versus finished cava wines found that the profile of the wines were more complex than that of the base wine [134]. Many studies have been conducted to show the effect of non-*Saccharomyces* yeast on still wines and their effects on the wines' sensorial properties. Results of these studies can be transferred to traditional method sparkling wines.

T. delbrueckii is one of the most widely studied non-*Saccharomyces* yeasts. It is used already on an industrial scale in wine production [51]. Tempranillo wines made from sequential fermentations of *T. delbrueckii* and *S. cerevisiae* were evaluated by descriptive sensory analysis [34]. The six panelists used a scale from 0 to 10 to rate the overall perception and aromatic quality of the wines. *T. delbrueckii* strains in sequential fermentations, performed better than *S. cerevisiae* alone. Significant differences in aromatic quality from sensory evaluation of several fermentations with *T. delbrueckii* were correlated with tgreater production of several VOCs measured including esters, diacetyl and 3-ethoxy propanol [34]. More recently, Belda et al. [37] compared Verdejo wines fermented by co-inoculation of *T. delbrueckii* with *S. cerevisiae* to wines made using only *S. cerevisiae*. A panel of ten experienced wine tasters (members of the staff of the Food Technology Department of the Polytechnic University of Madrid and Microbiology Department of the Complutense University of Madrid) assessed the wines. Following the generation of attributes, twelve were chosen to describe the wines using a 10 cm unstructured scale. The authors reported that wines produced using *T. delbrueckii*, had a higher aroma quality, intensity and fruity character than the other wines. This result was accredited to a significant increase in varietal thiols, especially 4-MSP and in 2-phenylethyl, along with the lower values of higher alcohols. Marcon et al. [38] also reported a positive effect on Moscato Branco wines fermented by co-inoculation of *T. delbrueckii* with *S. cerevisiae*. The descriptive sensory attributes included visual and olfactory terms (aroma intensity and gustatory), were scored with an intensity scale (0–5), while the general sensory quality was scored from 0 to 100. The positive contribution of co-inoculation of *T. delbrueckii* with *S. cerevisiae* was again related to the increase in ester concentrations, and the reduction in higher alcohols and volatile fatty acids [38]. These studies show the positive effects of *T. delbrueckii* on sensory properties of still wines, which could be of interest for sparkling wines, although their concentrations during aging and interaction with autolytic flavor compounds need to be monitored.

In the study of Benito et al. [29] the sequential fermentation using *S. cerevisiae* and three non-*Saccharomyces* yeasts (*P. kluyveri*, *L. thermotolerans*, or *M. pulcherrima*) was carried out using Riesling grapes. A sensory evaluation by a panel of thirteen participants (staff of the Department of Microbiology and Biochemistry of the Hochschule Geisenheim University, Germany) used 17 attributes and a ten-point scale. These wines produced from non-*Saccharomyces* yeast had higher scores for overall impression and fruitiness, while those with *S. cerevisiae* had the lowest score for aroma quality but highest scores for ethyl acetate, acetaldehyde and oxidation. The high scores for acetaldehyde in sensory evaluation were corroborated chemical data that confirmed high values of acetaldehyde in the wines.

5.3. Glycerol

As a by-product of fermentation, glycerol is one of the compounds, after water and ethanol, that is found at the highest concentrations in wine (5–20 g/L). The concentration of glycerol in wine may change depending on the vinification conditions such as temperature, aeration, sulphite level and yeast strains [58]. In still table wines, glycerol content has been found to contribute positively to mouthfeel and Jolly et al. [136] states that some non-*Saccharomyces* strains positively influence wine quality. In the case of sparkling wine, glycerol content affects viscosity, foaming and VOCs [3] with a high concentration of glycerol in base wines having a negative synergistic effect with ethanol that could retard completion of the second alcoholic fermentation [58].

Increases in glycerol concentrations were one of the first recognized effects of non-*Saccharomyces* yeast species in fermentation winemaking [18]. Borrull et al. [58] determined the effect of glycerol levels on the growth of yeast strains in the presence of ethanol. The effect of 0, 5 and 10 g/L of glycerol was studied in the basal growth medium with 0%, 10% and 15% (v/v) of ethanol. The results showed that the glycerol concentration of 5 g/L did not modify the behavior of yeast strains in the absence or presence of ethanol. However, 10 g/L of glycerol concentration significantly affected it, regardless of the ethanol concentration. This caused a lower maximum growth rate and the initiation of the growth stage was longer than usual. The study concluded that a high glycerol level in the base wine could impact the second alcoholic fermentation and may even cause stuck fermentations [58]. The yeasts that produce a high amount of glycerol during fermentation, such as *S. kudriavzevii* [137], would probably prevent a successful second fermentation [58]. Non-*Saccharomyces* yeast species that have been described as high glycerol producers include *T. delbruckii, Candida zeylanoides, Candida stellata, Starmerella bacillaris* (synonym *Candida zemplinina*) and *L. thermotolerans* [19,35–37,42,45,49,120,138].

Glycerol concentrations reported in wines produced from *S. cerevisiae* have been in the range of 4.5–9.9 g/L [57,139], while non-*Saccharomyces* yeast have been found to produce concentrations of 8.3–10.5 g/L for wines *(S. japonicas)* and 9–11.4 g/L for *S. pombe* strains [139].

Benito et al. [44] found similar results from different *S. pombe* strains. Although two of the strains showed the highest values of glycerol, the other strains produced similar results to *S. cerevisiae* (8.02–8.91 g/L). Additionally, *M. pulcherrima*, increased glycerol concentrations, without increasing volatile acidity and acetaldehyde in the final wine [49].

Sequential fermentations with non-*Saccharomyces* yeasts, including *Kluyveromyces thermotolerans, P. kluyveri* and *M. pulcherrima*, produced higher amount of glycerol in comparison to *S. cerevisiae* [29]. The levels of glycerol varied from 5.8 to 6.3 g/L [29]. *T. delbrueckii* is a yeast that produces lower levels of glycerol than other non-*Saccharomyces* yeasts [140,141]. Mixed fermentations with non-*Saccharomyces* yeasts, including *T. delbrueckii*, was studied by Comitini et al. [19]. The mixed fermentations with non-*Saccharomyces* yeasts produced high amounts of glycerol were reported. However, similar amounts of glycerol were produced from the mixed fermentation of *T. delbrueckii* (5.88 g/L to 6.29 g/L and the sole fermentation of *S. cerevisiae* (6.23 g/L to 6.65 g/L). Glycerol content of sparkling wines produced with *T. delbrueckii* showed significantly higher glycerol values in comparison to sparkling wines produced with *S. cerevisiae* in a study by González-Royo et al. [35]. White sparkling wines produced with *S. ludwigii* have been found to have significantly higher concentrations of glycerol (4.95 g/L) in comparison to those produced with *S. cerevisiae* and *S. pombe* (4.57 g/L and 4.67 g/L respectively). Interestingly, the glycerol content of red sparkling wines ranged between 4.89 g/L to 5.12 g/L without any significant differences [52]. It is apparent from these results that glycerol concentrations in sparkling wines made from non-*Saccharomyces* yeast differ depending on yeast species and strain, and whether the yeast is used alone, or in combination with another yeast species. Importantly, uncertainty surrounding the long-term effect on sparkling wine foam and mouthfeel remains due to negative perceptions associated with increased mouthfeel in high quality sparkling wines.

6. Conclusions and Further Research

The nutrient requirement differences of non-*Saccharomyces* yeasts from their preference for either ammonium and/or amino acids suggests an area for further research in combination with the nitrogen requirements of a co-inoculation fermentation for first and/or second fermentation. Some non-*Saccharomyces* yeasts have been used to decrease the biogenic amount of sparkling wines although, further studies are needed to study their effect on biogenic amine concentrations after several years of lees aging. Non-*Saccharomyces* yeasts can influence the aromas of sparkling wines through production of enzymes and metabolites during aging in contact with yeast lees. Non-*Saccharomyces* yeasts have shown significant differences in numerous VOCs between species and strains. The studies on sensory effects of non-*Saccharomyces* yeasts on sparkling wines have found that the use of yeasts as sole inoculations, or in mixed fermentations to obtain specific sensory attributes and distinctive characters is possible. However, the studies so far, conducted have used relatively short lees aging times (4, 6 and 12 months). The ability to reduce alcohol levels by some yeasts could be beneficial to warm climate sparkling wine producers. However, our knowledge of their effect on sparkling wine practical production stages (i.e., riddling, disgorging), foam stability, flavor and aroma in wines that have been aged for long periods of cellar aging is limited. A major challenge to overcome is their acceptance by sparkling winemakers and established brands. With further research these yeasts when combined, provide a point of difference for small sparkling wine producers. The related topic of interspecific hybridization, and encapsulation of non-*Saccharomyces* yeast have not been considered in our review but are both areas that necessitate consideration in sparkling wine research.

Author Contributions: The authors contributed equally to the writing of the paper.

Acknowledgments: The authors would like to thank Perennia Food and Agriculture Inc., Nova Scotia, Canada, Acadia University, Nova Scotia, Canada and the Cool Climate Oenology and Viticulture Institute (CCOVI), Brock University, Canada for their support. We would like to acknowledge the contribution of Jeff Graham and Mike Matyjewicz (Sparkling Winos) for their drawing of sparkling wine bottles that appears in the graphical abstract for this paper.

Conflicts of Interest: The authors and Perennia Food and Agriculture Inc. declare no conflict of interest.

References

1. Jackson, R.S. Specific and Distinctive Wine Styles. In *Wine Science: In Principles and Applications*; Academic Press/Elsevier: Hoboken, NJ, USA, 2014; pp. 677–759.
2. Vigentini, I.; Cardenas, S.B.; Valdetara, F.; Faccincani, M.; Panont, C.A.; Picozzi, C.; Foschino, R. Use of Native Yeast Strains for In-Bottle Fermentation to Face the Uniformity in Sparkling Wine Production. *Front. Microbiol.* **2017**, *8*, 1225. [CrossRef] [PubMed]
3. Kemp, B.; Alexandre, H.; Robillard, B.; Marchal, R. Effect of production phase on bottle-fermented sparkling wine quality. *J. Agric. Food Chem.* **2015**, *14*, 19–38. [CrossRef] [PubMed]
4. Pérez-Magariño, S.; Ortega-Herasa, M.; Bueno-Herrera, M.; Martínez-Lapuente, L.; Guadalupe, Z.; Ayestarán, B. Grape variety, aging on lees and aging in bottle after disgorging influence on volatile composition and foamability of sparkling wines. *LWT-Food Sci. Technol.* **2015**, *61*, 47–55. [CrossRef]
5. Martínez-García, R.; García-Martínez, T.; Puig-Pujol, A.; Mauricio, J.C.; Moreno, J. Changes in sparkling wine aroma during the second fermentation under CO_2 pressure in sealed bottle. *Food Chem.* **2017**, *237*, 1030–1040. [CrossRef] [PubMed]
6. Pozo-Bayón, M.A.; Martínez-Rodríguez, M.; Pueyo, E.; Moreno-Arribas, M.V. Chemical and biochemical features involved in sparkling wine production: From a traditional to an improved winemaking technology. *Trends Food Sci. Technol.* **2009**, *20*, 289–299. [CrossRef]
7. Torresi, S.; Frangipane, M.T.; Anelli, G. Biotechnologies in sparkling wine production. Interesting approaches for quality improvement: A review. *Food Chem.* **2011**, *129*, 1232–1241. [CrossRef] [PubMed]
8. Riu-Aumatell, M.; Torrens, J.; Buxaderas, S.; López-Tamames, E. Cava (Spanish sparkling wine) aroma: Composition and determination methods. In *Recent Advances in Pharmaceutical Sciences III*; Muñoz-Torrero, D., Cortés, A., Mariño, E.L., Eds.; Transworld Research Network: Kerala, India, 2013; pp. 45–60.
9. Caliari, V.; Burin, V.M.; Rosier, J.P.; BordignonLuiz, M.T. Aromatic profile of Brazilian sparkling wines produced with classical and innovative grape varieties. *Food Res. Int.* **2014**, *62*, 965–973. [CrossRef]
10. Mafata, M. The Effect of Grape Temperature on The Phenolic Extraction and Sensory Perception of Méthode Cap Classique Wines. Ph.D. Thesis, Stellenbosch University, Stellenbosch, South Africa, March 2017.
11. Buxaderas, S.; López-Tamames, E. Sparkling wines: Features and trends from tradition. *Adv. Food Nutr. Res.* **2012**, *66*, 1–45. [CrossRef] [PubMed]

12. Duteurtre, B. Assemblage. In *Le Champagne: De La Tradition à La Science*; Lavoisier/Tec & Doc: Paris, France, 2006; pp. 116–125.

13. Ribéreau-Gayon, P.; Dubourdieu, D.; Doneche, B.; Lonvaud, A. Other Winemaking Methods. In *Handbook of Enology, the Microbiology of Wine and Vinifications*, 2nd ed.; John Wiley and Sons Ltd.: Hoboken, NJ, USA, 2006; pp. 445–480.

14. Duteurtre, B. Degorgement et bouchage. In *Le Champagne: De La Tradition à La Science*, 2nd ed.; Lavoisier/Tec & Doc: Paris, France, 2006; pp. 205–232.

15. Masneuf-Pomarede, I.; Bely, M.; Marullo, P.; Albertin, W. The genetics of non-conventional wine yeasts: Current knowledge and future challenges. *Front. Microbiol.* **2016**, *6*, 1563. [CrossRef] [PubMed]

16. Romani, C.; Lencioni, L.; Gobbi, M.; Mannazzu, I.; Ciani, M.; Domizio, P. *Schizosaccharomyces japonicus*: A polysaccharide-overproducing yeast to be used in winemaking. *Fermentation* **2018**, *4*, 14. [CrossRef]

17. Domizio, P.; Romani, C.; Lencioni, L.; Comitini, F.; Gobbi, M.; Mannazzu, I.; Ciani, M. Outlining a future for non-*Saccharomyces* yeasts: Selection of putative spoilage wine strains to be used in association with *Saccharomyces cerevisiae* for grape juice fermentation. *Int. J. Food Microbiol.* **2011**, *147*, 170–180. [CrossRef] [PubMed]

18. Jolly, N.P.; Varela, C.; Pretorius, I.S. Not your ordinary yeast: Non-*Saccharomyces* yeasts in wine production uncovered. *FEMS Yeast Res.* **2014**, *14*, 215–237. [CrossRef] [PubMed]

19. Comitini, F.; Gobbi, M.; Domizio, P.; Romani, C.; Lencioni, L.; Mannazzu, I.; Ciani, M. Selected non-*Saccharomyces* wine yeasts in controlled multistarter fermentations with *Saccharomyces cerevisiae*. *Food Microbiol.* **2011**, *5*, 873–882. [CrossRef] [PubMed]

20. Varela, C.; Siebert, T.; Cozzolino, D.; Rose, L.; Mclean, H.; Henschke, P.A. Discovering a chemical basis for differentiating wines made by fermentation with 'wild' indigenous and inoculated yeasts: Role of yeast volatile compounds. *Aust. J. Grape Wine Res.* **2009**, *15*, 238–248. [CrossRef]

21. García, M.; Esteve-Zarzoso, B.; Arroyo, T. Non-*Saccharomyces* Yeasts: Biotechnological Role for Wine Production. In *Grape and Wine Biotechnology*; Morata, A., Ed.; IntechOpen: London, UK, 2016; pp. 249–271. [CrossRef]

22. Chasseriaud, L.; Coulon, J.; Marullo, P.; Albertin, W.; Bely, M. New oenological practice to promote non-*Saccharomyces* species of interest: Saturating grape juice with carbon dioxide. *Appl. Microbiol. Biotechnol.* **2018**, *102*, 3779–3791. [CrossRef] [PubMed]

23. Bagheri, B.; Bauer, F.F.; Setati, M.E. The impact of Saccharomyces cerevisiae on a wine yeast consortium in natural and inoculated fermentations. *Front. Microbiol.* **2017**, *8*, 1988. [CrossRef] [PubMed]

24. Padilla, B.; Gil, J.V.; Manzanares, P. Past and Future of Non-*Saccharomyces* Yeasts: From Spoilage Microorganisms to Biotechnological Tools for Improving Wine Aroma Complexity. *Front. Microbiol.* **2016**, *7*, 411. [CrossRef] [PubMed]

25. Gschaedler, A. Contribution of non-conventional yeasts in alcoholic beverages. *Curr. Opin. Food Sci.* **2017**, *13*, 73–77. [CrossRef]

26. Lleixà, J.; Martín, V.; Portillo, M.D.; Carrau, F.; Beltran, G.; Mas, A. Comparison of Fermentation and Wines Produced by Inoculation of *Hanseniaspora vineae* and *Saccharomyces cerevisiae*. *Front. Microbiol.* **2016**, *7*, 338. [CrossRef]

27. Comitini, F.; Capece, A.; Ciani, M.; Romano, P. New insights on the use of wine yeasts. *Curr. Opin. Food Sci.* **2017**, *13*, 44–49. [CrossRef]

28. Petruzzi, L.; Capozzi, V.; Berbegal, C.; Corbo, M.R.; Bevilacqua, A.; Spano, G.; Sinigaglia, M. Microbial Resources and Enological Significance: Opportunities and Benefits. *Front. Microbiol.* **2017**, *8*, 995. [CrossRef] [PubMed]

29. Benito, S.; Hofmann, T.; Laier, M.; Lochbühler, B.; Schüttler, A.; Ebert, K.; Fritsch, S.; Röcker, J.; Rauhut, D. Effect on quality and composition of Riesling wines fermented by sequential inoculation with non-*Saccharomyces* and *Saccharomyces cerevisiae*. *Eur. Food Res. Technol.* **2015**, *241*, 707–717. [CrossRef]

30. Varela, B.; Borneman, A.R. Yeasts found in vineyards and wineries. *Yeast* **2017**, *34*, 111–128. [CrossRef] [PubMed]

31. Tristezza, M.; Tufariello, M.; Capozzi, V.; Spano, G.; Mita, G.; Grieco, F. The Oenological Potential of *Hanseniaspora uvarum* in Simultaneous and Sequential Co-fermentation with *Saccharomyces cerevisiae* for Industrial Wine Production. *Front. Microbiol.* **2016**, *7*, 670. [CrossRef] [PubMed]

32. Hu, K.; Jin, G.J.; Xu, Y.H.; Tao, Y.S. Wine aroma response to different participation of selected *Hanseniaspora uvarum* in mixed fermentation with *Saccharomyces cerevisiae*. *Food Res. Int.* **2018**, *108*, 119–127. [CrossRef] [PubMed]

33. Medina, K.; Boido, E.; Fariña, L.; Gioia, O.; Gomez, M.E.; Barquet, M.; Gaggero, C.; Dellacassa, E.; Carrau, F. Increased flavour diversity of Chardonnay wines by spontaneous fermentation and co-fermentation with *Hanseniaspora vineae*. *Food Chem.* **2013**, *141*, 13–21. [CrossRef] [PubMed]

34. Loira, I.; Vejarano, R.; Banuelos, M.A.; Morata, A.; Tesfaye, W.; Uthurry, C.; Villa, A.; Cintora, I.; Suarez-Lepe, J.A. Influence of sequential fermentation with *Torulaspora delbrueckii* and *Saccharomyces cerevisiae* on wine quality. *LWT-Food Sci. Technol.* **2014**, *59*, 915–922. [CrossRef]

35. González-Royo, E.; Pascual, O.; Kontoudakis, N.; Esteruelas, M.; Esteve-Zarzoso, B.; Mas, A.; Canals, J.M.; Zamora, F. Oenological consequences of sequential inoculation with non-*Saccharomyces* yeasts (*Torulaspora delbrueckii* or *Metschnikowia pulcherrima*) and *Saccharomyces cerevisiae* in base wine for sparkling wine production. *Eur. Food Res. Technol.* **2015**, *240*, 999–1012. [CrossRef]

36. Medina-Trujillo, L.; González-Royo, E.; Sieczkowski, N.; Heras, J.; Fort, F.; Canals, J.M.; Zamora, F. Effect of sequential inoculation (*Torulaspora delbrueckii/Saccharomyces cerevisiae*) in the first fermentation on the foam properties of sparkling wine (Cava). *Eur. Food Res. Technol.* **2017**, *243*, 681–688. [CrossRef]

37. Belda, I.; Ruiz, J.; Beisert, B.; Navascués, E.; Marquina, D.; Calderon, F.; Rauhut, D.; Benito, S.; Santos, A. Influence of *Torulaspora delbrueckii* in varietal thiol (3-SH and 4-MSP) release in wine sequential fermentations. *Int. J. Food Microbiol.* **2017**, *257*, 183–191. [CrossRef] [PubMed]

38. Marcon, A.R.; Schwarz, L.V.; Dutra, S.V.; Moura, S.; Agostini, F.; Delamare, A.P.L.; Echeverrigaray, S. Contribution of a Brazilian *Torulaspora delbrueckii* isolate and a commercial *Saccharomyces cerevisiae* to the aroma profile and sensory characteristics of Moscato Branco wines. *Aust. J. Grape Wine Res.* **2018**. [CrossRef]

39. Varela, C.; Barker, A.; Tran, T.; Borneman, A.; Curtin, C. Sensory profile and volatile aroma composition of reduced alcohol Merlot wines fermented with *Metschnikowia pulcherrima* and *Saccharomyces uvarum*. *Int. J. Food Microbiol.* **2017**, *252*, 1–9. [CrossRef] [PubMed]

40. Chen, K.; Escott, C.; Loira, I.; Del Fresno, J.M.; Morata, M.; Tesfaye, W.; Calderon, F.; Suarez-Lepe, J.A.; Han, S.; Benito, S. Use of non-*Saccharomyces* yeasts and oenological tannin in red winemaking: Influence on colour, aroma and sensorial properties of young wines. *Food Microbiol.* **2018**, *69*, 51–63. [CrossRef] [PubMed]

41. Benucci, I.; Cerreti, M.; Liburdi, K.; Nardi, T.; Vagnolic, P.; Ortiz-Julien, A.; Esti, M. Pre-fermentative cold maceration in presence of non-*Saccharomyces* strains: Evolution of chromatic characteristics of Sangiovese red wine elaborated by sequential inoculation. *Food Res. Int.* **2018**, *107*, 257–266. [CrossRef] [PubMed]

42. Englezos, V.; Torchio, F.; Cravero, F.; Marengo, F.; Giacosa, S.; Gerbi, V.; Rantsiou, K.; Rolle, L.; Cocolin, L. Aroma profile and composition of Barbera wines obtained by mixed fermentations of *Starmerella bacillaris* (synonym *Candida zemplinina*) and *Saccharomyces cerevisiae*. *LWT-Food Sci. Technol.* **2016**, *73*, 567–575. [CrossRef]

43. Morata, A.; Benito, S.; Loira, I.; Palomero, F.; Gonzalez, M.C.; Suarez-Lepe, J.A. Formation of pyranoanthocyanins by *Schizosaccharomyces pombe* during the fermentation of red must. *Int. J. Food Microbiol.* **2012**, *159*, 47–53. [CrossRef] [PubMed]

44. Benito, A.; Jeffares, D.; Palomero, F.; Calderón, F.; Bai, F.-Y.; Bähler, J.; Benito, S. Selected *Schizosaccharomyces pombe* Strains Have Characteristics That Are Beneficial for Winemaking. *PLoS ONE* **2016**, *11*, e0151102. [CrossRef] [PubMed]

45. Gobbi, M.; Comitini, F.; Domizio, P.; Romani, C.; Lencioni, L.; Mannazzu, I.; Ciani, M. *Lachancea thermotolerans* and *Saccharomyces cerevisiae* in simultaneous and sequential co-fermentation: A strategy to enhance acidity and improve the overall quality of wine. *Food Microbiol.* **2013**, *33*, 271–281. [CrossRef] [PubMed]

46. Lencioni, L.; Romani, C.; Gobbi, M.; Comitini, F.; Ciani, M.; Domizio, P. Controlled mixed fermentation at winery scale using *Zygotorulaspora florentina* and *Saccharomyces cerevisiae*. *Int. J. Food Microbiol.* **2016**, *3*, 36–44. [CrossRef] [PubMed]

47. Albertin, W.; Zimmer, A.; Miot-Sertier, C.; Bernard, M.; Coulon, J.; Moine, V.; Colonna-Ceccaldi, B.; Bely, M.; Marullo, P.; Masneuf-Pomarede, I. Combined effect of the *Saccharomyces cerevisiae* lag phase and the non-*Saccharomyces* consortium to enhance wine fruitiness and complexity. *Appl. Microbiol. Biotechnol.* **2017**, *101*, 7603–7620. [CrossRef] [PubMed]

48. Garavaglia, J.; Schneider, R.D.; Mendes, S.D.; Welke, J.E.; Zini, C.A.; Caramao, E.B.; Valente, P. Evaluation of *Zygosaccharomyces bailii* BCV 08 as a co-starter in wine fermentation for the improvement of ethyl esters production. *Microbiol. Res.* **2015**, *173*, 59–65. [CrossRef] [PubMed]

49. Escribano, R.; González-Arenzana, L.; Portu, J.; Garijo, P.; López-Alfaro, I.; López, R.; Santamaría, P.; Gutiérrez, A.R. Wine aromatic compound production and fermentative behaviour within different non-*Saccharomyces* species and clones. *J. Appl. Microbiol.* **2018**, *124*, 1521–1531. [CrossRef] [PubMed]

50. Loira, I.; Morata, A.; Comuzzo, P.; Callejo, M.J.; González, C.; Calderón, F.; Suárez Lepe, J.A. Use of *Schizosaccharomyces pombe* and *Torulaspora delbrueckii* strains in mixed and sequential fermentations to improve red wine sensory quality. *Food Res. Int.* **2015**, *76*, 325–333. [CrossRef] [PubMed]

51. Canonico, L.; Comitini, F.; Ciani, M. *Torulaspora delbrueckii* for secondary fermentation in sparkling wine production. *Food Microbiol.* **2018**, *74*, 100–106. [CrossRef] [PubMed]

52. Ivit, N.N.; Loira, I.; Morata, A.; Benito, S.; Palomero, F.; Suarez-Lepe, J.A. Making natural sparkling wines with non-*Saccharomyces* yeasts. *Eur. Food Res. Technol.* **2018**, *244*, 925–935. [CrossRef]

53. Whitener, M.E.B.; Stanstrup, J.; Panzeri, V.; Carlin, S.; Divol, B.; Du Toit, M.; Vrhovsek, U. Untangling the wine metabolome by combining untargeted SPME–GCxGCTOF-MS and sensory analysis to profile Sauvignon blanc co-fermented with seven different yeasts. *Metabolomics* **2016**, *12*, 53. [CrossRef]

54. Balikci, E.K.; Tanguler, H.; Jolly, N.P.; Erten, H. Influence of *Lachancea thermotolerans* on cv. Emir wine fermentation. *Yeast* **2016**, *33*, 313–321. [CrossRef] [PubMed]

55. Martínez-Rodríguez, A.J.; Carrascosa, A.V.; Martin-Alvarez, P.J.; Moreno-Arribas, V.; Polo, M.C. Influence of the yeast strain on the changes of the amino acids, peptides and proteins during sparkling wine production by the traditional method. *J. Ind. Microbiol. Biotechnol.* **2002**, *29*, 314–322. [CrossRef] [PubMed]

56. Perpetuini, G.; Di Gianvito, P.; Arfelli, G.; Schirone, M.; Corsetti, A.; Tofalo, R.; Suzzi, G. Biodiversity of autolytic ability in flocculent Saccharomyces cerevisiae strains suitable for traditional sparkling wine fermentation. *Yeast* **2016**, *33*, 303–312. [CrossRef] [PubMed]

57. Di Gianvito, P.; Perpetuini, G.; Tittarelli, F.; Schirone, M.; Arfelli, G.; Piva, A.; Patrignani, F.; Lanciotti, R.; Olivastric, L.; Suzzi, G.; et al. Impact of *Saccharomyces cerevisiae* strains on traditional sparkling wines production. *Food Res. Int.* **2018**, *109*, 552–560. [CrossRef] [PubMed]

58. Borrull, A.; Poblet, M.; Rozes, N. New insights into the capacity of commercial wine yeasts to grow on sparkling wine media. Factor screening for improving wine yeast selection. *Food Microbiol.* **2015**, *48*, 41–48. [CrossRef] [PubMed]

59. Penacho, V.; Valero, E.; Gonzalez, R. Transcription profiling of sparkling wine second fermentation. *Int. J. Food Microbiol.* **2012**, *153*, 176–182. [CrossRef] [PubMed]

60. Garofalo, C.; Arena, M.P.; Laddomada, B.; Cappello, M.S.; Bleve, G.; Grieco, F.; Beneduce, L.; Berbegal, C.; Spano, G.; Capozzi, V. Starter Cultures for Sparkling Wine. *Fermentation* **2016**, *2*, 21. [CrossRef]

61. Suárez Lepe, J.A.; Iñigo Leal, B. Vinificaciones especiales desde el punto de vista microbiológico. In *Microbiología Enológica, Fundamentos de Vinificación*, 3rd ed.; Ediciones Mundi-Prensa: Madrid, Spain, 2004; pp. 607–673.

62. Jackson, R.S. Fermentation. In *Wine Science: In Principles and Applications*, 4th ed.; Academic Press/Elsevier: Hoboken, NJ, USA, 2014; pp. 427–534.

63. Martínez-Rodríguez, A.J.; Polo, M.C. Effect of the addition of bentonite to the tirage solution on the nitrogen composition and sensory quality of sparkling wines. *Food Chem.* **2003**, *81*, 383–388. [CrossRef]

64. Alexandre, H.; Guilloux-Benatier, M. Yeast autolysis in sparkling wine—A review. *Aust. J. Grape Wine Res.* **2006**, *12*, 119–127. [CrossRef]

65. Moreno-Arribas, M.V.; Polo, M.C.; Pozo-Bayón, M.A. Peptides. In *Wine Chemistry and Biochemistry*; Moreno-Arribas, M.V., Polo, M.C., Eds.; Springer Science Business Media LLC: Berlin, Germany, 2009; pp. 191–209.

66. Martí-Raga, M.; Sancho, M.; Guillamon, J.M.; Mas, A.; Beltran, G. The effect of nitrogen addition on the fermentative performance during sparkling wine production. *Food Res. Int.* **2015**, *67*, 126–135. [CrossRef]

67. Gobert, A.; Tourdot-Maréchal, T.; Morge, C.; Sparrow, C.; Liu, Y.; Quintanilla-Casas, B.; Vichi, S.; Alexandre, H. Non-*Saccharomyces* yeasts nitrogen source preferences: Impact on sequential fermentation and wine volatile compounds profile. *Front. Microbiol.* **2017**, *8*, 2175. [CrossRef] [PubMed]

68. Moreno-Arribas, V.; Pueyo, E.; Polo, M.C.; Martin-Alvarez, P.J. Changes in the Amino Acid Composition of the Different Nitrogenous Fractions during the Aging of Wine with Yeasts. *J. Agric. Food Chem.* **1998**, *46*, 4042–4051. [CrossRef]

69. Charpentier, C.; Feuillat, M. Yeast autolysis. In *Wine Microbiology and Biotechnology*; Fleet, G., Ed.; Harwood Academic Publishers: London, UK, 1993; pp. 225–242.

70. Borrull, A.; Lopez-Martínez, G.; Miro-Abella, E.; Salvado, Z.; Poblet, M.; Cordero-Otero, R.; Rozes, N. New insights into the physiological state of *Saccharomyces cerevisiae* during ethanol acclimation for producing sparkling wines. *Food Microbiol.* **2016**, *54*, 20–29. [CrossRef]

71. Leroy, M.J.; Charpentier, M.; Duteurtre, B.; Feuillat, M.; Charpentier, C. Yeast Autolysis during Champagne Aging. *Am. J. Enol. Vitic.* **1990**, *41*, 21–28.

72. Martí-Raga, M.; Marullo, P.; Beltran, G.; Mas, A. Nitrogen modulation of yeast fitness and viability during sparkling wine production. *Food Microbiol.* **2016**, *54*, 106–114. [CrossRef]

73. Feuillat, M.; Charpentier, C. Autolysis of yeasts in Champagne. *Am. J. Enol. Vitic.* **1982**, *33*, 6–13.

74. Ribéreau-Gayon, P.; Glories, Y.; Maujean, A.; Dubourdieu, D. Nitrogen Compounds. In *Handbook of Enology the Chemistry of Wine Stabilization and Treatments*, 2nd ed.; John Wiley and Sons Ltd.: Hoboken, NJ, USA, 2006; Volume 2, pp. 109–140.

75. Lambrechts, M.G.; Pretorius, I.S. Yeast and its Importance to Wine Aroma-A Review. *S. Afr. J. Enol. Vitic.* **2000**, *21*, 97–129.

76. Bozdogan, A.; Canbas, A. Influence of yeast strain, immobilisation and ageing time on the changes of free amino acids and amino acids in peptides in bottle-fermented sparkling wines obtained from *Vitis vinifera* cv. Emir. *Int. J. Food Sci. Technol.* **2011**, *46*, 1113–1121. [CrossRef]

77. Bozdogan, A.; Canbas, A. The effect of yeast strain, immobilisation, and ageing time on the amount of free amino acids and amino acids in peptides of sparkling wines obtained from cv. Dimrit grapes. *S. Afr. J. Enol. Vitic.* **2012**, *33*, 257–263. [CrossRef]

78. Puig-Deu, M.; Lopez-Tamames, E.; Buxaderas, S.; Torre-Boronat, M.C. Quality of base and sparkling wines as influenced by the type of fining agent added pre-fermentation. *Food Chem.* **1999**, *66*, 35–42. [CrossRef]

79. Lleixà, J.; Manzano, M.; Mas, A.; Portillo, M.C. *Saccharomyces* and non-*Saccharomyces* competition during microvinification under different sugar and nitrogen conditions. *Front. Microbiol.* **2016**, *7*, 1959. [CrossRef] [PubMed]

80. Palomero, F.; Morata, A.; Benito, S.; Calderón, F.; Suárez-Lepe, J.A. New genera of yeasts for over-lees aging of red wine. *Food Chem.* **2009**, *112*, 432–441. [CrossRef]

81. Marcobal, A.; Martin-Alvarez, P.J.; Polo, M.C.; Munoz, R.; Moreno-Arribas, M.V. Formation of Biogenic Amines throughout the Industrial Manufacture of Red Wine. *J. Food Prot.* **2006**, *69*, 397–404. [CrossRef] [PubMed]

82. Anli, R.E.; Bayram, M. Biogenic Amines in Wines. *Food Rev. Int.* **2008**, *25*, 86–102. [CrossRef]

83. Ke, R.; Weic, Z.; Bogdald, C.; Göktaşe, R.K.; Xiaoa, R. Profiling wines in China for the biogenic amines: A nationwide survey and pharmacokinetic fate modelling. *Food Chem.* **2018**, *250*, 268–275. [CrossRef] [PubMed]

84. Martuscelli, M.; Arfelli, G.; Manetta, A.C.; Suzzi, G. Biogenic amines content as a measure of the quality of wines of Abruzzo (Italy). *Food Chem.* **2013**, *140*, 590–597. [CrossRef] [PubMed]

85. Lorenzo, C.; Bordiga, M.; Pérez-Álvarez, E.P.; Travaglia, F.; Arlorio, M.; Salinasa, M.R.; Coïsson, J.D.; Garde-Cerdán, T. The impacts of temperature, alcoholic degree and amino acids content on biogenic amines and their precursor amino acids content in red wine. *Food Res. Int.* **2017**, *99*, 328–335. [CrossRef] [PubMed]

86. Marques, A.P.; Leitao, M.C.; San Romao, M.V. Biogenic amines in wines: Influence of oenological factor. *Food Chem.* **2008**, *107*, 853–860. [CrossRef]

87. Costantini, A.; Vaudano, E.; Del Prete, V.; Danei, M.; Garcia-Moruna, E. Biogenic Amine Production by Contaminating Bacteria Found in Starter Preparations Used in Winemaking. *J. Agric. Food Chem.* **2009**, *57*, 10664–10669. [CrossRef] [PubMed]

88. Garcia-Ruiz, A.; Gonzalez-Rompinelli, E.M.; Bartollome, B.; Moreno-Arribas, M.V. Potential of wine-associated lactic acid bacteria to degrade biogenic amines. *Int. J. Food Microbiol.* **2011**, *148*, 115–120. [CrossRef] [PubMed]

89. Granchi, L.; Romano, P.; Mangani, S.; Guerrini, S.; Vincenzini, M. Production of biogenic amines by wine microorganisms. *Bull. OIV* **2005**, *78*, 595–610.

90. Caruso, M.; Fiore, C.; Contursi, M.; Salzano, G.; Paparella, A.; Romano, P. Formation of biogenic amines as criteria for the selection of wine yeasts. *World J. Microbiol. Biotechnol.* **2002**, *18*, 159–163. [CrossRef]
91. Lehtonen, P. Determination of amines and amino acids in wine: A review. *Am. J. Enol. Vitic.* **1996**, *47*, 127–133.
92. Sanlibaba, P.; Uymaz, B. Biogenic Amine Formation in Fermented Foods: Cheese and Wine. *Eur. Int. J. Sci. Technol.* **2015**, *4*, 81–92.
93. Wantke, F.; Moritz, K.; Sesztak-Greinecker, G.; Gotz, M.; Hemmer, W. Histamine Content in Red and Sparkling Wine and Relationship with Wine Quality. *J. Allergy Clin. Immunol.* **2008**, *121*, 194. [CrossRef]
94. Konakovsky, V.; Focke, M.; Hoffmann-Sommergruber, K.; Schmid, R.; Scheiner, O.; Moser, P.; Hemmer, W.; Jarisch, R. Levels of Histamine and other Biogenic Amines in Red Wines and Sparkling Wines. *J. Allergy Clin. Immunol.* **2011**, *127*, AB242. [CrossRef]
95. Ancin-Azpilicueta, C.; Gonzalez-Marco, A.; Jimenez-Moreno, N. Current knowledge about the presence of amines in wine. *Crit. Rev. Food Sci. Nutr.* **2008**, *48*, 257–275. [CrossRef] [PubMed]
96. Pozo-Bayón, M.A.; Monagas, M.; Bartolome, B.; Moreno-Arribas, V. Wine futures related to safety and consumer health: An integrated perspective. *Crit. Rev. Food Sci. Nutr.* **2012**, *52*, 31–54. [CrossRef] [PubMed]
97. Rapp, A.; Mandery, H. Wine aroma. *Experentia* **1986**, *42*, 873–884. [CrossRef]
98. Azzolini, M.; Fedrizzi, B.; Tosi, E.; Finato, F.; Vagnoli, P.; Scrinzi, C.; Zapparoli, G. Effects of *Torulaspora delbrueckii* and *Saccharomyces cerevisiae* mixed cultures on fermentation and aroma of Amarone wine. *Eur. Food Res. Technol.* **2012**, *235*, 303–313. [CrossRef]
99. Varela, C. The impact of non-*Saccharomyces* yeasts in the production of alcoholic beverages. *Appl. Microbiol. Biotechnol.* **2016**, *100*, 9861–9874. [CrossRef] [PubMed]
100. Kemp, B.; Hogan, C.; Xu, S.; Dowling, L.; Inglis, D. The impact of wine style and sugar addition in liqueur d'expedition (*dosage*) solutions on traditional method sparkling wine composition. *Beverages* **2017**, *3*, 7. [CrossRef]
101. Gallardo-Chacon, J.; Vichi, S.; López-Tamames, E.; Buxaderas, S. Changes in the Sorption of Diverse Volatiles by *Saccharomyces cerevisiae* Lees during Sparkling Wine Aging. *J. Agric. Food Chem.* **2010**, *58*, 12426–12430. [CrossRef] [PubMed]
102. Mamede, M.E.O.; Cardello, H.M.A.B.; Pastore, G.M. Evaluation of an aroma similar to that of sparkling wine: Sensory and gas chromatography analyses of fermented grape musts. *Food Chem.* **2005**, *89*, 63–68. [CrossRef]
103. Kemp, B.; Condé, B.; Jégou, S.; Howell, K.; Vasserot, Y.; Marchal, R. Chemical Compounds and Mechanisms involved in the Formation and Stabilization of Foam in Sparkling Wines. *Crit. Rev. Food Sci. Nutr.* **2018**. [CrossRef] [PubMed]
104. Waters, E.J.; Wallace, W.; Williams, P.J. Identification of heat unstable wine proteins and their resistance to peptidases. *J. Agric. Food Chem.* **1992**, *40*, 1514–1519. [CrossRef]
105. Goncalves, F.; Heyraud, A.; de Pinho, M.N.; Rinaudo, M. Characterisation of white wine mannoproteins. *J. Agric. Food Chem.* **2002**, *50*, 6097–6101. [CrossRef] [PubMed]
106. Fleet, G.H. Cell wall. In *The Yeasts: Yeast Organelles*; Rose, A.H., Harrison, J.S., Eds.; Academic Press: London, UK, 1991; Volume 4, pp. 199–277.
107. Klis, F.M.; Boorsma, A.; De Grot, P.W.J. Cell wall construction in *Saccharomyces cerevisiae*. *Yeast* **2006**, *23*, 185–202. [CrossRef] [PubMed]
108. Giovani, G.; Rosi, I.; Bertuccioli, M. Quantification and characterization of cell wall polysaccharides released by non-*Saccharomyces* yeast during alcoholic fermentation. *Int. J. Food Microbiol.* **2012**, *160*, 113–118. [CrossRef] [PubMed]
109. Domizio, P.; Liu, Y.; Bission, L.F.; Barile, D. Use of non-*Saccharomyces* wine yeasts as novel sources of mannoproteins in wine. *Food Microbiol.* **2014**, *43*, 5–15. [CrossRef] [PubMed]
110. Benito, S.; Palomero, F.; Morata, A.; Calderón, F.; Suárez-Lepe, J.A. New applications for *Schizosaccharomyces pombe* in the alcoholic fermentation of red wines. *Int. Food J. Sci. Technol.* **2012**, *47*, 2101–2108. [CrossRef]
111. Bertuccioli, M.; Ferrari, S. Laboratory experience on the influence of yeast in mouthfeel. In Proceedings of the Les Entretiens Scientifiques De Lallemand, Montreal, QC, Canada, 25 September 1999.
112. Polo, M.C.; Gonzalez de Llano, M.D.; Ramos, M. Derivatization and liquid chromatographic separation of peptides. In *Food Analysis by HPLC*; Nollet, M.L., Ed.; Dekker: New York, NY, USA, 1992; pp. 117–140.

113. Contreras, A.; Hidalgo, C.; Schmidt, S.; Henshke, P.A.; Curtin, C.; Varela, C. The application of non-*Saccharomyces* yeast in fermentations with limited aeration as a strategy for the production of wine with reduced alcohol content. *Int. J. Food Microbiol.* **2015**, *105*, 7–15. [CrossRef] [PubMed]

114. Alonso-del-Rio, J.; Lairón-Peris, M.; Barrio, E.; Querol, A. Effect on the prevalence of *Saccharomyces* non *cerevisiae* species against *S. cerevisiae* wine strain on wine fermentation: Competition, physiological fitness, and influence in final wine composition. *Front. Microbiol.* **2017**, *8*, 150. [CrossRef]

115. Pérez-Torrado, R.; Barrio, E.; Querol, A. Alternative yeasts for winemaking: *Saccharomyces* non-*cerevisiae* and its hybrids. *Crit. Rev. Food Sci. Nutr.* **2018**, *58*, 1780–1790. [CrossRef] [PubMed]

116. Morata, A.; Loira, I.; Tesfaye, W.; Bañuelos, M.A.; González, C.; Suárez-Lepé, J.A. *Lachancea thermotolerans* applications in wine technology. *Fermentation* **2018**, *4*, 53. [CrossRef]

117. Kapsopoulou, K.; Kapaklis, A.; Spyropoulos, H. Growth and fermentation characteristics of a strain of the wine yeast *Kluyveromyces thermotolerans* isolated in Greece. *World J. Microbiol. Biotechnol.* **2005**, *21*, 1599–1602. [CrossRef]

118. Balmaseda, A.; Bordona, A.; Reguant, C.; Bauttista-Gallego, J. Non–*Saccharomyces* in wine: Effect upon *Oenococcus oeni* and malolactic fermentation. *Front. Microbiol.* **2018**, *9*, 534. [CrossRef] [PubMed]

119. Belda, I.; Navascués, E.; Marquina, D.; Santos, A.; Calderon, F.; Benito, S. Dynamic analysis of physiological properties of *Torulaspora delbrueckii* in wine fermentations and its incidence on wine quality. *Appl. Microbiol. Biotechnol.* **2015**, *99*, 1911–1922. [CrossRef] [PubMed]

120. Puertas, B.; Jiménez, M.J.; Cantos-Villar, E.; Cantoral, J.M.; Rodriguez, M.E. Use of *Torulaspora delbuckii* and *Saccharomyces cerevisiae* in semi-industrial inoculation to improve quality of Palomino and Chardonnay wines in warm countries. *J. Appl. Microbiol.* **2017**, *122*, 733–746. [CrossRef] [PubMed]

121. Benito, S. The impact of *Torulaspora delbrueckii* yeast in winemaking. *Appl. Microbiol. Biotechnol.* **2018**, *102*, 3081–3094. [CrossRef] [PubMed]

122. Maicas, S.; Ferrer, S.; Pardo, I. NAD(P)H regeneration is the key for heterolactic fermentation of hexoses in *Oenococcus oeni*. *Microbiology* **2002**, *148*, 325–332. [CrossRef] [PubMed]

123. Benito, S.; Palomero, P.; Calderón, F.; Palmero, D.; Suárez-Lepé, J.A. Selection of appropriate *Schizosaccharomyces* strains for winemaking. *Food Microbiol.* **2014**, *42*, 218–224. [CrossRef] [PubMed]

124. Zhang, X.K.; Lan, Y.B.; Zhu, B.Q.; Xiang, X.F.; Duana, C.Q.; Shia, Y. Changes in monosaccharides, organic acids and amino acids during Cabernet Sauvignon wine ageing based on a simultaneous analysis using gas chromatography–mass spectrometry. *J. Food Sci. Agric.* **2017**, *98*, 104–112. [CrossRef] [PubMed]

125. Vannier, A.; Brun, O.X.; Feinberg, M.H. Application of sensory analysis to champagne wine characterisation and discrimination. *Food Qual. Prefer.* **1999**, *10*, 101–107. [CrossRef]

126. Buxaderas, S.; López-Tamames, E. Managing the quality of sparkling wine. In *Managing Wine Quality, Viticulture and Wine Quality*; Reynolds, A., Ed.; Woodhead Food Series: Sawston, UK, 2010; Volume 2, pp. 553–588.

127. Martínez-Rodríguez, A.; Carrascosa, A.V.; Barcenilla, J.M.; Pozo-Bayón, M.A.; Polo, M.C. Autolytic capacity and foam analysis as additional criteria for the selection of yeast strains for sparkling wine production. *Food Microbiol.* **2001**, *18*, 183–191. [CrossRef]

128. Hidalgo, P.; Pueyo, E.; Pozo-Bayón, M.A.; Martínez-Rodríguez, A.J.; Martinez-Alvarez, P.; Polo, M.C. Sensory and Analytical Study of Rosé Sparkling Wines Manufactured by Second Fermentation in the Bottle. *J. Agric. Food Chem.* **2004**, *52*, 6640–6645. [CrossRef] [PubMed]

129. McMahon, K.M.; Diako, C.; Aplin, J.; Mattinson, D.S.; Culver, C.; Rossa, C.F. Trained and consumer panel evaluation of sparkling wines sweetened to brut or demi sec residual sugar levels with three different sugars. *Food Res. Int.* **2017**, *99*, 173–185. [CrossRef] [PubMed]

130. López de Lerma, N.; Peinado, R.A.; Puig-Pujol, A.; Mauricioc, J.C.; Moreno, J.; García-Martínez, T. Influence of two yeast strains in free, bioimmobilized or immobilized with alginate forms on the aromatic profile of long aged sparkling wines. *Food Chem.* **2018**, *250*, 22–29. [CrossRef] [PubMed]

131. Liger-Belair, G.; Polidoric, G.; Zéninaria, V. Unraveling the evolving nature of gaseous and dissolved carbon dioxide in champagne wines: A state-of-the-art review, from the bottle to the tasting glass. *Anal. Chim. Acta* **2012**, *732*, 1–15. [CrossRef] [PubMed]

132. Obiols, J.M.; De la Presa-Owens, C.; Buxaderas, S.; Bori, J.L.; De la Torre-Boronat, C. Protocolo de evaluación de la formación de la efervescencia y espuma en un vino espumoso. *ACE Revista de Enología* **1998**, *15*, 3–15.

133. Gallart, M.; Tomás, X.; Suberiola, G.; López-Tamames, E.; Buxaderas, S. Relationship between foam parameters obtained by the gas-sparging method and sensory evaluation of sparkling wines. *J. Sci. Food Agric.* **2004**, *84*, 127–133. [CrossRef]

134. Torrens, J.; Riu-Aumatell, M.; Vichi, S.; López-Tamames, E.; Buxaderas, S. Assessment of Volatile and Sensory Profiles between Base and Sparkling Wines. *J. Agric. Food Chem.* **2010**, *58*, 2455–2461. [CrossRef] [PubMed]

135. Duteurtre, B. L'organisation champenoise. In *Le Champagne: De la Tradition à la Science*, 2nd ed.; Lavoisier/Tec & Doc: Paris, France, 2006; pp. 19–32.

136. Jolly, N.P.; Augustyn, O.P.H.; Pretorius, I.S. The Role and Use of Non-*Saccharomyces* Yeasts in Wine Production. *S. Afr. J. Enol. Vitic.* **2006**, *27*, 15–39. [CrossRef]

137. Arroyo-Lopez, F.N.; Perez-Torrado, R.; Querol, A.; Barrio, E. Modulation of the glycerol and ethanol syntheses in the yeast *Saccharomyces kudriavzevii* differs from that exhibited by *Saccharomyces cerevisiae* and their hybrid. *Food Microbiol.* **2010**, *27*, 628–637. [CrossRef] [PubMed]

138. Soden, A.; Francis, I.L.; Oakey, H.; Henschke, P.A. Effects of co-fermentation with *Candida stellata* and *Saccharomyces cerevisiae* on the aroma and composition of Chardonnay wine. *Aust. J. Grape Wine Res.* **2000**, *6*, 21–30. [CrossRef]

139. Domizio, P.; Liu, Y.; Bisson, L.F.; Barile, D. Cell wall polysaccharides released during the alcoholic fermentation by *Schizosaccharomyces pombe* and *S. japonicus*: Quantification and characterization. *Food Microbiol.* **2017**, *61*, 136–149. [CrossRef] [PubMed]

140. Bely, M.; Stoeckle, P.; Masneuf-Pomarede, I.; Dubourdieu, D. Impact of mixed *Torulaspora delbrueckii Saccharomyces cerevisiae* culture on high-sugar fermentation. *Int. J. Food Microbiol.* **2008**, *122*, 312–320. [CrossRef] [PubMed]

141. Renault, P.; Miot-Sertier, C.; Marullo, P.; Hernandez-Orte, P.; Lagarrigue, L.; Lonvaud- Funel, A.; Bely, M. Genetic characterization and phenotypic variability in *Torulaspora delbrueckii* species: Potential applications in the wine industry. *Int. J. Food Microbiol.* **2009**, *134*, 201–210. [CrossRef] [PubMed]

![fermentation logo] *fermentation*

MDPI

Review

Challenges of the Non-Conventional Yeast *Wickerhamomyces anomalus* in Winemaking

Beatriz Padilla [1], Jose V. Gil [2,3] and Paloma Manzanares [2,*]

[1] INCLIVA Health Research Institute, 46010 Valencia, Spain; bpadilla@incliva.es
[2] Department of Biotechnology, Instituto de Agroquímica y Tecnología de Alimentos (IATA),
   Consejo Superior de Investigaciones Científicas (CSIC), Paterna, 46980 Valencia, Spain; giljv@uv.es
[3] Departamento de Medicina Preventiva y Salud Pública, Ciencias de la Alimentación,
   Toxicología y Medicina Legal, Facultad de Farmacia, Universitat de València, Burjassot, 46100 Valencia, Spain
* Correspondence: pmanz@iata.csic.es; Tel.: +34-96 390-0022

Received: 27 July 2018; Accepted: 18 August 2018; Published: 20 August 2018

Abstract: Nowadays it is widely accepted that non-*Saccharomyces* yeasts, which prevail during the early stages of alcoholic fermentation, contribute significantly to the character and quality of the final wine. Among these yeasts, *Wickerhamomyces anomalus* (formerly *Pichia anomala, Hansenula anomala, Candida pelliculosa*) has gained considerable importance for the wine industry since it exhibits interesting and potentially exploitable physiological and metabolic characteristics, although its growth along fermentation can still be seen as an uncontrollable risk. This species is widespread in nature and has been isolated from different environments including grapes and wines. Its use together with *Saccharomyces cerevisiae* in mixed culture fermentations has been proposed to increase wine particular characteristics. Here, we review the ability of *W. anomalus* to produce enzymes and metabolites of oenological relevance and we discuss its potential as a biocontrol agent in winemaking. Finally, biotechnological applications of *W. anomalus* beyond wine fermentation are briefly described.

Keywords: non-*Saccharomyces* yeasts; *Wickerhamomyces anomalus*; *Pichia anomala*; enzymes; glycosidases; acetate esters; biocontrol; mixed starters; wine

---

1. Introduction

Saccharomyces cerevisiae is the main microorganism involved in the alcoholic fermentation of grape must. Moreover, the use of selected *S. cerevisiae* strains has provided an improvement in the control and homogeneity of fermentations. However, winemaking is a non-sterile process, and many other species of yeasts belonging to various non-*Saccharomyces* genera prevail during the early stages of alcoholic fermentation and contribute significantly to the character and quality of the final wine [1].

In the past, non-*Saccharomyces* yeasts were considered of secondary significance or as undesirable spoilage yeasts. Nowadays, the role of non-*Saccharomyces* has been re-evaluated, and it is widely accepted that selected strains can positively influence the winemaking process [2]. Beyond the contribution of non-*Saccharomyces* yeasts to wine aroma complexity [3], these yeasts can help address some of the modern challenges in winemaking, including the reduction of the ethanol content of wine [4–7] or the control of wine spoilage [8,9].

Ecological studies have shown that species of mainly *Hanseniaspora* (*Kloeckera*), *Candida*, and *Metschnikowia* initiate the fermentation together with species of *Pichia*, *Issatchenkia*, and *Kluyveromyces*. Occasionally, representatives of *Brettanomyces, Schizosaccharomyces, Torulaspora, Rhodotorula, Zygosaccharomyces*, and *Cryptococcus* genera are also present. These yeasts decline by mid-fermentation, and then, *S. cerevisiae* becomes predominant and continues the fermentation [10]. Based on the capability of some of these non-*Saccharomyces* yeasts to produce flavor-enhancing enzymes or to modify the concentration of secondary metabolites, different mixed starters have been designed and proposed as a tool to enhance

Fermentation **2018**, 4, 68

wine quality [3,11]. Moreover, several species including *Lachancea thermotolerans, Metschnikowia pulcherrima, Torulaspora delbrueckii, Pichia kluyveri*, and *Schizosaccharomyces pombe* are already commercially available.

Wickerhamomyces anomalus, formerly known as *Pichia anomala, Hansenula anomala, Candida pelliculosa* was recently assigned to the genus *Wickerhamomyces* based on phylogenetic analysis of gene sequences, which has caused major changes in the classification of yeasts. [12]. This species has been frequently isolated from grapes and wines. Although traditionally *W. anomalus* is associated with excessive production of ethyl acetate, which represents a serious handicap for their use in winemaking, this species has gained considerable importance for the wine industry since it exhibits interesting and potentially exploitable physiological and metabolic characteristics as summarized in Figure 1. Here, we revisit the contribution of *W. anomalus* in wine production. First, we review the ecology and prevalence of this yeast in winemaking, and we discuss its ability to produce enzymes, killer toxins, and metabolites of enological relevance. Second, we review the design of mixed starters of *W. anomalus* with *S. cerevisiae* to improve wine aroma complexity. Finally, we discuss biotechnological applications of *W. anomalus* beyond wine fermentation. When citing older literature, the original yeast species name will be kept.

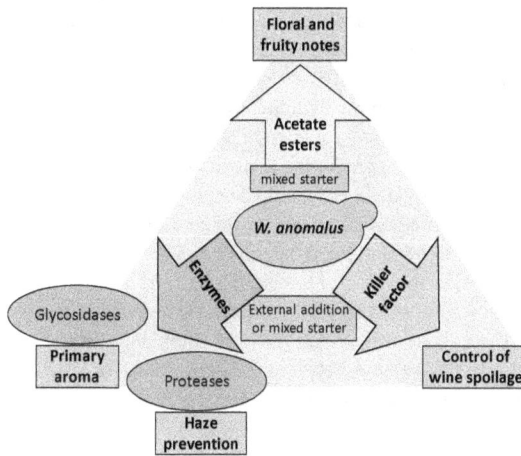

Figure 1. Benefits of *Wickerhamomyces anomalus* in winemaking.

2. *W. anomalus* Is a Ubiquitous Yeast Generally Associated with Winemaking

W. anomalus is a heterothallic, ascomycetous yeast, forming one to four hat-shaped ascospores [13, 14]. The placement of *P. anomala* in the genus *Wickerhamomyces* was due to multigene phylogenetic analysis [11]. *W. anomalus* is a widely used name and a proposal to conserve the species name anomala (-us). *W. anomalus* is a biotechnologically relevant yeast species with food, environmental, industrial, and medical applications. Natural habitats of *W. anomalus* are very diverse and include tree exudates, plants and fruit skins, insects, human tissues, and faeces, and also wastewaters and marine environments. The versatility of this species is encouraged by its ability to tolerate extreme environmental conditions like oxidative, salt, and osmotic stress, as well as pH and temperature shocks [15]. Due to these characteristics, this yeast can be a spoilage organism, for instance, in high-sugar food products [16,17] or silage [18]. Its genome sequence is already available, providing the basis to analyse metabolic capabilities, phylogenetic relationships, and biotechnologically important features [19,20]. The main physiological and genetics features of *W. anomalus* are reviewed in Reference [14].

In winemaking, *W. anomalus* is a ubiquitous yeast which has been previously associated with grape, must, wine, and winery facilities. It was shown that *H. anomala* is present during the early stages of red wine fermentation even when the must is inoculated with 10^5 to 10^7 cells of *S. cerevisiae* per mL, thus, making a significant contribution to fermentation [21]. Different studies described that *W. anomalus* isolated from grape must was able to persist until the end of fermentation [22,23]. Some strains of *W. anomalus* can tolerate up to 12.5% (*v*/*v*) ethanol and are known to produce killer toxins [15,24], allowing this species to compete against other yeasts in the same environment. *W. anomalus* is able to grow abundantly in wine due to its fully aerobic or weakly fermentative metabolism, and it is known for film formation on the surface of bulk wines in unfilled containers and with insufficient sulphite levels to prevent their growth [25].

Grapes are a primary source of non-*Saccharomyces* yeasts including several *Pichia* species [26]. *P. anomala* was found throughout different vineyards over a period of three years in conventional and organic vineyards, representing approximately 20% and 25% of yeast species isolated from musts obtained from Grenache and Shiraz varieties [27]. In a similar study, *W. anomalus* was the second dominant yeast after *Hanseniaspora uvarum* in Cabernet Sauvignon grape must derived from integrated vineyards [28]. However, it was observed that the cell concentration of *W. anomalus* only increased marginally throughout fermentation, suggesting that its growth is severely hampered by the lack of oxygen [28]. This yeast generally shows low growth rates and biomass yields under anaerobic conditions [15]. Yeast isolations from Malvar grape musts pointed out *W. anomalus* as one of the most frequent non-*Saccharomyces* species, and in addition, the yeast was a good producer of extracellular enzymes which may be beneficial in winemaking [29]. Recently, the dynamics of several non-*Saccharomyces* species were evaluated in synthetic must in the presence or absence of *S. cerevisiae* [30]. The study showed that the behaviour of the non-*Saccharomyces* species was differentially influenced by the presence of *S. cerevisiae*. Interestingly, in the absence of *S. cerevisiae*, *W. anomalus* suppressed the rest of non-*Saccharomyces* species suggesting that the yeast can survive in the early stages of the fermentation better than the other yeast species and may utilize the nitrogen released by dead cells. However, in the presence of *S. cerevisiae*, *W. anomalus* specifically declined early in fermentation, suggesting an antagonistic interaction between both yeasts [30]. This interaction has also been proposed in apple cider fermentations [31].

The prevalence of *P. anomala* in cellar equipment has been described in several Spanish wineries, and it was the only species among all detected that it was present in all four wineries evaluated [32]. Previously, it was found that besides *S. cerevisiae*, the most commonly detected species were *P. anomala*, *Pichia membranifaciens*, *Candida* spp. and *Cryptococcus* spp. [33]. Finally, *Pichia* spp. accounted for 83% of non-*Saccharomyces* yeasts present in winery surfaces, such as floor, pumps, and empty tanks, whereas *Hanseniaspora* spp. accounted for the remaining 17% [34].

Ecological studies in different wine regions of the world have also identified other *Pichia* species. Although at lower levels, several species such as *Pichia terricola*, *Pichia kudriavzevii*, and *P. kluyvery* were present in freshly extracted grape musts from Bordeaux region, although they rapidly disappeared from fermenting musts [35]. However, *P. membranifaciens* and *Pichia fermentans* appeared after the starting of the malolactic fermentation, and the former was present in samples of red and white Bordeaux wines examined at 1- and 2-month intervals after fermentation [35]. *P. membranifaciens* was also identified in spontaneous fermentations of musts from La Mancha, Spain [36] and in grape varieties used in India for winemaking [37]. As a minor species, *P. membranifaciens* was described as part of the indigenous population during spontaneous fermentations of wines in Mendoza, Argentina [38]. Yeast diversity studies of grape varieties from vine-growing regions of China identified *P. fermentans* and *Pichia guilliermondii* [39], and the former was also isolated from a Southern Italian autochthonous grape cultivar [40]. *P. kluyveri* and *Pichia farinosa* were found in vineyards and grape musts from four production regions of South Africa, although they were not the predominant species [41]. By contrast, significant amounts of *P. kluyveri* and *P. kudriavzevii* were isolated from grapes varieties of the Strekov winegrowing region in Slovakia, and they were more associated with damaged than with intact berries [42]. Both species are considered as indicators of mould-damaged grapes [26,43]. Recently the species *Pichia galeiformis* was identified for the first time on grape berries by FT-IR spectroscopy [44].

3. *W. anomalus* Is a Good Producer of Relevant Enzymes for Winemaking

W. anomalus strains isolated from enological ecosystems have been reported as an interesting source of different enzymes which could be used in the winemaking industry [45]. Aroma is one of the most appreciable characteristics influencing the overall quality of wine. Non-*Saccharomyces* yeasts may have an impact on both the primary and secondary aroma through the production of enzymes and metabolites, respectively. Strains identified as *W. anomalus* or its former names have been reported to produce glycosidases such as β-D-glucosidase, α-L-arabinofuranosidase, α-L-rhamnosidase, and β-D-xylosidase, which are involved in the release of aroma compounds from grape precursors [3].

Several authors have explored the enzymatic potential of non-*Saccharomyces* isolates with the aim of identifying good producer strains. A study conducted on 20 different yeast species showed that all tested strains of *H. anomala* presented β-glucosidase activity [46]. Results from other screenings concluded that *P. anomala* strains exhibited higher β-glucosidase activity when compared with yeast species belonging to other genera such as *Candida, Dekkera* or *Torulaspora* [47,48].

Besides showing β-D-glucosidase activity, other *P. anomala/W. anomalus* strains exhibited α-L-arabinofuranosidase or β-D-xylosidase activity in screenings as well, including more than 300 and 100 wine yeast isolates, respectively [49,50]. Similarly, *P. anomala* produced β-D-xylosidase with activity at pH, temperature, and concentrations of glucose and ethanol usually found during wine fermentation [51]. Interestingly, selected *P. anomala* strains are able to produce several glycosidase activities, for instance one *P. anomala* strain was described as producer of the four glycosidase activities [52], and a *W. anomalus* strain producing β-D-glucosidase, also showed α-L-arabinofuranosidase and β-D-xylosidase activities [24].

Despite the potential of *W. anomalus* to produce glycosidases, the effect of purified *W. anomalus* enzymes on the releasing of wine volatile compounds has been scarcely explored. Terpene production was observed in Muscat-type grape juice and wine treated with β-D-glucosidase from *P. anomala* MDD24 [53,54]. This glucosidase was efficient in releasing desirable aromas particularly during the final stage of alcoholic fermentation due to its tolerance to high concentrations of ethanol. Furthermore, isolates of *W. anomalus* showing β-D-glucosidase activity provoked a moderated overall terpene increase when inoculated to final wines [49]. However, the effectiveness of *W. anomalus* α-L-arabinofuranosidase, α-L-rhamnosidase or β-D-xylosidase for aromatic compounds releasing during winemaking has not been explored yet.

The strain *W. anomalus* AS1 was selected by the capability of its cells to hydrolyze different synthetic and natural glycosides under wine related conditions [24]. Afterwards, the enzyme was purified from the culture supernatant of AS1 and characterized as a multifunctional exo-β-1,3-glucanase active under typical oenological conditions [55]. Thus, the enzyme might have multiple applications in winemaking such as increasing concentrations of sensory and bioactive compounds by splitting glycosylated precursors or to reduce viscosity by hydrolysis of glycan slimes. The role of exo-β-1,3-glucanases in increasing wine aroma through the release of glycosidic precursors has been previously discussed [56].

Besides the contribution to the aromatic profile of wines, other relevant enzymes for winemaking are also produced by *W. anomalus* strains. Degradation of haze forming-proteins by enzymes is an attractive alternative to bentonite fining because it would minimize losses of wine volume and aroma [57]. In fact, wine yeasts secreting proteolytic enzymes are of high biotechnological interest for protein haze prevention because they could be directly added as starter cultures to the grape must. This is the case of *W. anomalus* 227 which secretes the aspartic protease WaAPR1 in white grape juice, suggesting its suitability for reducing grape must protein content [58].

4. *W. anomalus* Is a Good Producer of Acetate Esters

Non-*Saccharomyces* yeast, were traditionally considered as spoilage wine microorganisms due to high ethyl acetate production. In particular, *P. anomala* is a major ethyl acetate producer, and some strains show levels of ethyl acetate higher than 150 mg/L [59], close to the concentration at which this acetate ester can impart spoilage character to wine (150–200 mg/L) [60]. However, *P. anomala* is also a good producer of fruity acetate esters and other volatiles with a positive impact on wine aroma. The ability of 37 strains of non-*Saccharomyces* yeasts, including seven *P. anomala* strains, to produce the main wine acetate esters; ethyl acetate, isoamyl acetate, and 2-phenylethyl acetate was examined in Reference [59] (Figure 2). Among the genera evaluated, *Pichia* and *Hanseniaspora* stood out as the best producers of acetate esters, although significant differences among strains were found in the production of the three esters, highlighting the convenience of carrying out adequate screenings for selection of the appropriate strains. All the seven *P. anomala* strains included in the study were good isoamyl acetate producers, and interestingly, five of them produced a level of ethyl acetate lower than 200 mg/L. None of them was able to produce 2-phenylethyl acetate [59]. A similar screening studied the main oenological characteristics of 55 non-*Saccharomyces* yeast strains, 14 of them belonging to the *Pichia* genus. Levels of ethyl acetate production of these *Pichia* species ranged between 0.35 and 272 mg/L, whereas the only strain of *P. anomala* evaluated produced around 100 mg/L. This *P. anomala* strain was selected to be included in a mixed starter due to its fermentative characteristics and its ethanol and polysaccharides production [61].

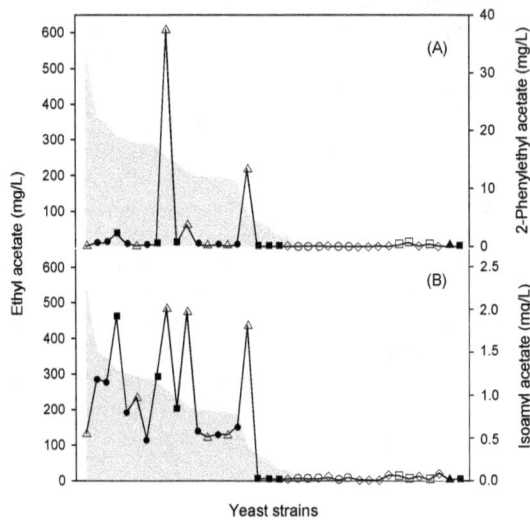

Figure 2. Production of acetate esters by non-*Saccharomyces* yeast strains. (**A**) 2-Phenylethyl acetate (symbols; right axis). (**B**) Isoamyl acetate (symbols; right axis). Ethyl acetate is represented in both panels as an area plot (left axis). Strains: ●*Pichia anomala* (seven different strains), ■ *P. membranifaciens* (seven different strains), ▲ *P. fermentans* (one strain), ◇ *Candida* spp., △ *Hanseniaspora* spp., ○ *Torulaspora* spp., □ *Zygosaccharomyces* spp. Adapted from Reference [59].

5. *W. anomalus* Produces Killer Toxins of Broad Spectrum

After its initial discovery in *S. cerevisiae*, the killer phenotype was described in non-*Saccharomyces* yeasts [62]. Killer toxins represent a biocontrol strategy alternative to the use of chemical preservatives or physical methodologies during the winemaking process [63]. In this context, *W. anomalus* killer proteins have been reported as antimicrobial agents against undesired microorganisms present in different food and beverages [14,64].

In the oenological environment, *W. anomalus* killer toxins are mainly tested against the prevailing wine spoilage microorganism *Dekkera/Brettanomyces* [8,65]. Nevertheless, the antimicrobial activity of *W. anomalus* towards other minor yeast species present during the early stages of grape fermentation such as *P. guilliermondii* or *P. membranifaciens* has also been reported [66,67]. Moreover, killer cultures belonging to *P. anomala* showed a broad killer spectrum against regionally relevant spoilage yeast and *Dekkera bruxellensis* collection strains [68]. The killer toxin Pikt produced by the *P. anomala* DBVPG 3003 strain was active on 15 isolates belonging to the genus *Dekkera/Brettanomyces*, and its fungicidal effect in wine was maintained during at least 10 days [69]. Further research revealed that this toxin presented a ubiquitin-like peptide structure with a molecular mass of approximately 8 kDa, and that it selectively interacts with β-1,6 glucans, which are the putative binding sites for Pikt on the cell wall of the sensitive targets [70]. Recently, the killer toxin KTCf20 secreted by the strain *W. anomalus* Cf20 was also suggested to bind to β-1,3 and β-1,6 glucans of the cell wall of sensitive strains. Moreover, the toxin was produced and showed to be stable and highly active at physicochemical conditions suitable for the winemaking process [66]. Finally, the potential use of Pikt from *W. anomalus* D2 as an alternative to sulphur dioxide (SO_2) has been proposed, since differently to SO_2, Pikt produced irreversible damage on sensitive yeasts, ensuring the complete control of spoilage *Brettanomyces* yeasts [71].

Beside the biocontrol effect of *W. anomalus* on non-*Saccharomyces* spoilage yeasts, it has also been reported that some isolates showed killer activity against *S. cerevisiae* strains [66,72]. Thus, the compatibility of selected killer *W. anomalus* strains with the main microbial agents involved in wine production needs to be tested during the selection procedure to avoid technological problems due to sluggish or incomplete alcoholic fermentations.

6. *W. anomalus* in Mixed Starters with *S. cerevisiae*

In recent years, the possibility to improve the fermentation process and the aromatic complexity of wine using selected non-*Saccharomyces* strains in mixed starters with *S. cerevisiae* has been investigated by many authors. This practice is proposed as a way to avoid stuck fermentations, control the ecological balance, achieve unique and distinctive aromatic characteristics, and control some specific oenological aspects, such as acidity, ethanol, or glycerol content [3,61,73–79]. Screening studies are useful to select appropriate non-*Saccharomyces* strains that, based on their quality profiles, could be good candidates to be part of a mixed starter. However, the behaviour of the selected strains could be modified by the presence of *S. cerevisiae* in the mixed starter [61,80]. Moreover, the appropriate modality (sequential or simultaneous) and inoculation time, the proportion of yeasts in the culture, and the potential microorganism interactions should be taken into account [61,81,82].

Different mixed starters containing *P. anomala* and other species of the *Pichia* genus have been proposed to improve wine quality (Table 1). Wines obtained with mixed cultures *P. anomala/S. cerevisiae* are characterized by higher concentrations of acetate esters, particularly ethyl acetate and isoamyl acetate [61,79,83–85]. Some of these wines showed levels of ethyl acetate higher than 150 mg/L [61,79], close to the concentration at which this acetate ester can impart spoilage character to wine (150–200 mg/L) [60]. However, wines with the highest concentrations of ethyl acetate were fermented in small volumes (less than 140 mL) where excessive aeration could promote the production of ethyl acetate. In contrast, experimental wines produced in 100 L tanks showed ethyl acetate levels less than 45 mg/L [83,84].

With the aim of reducing the production of ethyl acetate due to *P. anomala* in mixed cultures, the efficacy of a *petite P. anomala* mutant with low respiratory activity was investigated. In mixed cultures with *S. cerevisiae*, the *P. anomala* mutant died quicker and produced lower amounts of ethyl acetate than the wild type. Moreover, wines fermented by mixed cultures with the *petite* mutant strain of *P. anomala* and *S. cerevisiae* presented 100 mg/L of ethyl acetate, half the amount detected using the *P. anomala* wild-type strain, and had a better flavour profile [85]. Increases in acetate esters in wines fermented with *P. anomala* mixed cultures have been correlated with high scores in sensory preference tests, mainly in terms of floral and fruity notes [83,84]. In addition, herbaceous notes were related to higher levels of lineal alcohols in wines fermented with mixed cultures [83].

Table 1. Mixed starters of *Pichia* species and their main impact on wine quality.

Mixed Starter	Impact on Wine	Inoculation	Must	Ref.
P. anomala/S. cerevisiae	Isoamyl acetate increase	Co-inoculation	Bobal	[79]
	Isoamyl acetate increase	Co-inoculation	Commercial	[61]
	Acetate ester increase	Co-inoculation	Synthetic	[85]
	Acetate ester increase and alcohol decrease	Sequential	Airén	[84]
	Acetate and ethyl ester increase	Sequential	Mazuela	[83]
P. kudriavzevii/S. cerevisiae	Isoamyl acetate increase	Co-inoculation	Cabernet Sauvignon	[86]
P. membranifaciens/S. cerevisiae	Isoamyl and 2-phenetyl acetate	Sequential	Muscat	[87]
P. burtonii/S. cerevisiae	Ethyl ester increase	Sequential	Synthetic	[88]
P. kluyveri/S. cerevisiae	3-Mercaptohexyl acetate increase	Co-inoculation	Sauvignon Blanc	[81]
	3-Sulfanylhexan-1-ol increase	Co-inoculation	Sauvignon Blanc	[89]
	Off-flavor formation	Sequential	Sauvignon Blanc	[90]
P. fermentans/S. cerevisiae	Polysaccharide increase	Co-inoculation	Commercial	[91]

Other species of the *Pichia* genus have been included in mixed starter cultures together with *S. cerevisiae*. Some examples are summarized in Table 1. Wines produced with mixed cultures of *P. kudriavzevii* [86] and *P. membranifaciens* [87] presented increases in acetate esters as described for *P. anomala* strains. Wine fermented with a mixed starter of *Pichia burtonii/S. cerevisiae* contained higher amounts of ethyl esters [88], whilst co-inoculation of *P. kluyveri* and *S. cerevisiae* increased varietal aromas, mainly 3-mercaptohexyl acetate (3MHA) and 3-sulfanylhexan-1-ol in Sauvignon Blanc wines [81,89]. By contrast, a different *P. kluyvery* isolate did not show a sensorial significant increase in the tropical fruity aromas characterized by 3MHA, and the production of 3-methyl-butanoic acid was associated with an off-putting sour, sweaty, and cheesy aroma that is considered a wine fault [90]. Interestingly, the association of *P. fermentans* with *S. cerevisiae* in mixed cultures produced significant increases in the production of polysaccharides, which improve wine taste and body and exert positive effects on aroma persistence and protein and tartrate stability [91].

7. Applications of *W. anomalus* beyond Wine Fermentation

Different biotechnological applications of *W. anomalus* beyond winemaking are summarized in Table 2. Similar to wine fermentations, the application of non-*Saccharomyces* yeasts in the production of other alcoholic beverages and in bread fermentation is being explored. Besides the use of *Dekkera/Brettanomyces* for the production of sour beers, *W. anomalus* stands out as a promising yeast in brewing fermentations mainly due to its diversified enzymatic activities and bioconversion abilities [92]. The fermentation of cider by sequentially mixed cultures of *W. anomalus* and *S. cerevisiae* improved the final quality of cider as a result of a greater variety and amount of esters, higher alcohols, aldehydes, and ketones [31]. *P. anomala* mixed starters have also been proposed to improve the sensorial quality of the sugar cane spirit cachaça since co-inoculation with *S. cerevisiae* led to increases in acetate esters and other volatile compounds associated to good sensory descriptors [93]. Recently a mixed culture of *W. anomalus* with *S. cerevisiae* has been proposed for Chinese Baijiu making due to its positive effects on the end flavor of the beverage [94]. In addition, co-cultures of *S. cerevisiae*, *T. delbrueckii* and *P. anomala* as leavening agents for bread resulted in a higher abundance of volatile organic compounds and in higher sensorial ratings [95]. Finally, the dietary inclusion of *W. anomalus* as single cell protein in aquaculture showed positive effects on rainbow trout gut microbiota abundance and composition [96].

Table 2. Biotechnological applications of *W. anomalus* beyond wine fermentation.

Application	Yeast Strain	Reference
Food and beverage production		
beer	*W. anomalus* [1]	[92]
cider	*W. anomalus* YN6	[31]
cachaça	*P. anomala* UFLA CAF70 and CAF119	[93]
Chinese Baijiu	*W. anomalus* GZ3	[94]
bread	*P. anomala* JK04	[95]
Aquaculture	*W. anomalus* [1]	[96]
Biocontrol		
cereal grain preservation	*P. anomala* J121	[97,98]
antimycotic agent	*P. anomala* C33, C85, Di8, Di28, DBVPG3649	[99]
	P. anomala CMGB88	[72]
Production of fuels and chemicals		
bioethanol	*P. anomala* CBS132101	[100]
	W. anomalus [1]	[101]
ethyl acetate	*W. anomalus* NCYC16	[102]
	W. anomalus DSM 6766	[103]

[1] Strain not specified.

Regarding biocontrol capacity, the positive role of *P. anomala* in grain biopreservation is well established [97]. The yeast improved feed hygiene during storage of moist crimped barley grain by reduction of moulds and *Enterobacteriaceae*. Moreover, *P. anomala* enhanced the nutritional quality of the feed by increasing protein content and reducing the concentration of the antinutritional compound phytate [98]. The killer activity of *P. anomala* is also of interest in biomedical applications due to its activity against potential pathogenic yeast species, which may lead to the development of new antimycotic agents [72,99].

Production of fuels and chemicals is another area of potential application of *W. anomalus*. Bioethanol production exposes yeasts to complex fermentation medium with specific inhibitors and sugar mixtures. *W. anomalus* is able to produce ethanol in multiple biomass hydrolysates with different toxicity levels, is capable of utilizing xylose for growth when supplied with air, and can use nitrate as nitrogen source, making this species a potential ethanol producer using lignocellulosic biomass as a feedstock [100]. Moreover, other studies have identified *W. anomalus* strains that have a comparable ethanol yield to *S. cerevisiae*, although longer fermentation time was needed [101]. In addition to ethanol, *W. anomalus* has the potential to produce the industrially-relevant chemical ethyl acetate from numerous different carbon sources [102]. Ethyl acetate can be used as a microbiologically degradable and environmentally friendly solvent in the manufacture of food, glues, inks, and perfumes, and *W. anomalus* can be an alternative to the chemical processes. Recently, the identification of a novel enzyme Eat1 from *W. anomalus* resulted in high ethyl acetate production when expressed in *S. cerevisiae* and *Escherichia coli*, opening new possibilities for the production of biobased ethyl acetate [103].

8. Final Considerations

Based on the studies reviewed here, the potential positive influence of *W. anomalous* in winemaking seems clear. Indeed, mixed starters with selected *W. anomalus* strains and *S. cerevisiae* can enhance wine aroma, but also control spoilage wine microorganisms. Moreover, *W. anomalus* can exert positive effects in other fermentation processes.

Considering that *W. anomalus* is still seen as a spoilage yeast by winemakers, the commercial application of this yeast seems distant. Since *W. anomalous* is a ubiquitous yeast in the winemaking environment, smart strain screenings will provide appropriate candidates to be included as part of commercial mixed starters. These new strains will allow to exploit positive features of *W. anomalus* while minimizing negative aspects. Undoubtedly, further studies must test the feasibility of *W. anomalus* in different grape musts at industrial or semi-industrial scales, considering the impact of common oenological practices on the dynamics of this yeast. Finally, interactions among wine yeasts should be considered, taking into account that these interactions seem to be strain-dependent for both non-*Saccharomyces* and *S. cerevisiae* strains.

Author Contributions: B.P., J.V.G., and P.M. contributed to conception and design of the review. B.P. and J.V.G. drafted the manuscript. P.M. supervised and edited the manuscript. All authors commented on the manuscript at all stages.

Funding: This work was funded by grant BIO2015-68790-C2-1-R from the "Ministerio de Economía y Competitividad" (Spain) (MINECO/FEDER Funds). B.P. is the recipient of a Juan de la Cierva Formación post-doctoral research contract (FJCI-2015-23701) from "Ministerio de Economía, Industria y Competitividad" (Spain).

Acknowledgments: Authors thank Sandra Garrigues for critical reading of the manuscript.

Conflicts of Interest: The authors declare no conflict of interest.

References

1. Fleet, G.H. Wine yeasts for the future. *FEMS Yeast Res.* **2008**, *8*, 979–995. [CrossRef] [PubMed]
2. Comitini, F.; Capece, A.; Ciani, M.; Romano, P. New insights on the use of wine yeasts. *Curr. Opin. Food Sci.* **2017**, *13*, 44–49. [CrossRef]
3. Padilla, B.; Gil, J.V.; Manzanares, P. Past and future of non-*Saccharomyces* yeasts: From spoilage microorganisms to biotechnological tools for improving wine aroma complexity. *Front. Microbiol.* **2016**, *7*, 411. [CrossRef] [PubMed]
4. Ciani, M.; Morales, P.; Comitini, F.; Tronchoni, J.; Canonico, L.; Curiel, J.A.; Oro, L.; Rodrigues, A.J.; Gonzalez, R. Non-conventional yeast species for lowering ethanol content of wines. *Front. Microbiol.* **2016**, *7*, 642. [CrossRef] [PubMed]
5. Contreras, A.; Hidalgo, C.; Schmidt, S.; Henschke, P.A.; Curtin, C.; Varela, C. The application of non-*Saccharomyces* yeast in fermentations with limited aeration as a strategy for the production of wine with reduced alcohol content. *Int. J. Food Microbiol.* **2015**, *205*, 7–15. [CrossRef] [PubMed]
6. Englezos, V.; Rantsiou, K.; Cravero, F.; Torchio, F.; Ortiz-Julien, A.; Gerbi, V.; Rolle, L.; Cocolin, L. *Starmerella bacillaris* and *Saccharomyces cerevisiae* mixed fermentations to reduce ethanol content in wine. *Appl. Microbiol. Biotechnol.* **2016**, *100*, 5515–5526. [CrossRef] [PubMed]
7. Gonzalez, R.; Quiros, M.; Morales, P. Yeast respiration of sugars by non-*Saccharomyces* yeast species: A promising and barely explored approach to lowering alcohol content of wines. *Trends Food Sci. Technol.* **2013**, *29*, 55–61. [CrossRef]
8. Berbegal, C.; Spano, G.; Fragasso, M.; Grieco, F.; Russo, P.; Capozzi, V. Starter cultures as biocontrol strategy to prevent *Brettanomyces bruxellensis* proliferation in wine. *Appl. Microbiol. Biotechnol.* **2018**, *102*, 569–576. [CrossRef] [PubMed]
9. Oro, L.; Ciani, M.; Comitini, F. Antimicrobial activity of *Metschnikowia pulcherrima* on wine yeasts. *J. Appl. Microbiol.* **2014**, *116*, 1209–1217. [CrossRef] [PubMed]
10. Fleet, G.H.; Heard, G.M. Yeast-growth during winemaking. In *Wine Microbiology and Biotechnology*; Fleet, G.H., Ed.; Harwood Academic Publishers: Chur, Switzerland, 1993; pp. 27–54.
11. Petruzzi, L.; Capozzi, V.; Berbegal, C.; Corbo, M.R.; Bevilacqua, A.; Spano, G.; Sinigaglia, M. Microbial resources and enological significance: Opportunities and benefits. *Front. Microbiol.* **2017**, *8*, 13. [CrossRef] [PubMed]
12. Kurtzman, C.P. Phylogeny of the ascomycetous yeasts and the renaming of *Pichia anomala* to *Wickerhamomyces anomalus*. *Antonie van Leeuwenhoek* **2011**, *99*, 13–23. [CrossRef] [PubMed]
13. Kurtzman, C.P. *Pichia* e.C. Hansen emend. Kurtzman. In *The Yeasts*, 4th ed.; Kurtzman, C.P., Fell, J.W., Eds.; Elsevier: Amsterdam, The Netherlands, 1998; pp. 273–352.
14. Passoth, V.; Fredlund, E.; Druvefors, U.A.; Schnurer, J. Biotechnology, physiology and genetics of the yeast *Pichia anomala*. *FEMS Yeast Res.* **2006**, *6*, 3–13. [CrossRef] [PubMed]
15. Walker, G.M. *Pichia anomala*: Cell physiology and biotechnology relative to other yeasts. *Antonie van Leeuwenhoek* **2011**, *99*, 25–34. [CrossRef] [PubMed]
16. Lanciotti, R.; Sinigaglia, M.; Gardini, F.; Elisabetta Guerzoni, M. *Hansenula anomala* as spoilage agent of cream-filled cakes. *Microbiol. Res.* **1998**, *153*, 145–148. [CrossRef]
17. Tokuoka, K.; Ishitani, T.; Goto, S.; Komagata, K. Identification of yeasts isolated from high-sugar foods. *J. Gen. Appl. Microbiol.* **1985**, *31*, 411–427. [CrossRef]

18. Jonsson, A.; Pahlow, G. Systematic classification and biochemical characterization of yeasts growing in grass silage inoculated with *Lactobacillus* cultures. *Anim. Res. Dev.* **1984**, *20*, 7–22.

19. Fletcher, E.; Feizi, A.; Kim, S.; Siewers, V.; Nielsen, J. RNA-seq analysis of *Pichia anomala* reveals important mechanisms required for survival at low pH. *Microb. Cell. Fact.* **2015**, *14*, 11. [CrossRef] [PubMed]

20. Schneider, J.; Rupp, O.; Trost, E.; Jaenicke, S.; Passoth, V.; Goesmann, A.; Tauch, A.; Brinkrolf, K. Genome sequence of *Wickerhamomyces anomalus* DSM 6766 reveals genetic basis of biotechnologically important antimicrobial activities. *FEMS Yeast Res.* **2012**, *12*, 382–386. [CrossRef] [PubMed]

21. Heard, G.M.; Fleet, G.H. Growth of natural yeast flora during the fermentation of inoculated wines. *Appl. Environ. Microbiol.* **1985**, *50*, 727–728. [PubMed]

22. Díaz, C.; Molina, A.M.; Nähring, J.; Fischer, R. Characterization and dynamic behavior of wild yeast during spontaneous wine fermentation in steel tanks and amphorae. *BioMed Res. Int.* **2013**, *2013*, 540465. [CrossRef] [PubMed]

23. Renouf, V.; Claisse, O.; Lonvaud-Funel, A. Inventory and monitoring of wine microbial consortia. *Appl. Microbiol. Biotechnol.* **2007**, *75*, 149–164. [CrossRef] [PubMed]

24. Sabel, A.; Martens, S.; Petri, A.; Konig, H.; Claus, H. *Wickerhamomyces anomalus* AS1: A new strain with potential to improve wine aroma. *Ann. Microbiol.* **2014**, *64*, 483–491. [CrossRef]

25. Du Toit, M.; Pretorius, I.S. Microbial spoilage and preservation of wine: Using weapons from nature's own arsenal—A review. *S. Afr. J. Enol. Vitic.* **2000**, *21*, 74–92.

26. Barata, A.; Malfeito-Ferreira, M.; Loureiro, V. The microbial ecology of wine grape berries. *Int. J. Food Microbiol.* **2012**, *153*, 243–259. [CrossRef] [PubMed]

27. Cordero-Bueso, G.; Arroyo, T.; Serrano, A.; Tello, J.; Aporta, I.; Vélez, M.D.; Valero, E. Influence of the farming system and vine variety on yeast communities associated with grape berries. *Int. J. Food Microbiol.* **2011**, *145*, 132–139. [CrossRef] [PubMed]

28. Bagheri, B.; Bauer, F.F.; Setati, M.E. The diversity and dynamics of indigenous yeast communities in grape must from vineyards employing different agronomic practices and their influence on wine fermentation. *S. Afr. J. Enol. Vitic.* **2015**, *36*, 243–251. [CrossRef]

29. Cordero-Bueso, G.; Esteve-Zarzoso, B.; Cabellos, J.M.; Gil-Diaz, M.; Arroyo, T. Biotechnological potential of non-*Saccharomyces* yeasts isolated during spontaneous fermentations of Malvar (*Vitis vinifera* cv. L.). *Eur. Food Res. Technol.* **2013**, *236*, 193–207. [CrossRef]

30. Bagheri, B.; Bauer, F.F.; Setati, M.E. The impact of *Saccharomyces cerevisiae* on a wine yeast consortium in natural and inoculated fermentations. *Front. Microbiol.* **2017**, *8*, 13. [CrossRef] [PubMed]

31. Ye, M.; Yue, T.; Yuan, Y. Effects of sequential mixed cultures of *Wickerhamomyces anomalus* and *Saccharomyces cerevisiae* on apple cider fermentation. *FEMS Yeast Res.* **2014**, *14*, 873–882. [CrossRef] [PubMed]

32. Ocón, E.; Gutiérrez, A.R.; Garijo, P.; López, R.; Santamaría, P. Presence of non-*Saccharomyces* yeasts in cellar equipment and grape juice during harvest time. *Food Microbiol.* **2010**, *27*, 1023–1027. [CrossRef] [PubMed]

33. Martini, A. Origin and domestication of the wine yeast *Saccharomyces cerevisiae*. *J. Wine Res.* **1993**, *4*, 165–176. [CrossRef]

34. Ciani, M.; Mannazzu, I.; Marinangeli, P.; Clementi, F.; Martini, A. Contribution of winery-resident *Saccharomyces cerevisiae* strains to spontaneous grape must fermentation. *Antonie van Leeuwenhoek* **2004**, *85*, 159–164. [CrossRef] [PubMed]

35. Fleet, G.H.; Lafon-Lafourcade, S.; Ribéreau-Gayon, P. Evolution of yeasts and lactic acid bacteria during fermentation and storage of Bordeaux wines. *Appl. Environ. Microbiol.* **1984**, *48*, 1034–1038. [PubMed]

36. Fernández, M.T.; Ubeda, J.F.; Briones, A.I. Comparative study of non-*Saccharomyces* microflora of musts in fermentation by physiological and molecular methods. *FEMS Microbiol. Lett.* **1999**, *173*, 223–229. [CrossRef]

37. Chavan, P.; Mane, S.; Kulkarni, G.; Shaikh, S.; Ghormade, V.; Nerkar, D.P.; Shouche, Y.; Deshpande, M.V. Natural yeast flora of different varieties of grapes used for wine making in India. *Food Microbiol.* **2009**, *26*, 801–808. [CrossRef] [PubMed]

38. Combina, M.; Elia, A.; Mercado, L.; Catania, C.; Ganga, A.; Martinez, C. Dynamics of indigenous yeast populations during spontaneous fermentation of wines from Mendoza, Argentina. *Int. J. Food Microbiol.* **2005**, *99*, 237–243. [CrossRef] [PubMed]

39. Li, S.-S.; Cheng, C.; Li, Z.; Chen, J.-Y.; Yan, B.; Han, B.-Z.; Reeves, M. Yeast species associated with wine grapes in China. *Int. J. Food Microbiol.* **2010**, *138*, 85–90. [CrossRef] [PubMed]

40. Garofalo, C.; Tristezza, M.; Grieco, F.; Spano, G.; Capozzi, V. From grape berries to wine: Population dynamics of cultivable yeasts associated to "Nero di Troia" autochthonous grape cultivar. *World J. Microbiol. Biotechnol.* **2016**, *32*, 10. [CrossRef] [PubMed]

41. Jolly, N.P.; Pretorius, I.S. The effect of non-*Saccharomyces* yeasts on fermentation and wine quality. *S. Afr. J. Enol. Vitic.* **2003**, *24*, 55–62. [CrossRef]

42. Nemcová, K.; Breierová, E.; Vadkertiová, R.; Molnarová, J. The diversity of yeasts associated with grapes and musts of the Strekov winegrowing region, Slovakia. *Folia Microbiol.* **2015**, *60*, 103–109. [CrossRef] [PubMed]

43. Mills, D.A.; Johannsen, E.A.; Cocolin, L. Yeast diversity and persistence in *Botrytis*-affected wine fermentations. *Appl. Environ. Microbiol.* **2002**, *68*, 4884–4893. [CrossRef] [PubMed]

44. Grangeteau, C.; Gerhards, D.; Terrat, S.; Dequiedt, S.; Alexandre, H.; Guilloux-Benatier, M.; von Wallbrunn, C.; Rousseaux, S. FT-IR spectroscopy: A powerful tool for studying the inter- and intraspecific biodiversity of cultivable non-*Saccharomyces* yeasts isolated from grape must. *J. Microbiol. Methods* **2016**, *121*, 50–58. [CrossRef] [PubMed]

45. Madrigal, T.; Maicas, S.; Tolosa, J.J.M. Glucose and ethanol tolerant enzymes produced by *Pichia* (*Wickerhamomyces*) isolates from enological ecosystems. *Am. J. Enol. Vitic.* **2013**, *64*, 126–133. [CrossRef]

46. Rosi, I.; Vinella, M.; Domizio, P. Characterization of β-glucosidase activity in yeasts of oenological origin. *J. Appl. Microbiol.* **1994**, *77*, 519–527. [CrossRef]

47. Charoenchai, C.; Fleet, G.H.; Henschke, P.A.; Todd, B.E.N. Screening of non-*Saccharomyces* wine yeasts for the presence of extracellular hydrolytic enzymes. *Aust. J. Grape Wine Res.* **1997**, *3*, 2–8. [CrossRef]

48. Manzanares, P.; Rojas, V.; Genovés, S.; Vallés, S. A preliminary search for anthocyanin-β-D-glucosidase activity in non-*Saccharomyces* wine yeasts. *Int. J. Food Sci. Tech.* **2000**, *35*, 95–103. [CrossRef]

49. Lopez, M.C.; Mateo, J.J.; Maicas, S. Screening of β-glucosidase and β-xylosidase activities in four non-*Saccharomyces* yeast isolates. *J. Food Sci.* **2015**, *80*, C1696–C1704. [CrossRef] [PubMed]

50. Spagna, G.; Barbagallo, R.N.; Palmeri, R.; Restuccia, C.; Giudici, P. Properties of endogenous β-glucosidase of a *Pichia anomala* strain isolated from Sicilian musts and wines. *Enzym. Microb. Technol.* **2002**, *31*, 1036–1041. [CrossRef]

51. Manzanares, P.; Ramón, D.; Querol, A. Screening of non-*Saccharomyces* wine yeasts for the production of β-D-xylosidase activity. *Int. J. Food Microbiol.* **1999**, *46*, 105–112. [CrossRef]

52. Mateo, J.J.; Peris, L.; Ibañez, C.; Maicas, S. Characterization of glycolytic activities from non-*Saccharomyces* yeasts isolated from Bobal musts. *J. Ind. Microbiol. Biotechnol.* **2011**, *38*, 347–354. [CrossRef] [PubMed]

53. Swangkeaw, J.; Vichitphan, S.; Butzke, C.E.; Vichitphan, K. The characterisation of a novel *Pichia anomala* β-glucosidase with potentially aroma-enhancing capabilities in wine. *Ann. Microbiol.* **2009**, *59*, 335. [CrossRef]

54. Swangkeaw, J.; Vichitphan, S.; Butzke, C.E.; Vichitphan, K. Characterization of β-glucosidases from *Hanseniaspora* sp. and *Pichia anomala* with potentially aroma-enhancing capabilities in juice and wine. *World J. Microbiol. Biotechnol.* **2011**, *27*, 423–430. [CrossRef]

55. Schwentke, J.; Sabel, A.; Petri, A.; Konig, H.; Claus, H. The yeast *Wickerhamomyces anomalus* AS1 secretes a multifunctional exo-β-1,3-glucanase with implications for winemaking. *Yeast* **2014**, *31*, 349–359. [CrossRef] [PubMed]

56. Gil, J.V.; Manzanares, P.; Genovés, S.; Vallés, S.; González-Candelas, L. Over-production of the major exoglucanase of *Saccharomyces cerevisiae* leads to an increase in the aroma of wine. *Int. J. Food Microbiol.* **2005**, *103*, 57–68. [CrossRef] [PubMed]

57. Van Rensburg, P.; Pretorius, I.S. Enzymes in winemaking: Harnessing natural catalysts for efficient biotransformations—A review. *S. Afr. J. Enol. Vitic.* **2000**, *21*, 52–73.

58. Schlander, M.; Distler, U.; Tenzer, S.; Thines, E.; Claus, H. Purification and properties of yeast proteases secreted by *Wickerhamomyces anomalus* 227 and *Metschnikovia pulcherrima* 446 during growth in a white grape juice. *Fermentation* **2017**, *3*, 2. [CrossRef]

59. Viana, F.; Gil, J.V.; Genoves, S.; Valles, S.; Manzanares, P. Rational selection of non-*Saccharomyces* wine yeasts for mixed starters based on ester formation and enological traits. *Food Microbiol.* **2008**, *25*, 778–785. [CrossRef] [PubMed]

60. Lambretchts, M.G.; Pretorius, I.S. Yeast and its importance to wine aroma—A review. *S. Afr. J. Enol. Vitic.* **2000**, *21*, 97–129.

61. Domizio, P.; Romani, C.; Lencioni, L.; Comitini, F.; Gobbi, M.; Mannazzu, I.; Ciani, M. Outlining a future for non-*Saccharomyces* yeasts: Selection of putative spoilage wine strains to be used in association with *Saccharomyces cerevisiae* for grape juice fermentation. *Int. J. Food Microbiol.* **2011**, *147*, 170–180. [CrossRef] [PubMed]

62. Liu, G.L.; Chi, Z.; Wang, G.Y.; Wang, Z.P.; Li, Y.; Chi, Z.M. Yeast killer toxins, molecular mechanisms of their action and their applications. *Crit. Rev. Biotechnol.* **2015**, *35*, 222–234. [CrossRef] [PubMed]

63. Mehlomakulu, N.N.; Setati, M.E.; Divol, B. Non-*Saccharomyces* killer toxins: Possible biocontrol agents against *Brettanomyces* in wine? *S. Afr. J. Enol. Vitic.* **2015**, *36*, 94–104. [CrossRef]

64. Passoth, V.; Olstorpe, M.; Schnürer, J. Past, present and future research directions with *Pichia anomala*. *Antonie van Leeuwenhoek* **2011**, *99*, 121–125. [CrossRef] [PubMed]

65. Ciani, M.; Comitini, F. Non-*Saccharomyces* wine yeasts have a promising role in biotechnological approaches to winemaking. *Ann. Microbiol.* **2011**, *61*, 25–32. [CrossRef]

66. Fernández de Ullivarri, M.; Mendoza, L.M.; Raya, R.R. Characterization of the killer toxin KTCf20 from *Wickerhamomyces anomalus*, a potential biocontrol agent against wine spoilage yeasts. *Biol. Control* **2018**, *121*, 223–228. [CrossRef]

67. Lopes, C.A.; Sáez, J.S.; Sangorrín, M.P. Differential response of *Pichia guilliermondii* spoilage isolates to biological and physico-chemical factors prevailing in Patagonian wine fermentations. *Can. J. Microbiol.* **2009**, *55*, 801–809. [CrossRef] [PubMed]

68. Sangorrín, M.P.; Lopes, C.A.; Jofré, V.; Querol, A.; Caballero, A.C. Spoilage yeasts from Patagonian cellars: Characterization and potential biocontrol based on killer interactions. *World J. Microbiol. Biotechnol.* **2008**, *24*, 945–953. [CrossRef]

69. Comitini, F.; De, J.I.; Pepe, L.; Mannazzu, I.; Ciani, M. *Pichia anomala* and *Kluyveromyces wickerhamii* killer toxins as new tools against *Dekkera/Brettanomyces* spoilage yeasts. *FEMS Microbiol. Lett.* **2004**, *238*, 235–240. [CrossRef] [PubMed]

70. De Ingeniis, J.; Raffaelli, N.; Ciani, M.; Mannazzu, I. *Pichia anomala* DBVPG 3003 secretes a ubiquitin-like protein that has antimicrobial activity. *Appl. Environ. Microbiol.* **2009**, *75*, 1129–1134. [CrossRef] [PubMed]

71. Oro, L.; Ciani, M.; Bizzaro, D.; Comitini, F. Evaluation of damage induced by Kwkt and Pikt zymocins against *Brettanomyces/Dekkera* spoilage yeast, as compared to sulphur dioxide. *J. Appl. Microbiol.* **2016**, *121*, 207–214. [CrossRef] [PubMed]

72. Csutak, O.; Vassu, T.; Corbu, V.; Cirpici, I.; Ionescu, R. Killer activity of *Pichia anomala* CMGB 88. *Biointerface Res. Appl. Chem.* **2017**, *7*, 2085–2089.

73. Andorrà, I.; Berradre, M.; Mas, A.; Esteve-Zarzoso, B.; Guillamón, J.M. Effect of mixed culture fermentations on yeast populations and aroma profile. *LWT-Food Sci. Technol.* **2012**, *49*, 8–13. [CrossRef]

74. Gil, J.V.; Mateo, J.J.; Jiménez, M.; Pastor, A.; Huerta, T. Aroma compounds in wine as influenced by apiculate yeasts. *J. Food Sci.* **1996**, *61*, 1247–1249. [CrossRef]

75. Gobbi, M.; Comitini, F.; Domizio, P.; Romani, C.; Lencioni, L.; Mannazzu, I.; Ciani, M. *Lachancea thermotolerans* and *Saccharomyces cerevisiae* in simultaneous and sequential co-fermentation: A strategy to enhance acidity and improve the overall quality of wine. *Food Microbiol.* **2013**, *33*, 271–281. [CrossRef] [PubMed]

76. Quirós, M.; Rojas, V.; Gonzalez, R.; Morales, P. Selection of non-*Saccharomyces* yeast strains for reducing alcohol levels in wine by sugar respiration. *Int. J. Food Microbiol.* **2014**, *181*, 85–91. [CrossRef] [PubMed]

77. Canonico, L.; Agarbatu, A.; Comitini, F.; Ciani, M. *Torulaspora delbrueckii* in the brewing process: A new approach to enhance bioflavour and to reduce ethanol content. *Food Microbiol.* **2016**, *56*, 45–51. [CrossRef] [PubMed]

78. Contreras, A.; Hidalgo, C.; Henschke, P.A.; Chambers, P.J.; Curtin, C.; Varela, C. Evaluation of non-*Saccharomyces* yeasts for the reduction of alcohol content in wine. *Appl. Environ. Microbiol.* **2011**, *80*, 1670–1678. [CrossRef] [PubMed]

79. Rojas, V.; Gil, J.V.; Piñaga, F.; Manzanares, P. Acetate ester formation in wine by mixed cultures in laboratory fermentations. *Int. J. Food Microbiol.* **2003**, *86*, 181–188. [CrossRef]

80. Bely, M.; Stoeckle, P.; Masneuf-Pomarede, I.; Dubourdieu, D. Impact of mixed *Torulaspora delbrueckii-Saccharomyces cerevisiae* culture on high-sugar fermentation. *Int. J. Food Microbiol.* **2008**, *122*, 312–320. [CrossRef] [PubMed]

81. Anfang, N.; Brajkovich, M.; Goddard, M.R. Co-fermentation with *Pichia kluyveri* increases varietal thiol concentrations in Sauvignon blanc. *Aust. J. Grape Wine Res.* **2009**, *15*, 1–8. [CrossRef]

82. Viana, F.; Gil, J.V.; Valles, S.; Manzanares, P. Increasing the levels of 2-phenylethyl acetate in wine through the use of a mixed culture of *Hanseniaspora osmophila* and *Saccharomyces cerevisiae*. *Int. J. Food Microbiol.* **2009**, *135*, 68–74. [CrossRef] [PubMed]

83. Izquierdo Cañas, P.; García-Romero, E.; Heras Manso, J.; Fernández-González, M. Influence of sequential inoculation of *Wickerhamomyces anomalus* and *Saccharomyces cerevisiae* in the quality of red wines. *Eur. Food Res. Technol.* **2014**, *239*, 279–286. [CrossRef]

84. Izquierdo Cañas, P.M.I.; Garcia, A.T.P.; Romero, E.G. Enhancement of flavour properties in wines using sequential inoculations of non-*Saccharomyces* (*Hansenula* and *Torulaspora*) and *Saccharomyces* yeast starter. *Vitis* **2011**, *50*, 177–182.

85. Kurita, O. Increase of acetate ester-hydrolysing esterase activity in mixed cultures of *Saccharomyces cerevisiae* and *Pichia anomala*. *J. Appl. Microbiol.* **2008**, *104*, 1051–1058. [CrossRef] [PubMed]

86. Luan, Y.; Zhang, B.Q.; Duan, C.Q.; Yan, G.L. Effects of different pre-fermentation cold maceration time on aroma compounds of *Saccharomyces cerevisiae* co-fermentation with *Hanseniaspora opuntiae* or *Pichia kudriavzevii*. *LWT-Food Sci. Technol.* **2018**, *92*, 177–186. [CrossRef]

87. Gobert, A.; Tourdot-Marechal, R.; Morge, C.; Sparrow, C.; Liu, Y.Z.; Quintanilla-Casas, B.; Vichi, S.; Alexandre, H. Non-*Saccharomyces* yeasts nitrogen source preferences: Impact on sequential fermentation and wine volatile compounds profile. *Front. Microbiol.* **2017**, *8*, 2175. [CrossRef] [PubMed]

88. Rollero, S.; Bloem, A.; Ortiz-Julien, A.; Camarasa, C.; Divol, B. Altered fermentation performances, growth, and metabolic footprints reveal competition for nutrients between yeast species inoculated in synthetic grape juice-like medium. *Front. Microbiol.* **2018**, *9*, 196. [CrossRef] [PubMed]

89. Chasseriaud, L.; Coulon, J.; Marullo, P.; Albertin, W.; Bely, M. New oenological practice to promote non-*Saccharomyces* species of interest: Saturating grape juice with carbon dioxide. *Appl. Microbiol. Biotechnol.* **2018**, *102*, 3779–3791. [CrossRef] [PubMed]

90. Beckner Whitener, M.E.; Stanstrup, J.; Panzeri, V.; Carlin, S.; Divol, B.; Du Toit, M.; Vrhovsek, U. Untangling the wine metabolome by combining untargeted SPME–GCxGC-TOF-MS and sensory analysis to profile Sauvignon blanc co-fermented with seven different yeasts. *Metabolomics* **2016**, *12*, 53. [CrossRef]

91. Domizio, P.; Romani, C.; Comitini, F.; Gobi, M.; Lencioni, L.; Mannazzu, I.; Ciani, M. Potential spoilage non-*Saccharomyces* yeasts in mixed cultures with *Saccharomyces cerevisiae*. *Ann. Microbiol.* **2011**, *61*, 137–144. [CrossRef]

92. Basso, R.F.; Alcarde, A.R.; Portugal, C.B. Could non-*Saccharomyces* yeasts contribute on innovative brewing fermentations? *Food Res. Int.* **2016**, *86*, 112–120. [CrossRef]

93. Duarte, W.F.; Amorim, J.C.; Schwan, R.F. The effects of co-culturing non-*Saccharomyces* yeasts with *S. cerevisiae* on the sugar cane spirit (cachaça) fermentation process. *Antonie van Leeuwenhoek* **2013**, *103*, 175–194. [CrossRef] [PubMed]

94. Zha, M.; Sun, B.; Wu, Y.; Yin, S.; Wang, C. Improving flavor metabolism of *Saccharomyces cerevisiae* by mixed culture with *Wickerhamomyces anomalus* for Chinese Baijiu making. *J. Biosci. Bioeng.* **2018**, *126*, 189–195. [CrossRef] [PubMed]

95. Wahyono, A.; Lee, S.B.; Kang, W.W.; Park, H.D. Improving bread quality using co-cultures of *Saccharomyces cerevisiae*, *Torulaspora delbrueckii* JK08, and *Pichia anomala* JK04. *Ital. J. Food Sci.* **2016**, *28*, 298–313.

96. Huyben, D.; Nyman, A.; Vidakovic, A.; Passoth, V.; Moccia, R.; Kiessling, A.; Dicksved, J.; Lundh, T. Effects of dietary inclusion of the yeasts *Saccharomyces cerevisiae* and *Wickerhamomyces anomalus* on gut microbiota of rainbow trout. *Aquaculture* **2017**, *473*, 528–537. [CrossRef]

97. Olstorpe, M.; Passoth, V. *Pichia anomala* in grain biopreservation. *Antonie van Leeuwenhoek* **2011**, *99*, 57–62. [CrossRef] [PubMed]

98. Olstorpe, M.; Borling, J.; Schnürer, J.; Passoth, V. *Pichia anomala* yeast improves feed hygiene during storage of moist crimped barley grain under swedish farm conditions. *Anim. Feed Sci. Technol.* **2010**, *156*, 47–56. [CrossRef]

99. Buzzini, P.; Martini, A. Large-scale screening of selected *Candida maltosa*, *Debaryomyces hansenii* and *Pichia anomala* killer toxin activity against pathogenic yeasts. *Med. Mycol.* **2001**, *39*, 479–482. [CrossRef] [PubMed]

100. Zha, Y.; Hossain, A.H.; Tobola, F.; Sedee, N.; Havekes, M.; Punt, P.J. *Pichia anomala* 29X: A resistant strain for lignocellulosic biomass hydrolysate fermentation. *FEMS Yeast Res.* **2013**, *13*, 609–617. [CrossRef] [PubMed]

101. Ruyters, S.; Mukherjee, V.; Verstrepen, K.J.; Thevelein, J.M.; Willems, K.A.; Lievens, B. Assessing the potential of wild yeasts for bioethanol production. *J. Ind. Microbiol. Biotechnol.* **2015**, *42*, 39–48. [CrossRef] [PubMed]

102. Wu, J.; Elliston, A.; Le Gall, G.; Colquhoun, I.J.; Collins, S.R.A.; Dicks, J.; Roberts, I.N.; Waldron, K.W. Yeast diversity in relation to the production of fuels and chemicals. *Sci. Rep.* **2017**, *7*, 11. [CrossRef] [PubMed]

103. Kruis, A.J.; Levisson, M.; Mars, A.E.; van der Ploeg, M.; Daza, F.G.; Ellena, V.; Kengen, S.W.M.; van der Oost, J.; Weusthuis, R.A. Ethyl acetate production by the elusive alcohol acetyltransferase from yeast. *Metab. Eng.* **2017**, *41*, 92–101. [CrossRef] [PubMed]

![fermentation logo] *fermentation*

MDPI

Review

Oenological Impact of the *Hanseniaspora/Kloeckera* Yeast Genus on Wines—A Review

Valentina Martin, Maria Jose Valera, Karina Medina, Eduardo Boido and Francisco Carrau *

Enology and Fermentation Biotechnology Area, Food Science and Technology Department, Facultad de Quimica, Universidad de la Republica, Montevideo 11800, Uruguay; vmartin@fq.edu.uy (V.M.); mariajose_valera_martinez@hotmail.com (M.J.V.); kmedina@fq.edu.uy (K.M.); eboido@fq.edu.uy (E.B.)
* Correspondence: fcarrau@fq.edu.uy; Tel.: +598-292-481-94

Received: 7 August 2018; Accepted: 5 September 2018; Published: 10 September 2018

Abstract: Apiculate yeasts of the genus *Hanseniaspora/Kloeckera* are the main species present on mature grapes and play a significant role at the beginning of fermentation, producing enzymes and aroma compounds that expand the diversity of wine color and flavor. Ten species of the genus *Hanseniaspora* have been recovered from grapes and are associated in two groups: *H. valbyensis*, *H. guilliermondii*, *H. uvarum*, *H. opuntiae*, *H. thailandica*, *H. meyeri*, and *H. clermontiae*; and *H. vineae*, *H. osmophila*, and *H. occidentalis*. This review focuses on the application of some strains belonging to this genus in co-fermentation with *Saccharomyces cerevisiae* that demonstrates their positive contribution to winemaking. Some consistent results have shown more intense flavors and complex, full-bodied wines, compared with wines produced by the use of *S. cerevisiae* alone. Recent genetic and physiologic studies have improved the knowledge of the *Hanseniaspora/Kloeckera* species. Significant increases in acetyl esters, benzenoids, and sesquiterpene flavor compounds, and relative decreases in alcohols and acids have been reported, due to different fermentation pathways compared to conventional wine yeasts.

Keywords: non-*Saccharomyces*; genome; aroma compounds; anthocyanin; mixed cultures fermentation; flavor complexity

---

1. Introduction

Non-*Saccharomyces* (NS) yeasts were considered unattractive in traditional winemaking, and sulphites addition was the way to prevent the risk of their growth at the beginning of the vinification process. However, today's increased knowledge about yeast diversity has demonstrated that there are many NS yeasts with beneficial properties that contribute to increasing the sensory complexity of wines [1–3]. The main NS yeasts associated with grapes are the apiculate group with bipolar budding, more precisely, the genus *Hanseniaspora* and its asexual anamorph *Kloeckera* [4,5]. In Figure 1, the plating and microscopy characteristics of two species of the genus are shown.

The *Hanseniaspora/Kloeckera* (H/K) group is currently composed of 10 recognized species associated with grapes [6–8]. One of the main characteristics of these species is the weak fermentation capacity compared to *Saccharomyces cerevisiae* (SC). However, some species, such as *"vineae"*, might reach about 10% of the alcohol by volume of fermentative capacity under winemaking conditions. Furthermore, these species are important in the production of an increased diversity of volatile compounds in wine, and it was demonstrated the chemical composition of wines made with H/K in combination with SC differ from reference wines [9–12]. During these early studies about apiculate yeasts, some authors [13–15] showed that not all H/K strains formed high levels of volatile acidity and many of them produced similar levels to SC in this regard. These results indicate that although some strains of H/K can provide higher levels of ethanol than other strains, the main characteristic of many of these known strains is the increased formation of some acetate esters. The production of other secondary metabolites—i.e., glycerol, acetaldehyde, and hydrogen sulphide—also differed between strains [16]. Thus, differences in chemical analyses of the wines were noted.

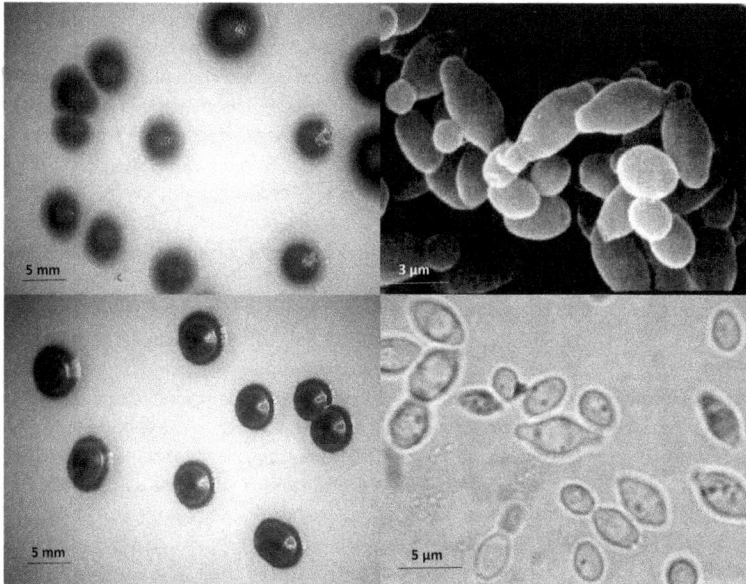

Figure 1. On the left-hand side are the typical colony color and morphology for *Hanseniaspora/Kloeckera* strains that are readily differentiated from other yeast genera in WL nutrient agar medium. However, is more difficult to distinguish between species, although some slight differences between *H. uvarum* (**upper**) colonies and *H. vineae* (**lower**) colonies can be appreciated in these photos. On the right-hand side, *H. uvarum* is visualized by electron microscopy and *H. vineae* is visualized with blue methylene stain.

The initial growth of H/K had a retarding effect on the subsequent growth of SC, as also shown for other NS species in mixed cultures [17]. Therefore, when considering the use of H/K strains at winemaking, grape must nutrient composition and competition for assimilable nitrogen by mixed cultures should be understood to prevent sluggish fermentations [18]. Some other cell interactions between H/K strains and SC that inhibit their growth were reported [19], however, H/K strains are intense removers of some vitamins, such as thiamine [20] or calcium pantothenate [21]. Medina et al. [17] found these two vitamins, in combination with ammonium salts, improved the development of SC strains to complete fermentation. Addition of yeast extract at 2 g/L was demonstrated to be more effective for *H. vineae* utilisation than ammonium salts in agave juice for tequila [22]. In white wine production, a *K. apiculata* isolate was used with SC at laboratory scale [23]. Inoculation of SC occurred 1 h after *K. apiculata*, and a dry wine of 13% by volume of ethanol was produced. A positive sensory evaluation of the Sauvignon blanc wines was obtained at 5 and 18 months after production. The production of β-phenylethyl acetate and ethyl acetate by the apiculate yeast *H. guilliermondii* has been investigated in laboratory fermentations [24]. The β-phenylethyl acetate ester contributes to 'rose', 'honey', 'fruity', and 'flowery' aroma nuances, and is formed to a greater or lesser extent by yeasts. As part of the 'fermentation bouquet', it can contribute to the overall flavor of a young wine.

The positive oenological characteristics that this H/K confers to wine have been broadly reported [18,25,26]. In addition, some studies have shown that several H/K strains have potential as biocontrol agents against fungi, such as *Botrytis*. The competition for nutrients is the action mechanism of protection used by *H. uvarum* against fungi in grapes and apples [27,28].

2. Genetics Context of *Hanseniaspora* Species

2.1. Application of Molecular Techniques for Taxonomy and Whole Genome Analysis

Classical microbiology techniques have been extensively used in oenology to select and inoculate the best yeast strains for obtaining enhanced positive characteristics in final wines. During the last decades, the use of molecular techniques for the identification and selection of specific strains has proven invaluable for the winemaking industry. Polymerase chain reaction-based methods allow the identification of distinct species and, also, strain genotyping, resulting in a more accurate strain selection [29,30]. Based on this knowledge, culture-independent techniques have been developed to detect microorganisms present during fermentation that are not cultivable by conventional methods [31–33]. Among these culture-independent techniques, the development of next-generation sequencing permitted the description of the whole microbiota present in a specific environment, even in complex communities, such as those found throughout the wine fermentation process [34,35].

Hanseniaspora species have been widely detected in various wine-related environments, especially from soil and grapes to the early stages of vinification [36,37]. This genus is part of the apiculate group of yeast formed by *Hanseniaspora*, *Saccharomycodes*, and *Nadsonia*. The genus *Hanseniaspora* presents heterogeneous morphological, serological, and chemotaxonomic features [38]. Ten species of *Hanseniaspora* have been recovered from grapes or wines, which are taxonomically associated in two groups (Figure 2): *H. valbyensis*, *H. guilliermondii*, *H. uvarum*, *H. opuntiae*, *H. thailandica*, *H. meyeri*, and *H. clermontiae* in one cluster; and *H. vineae*, *H. osmophila* and *H. occidentalis* in the other, as revealed by partial sequence alignment of the *26S rRNA* gene. The favorable oenological characteristics that this genus confers to wine have been broadly reported [18,25,39]. However, the biotechnological potential of these species is still under evaluation at the industrial level, as compared to the traditional SC conventional strains. In this context, the development of molecular techniques and the recent identification of the whole genome sequences from *Hanseniaspora* species related to wine have created the possibility to understand and applied them from a novel, precise oenological perspective.

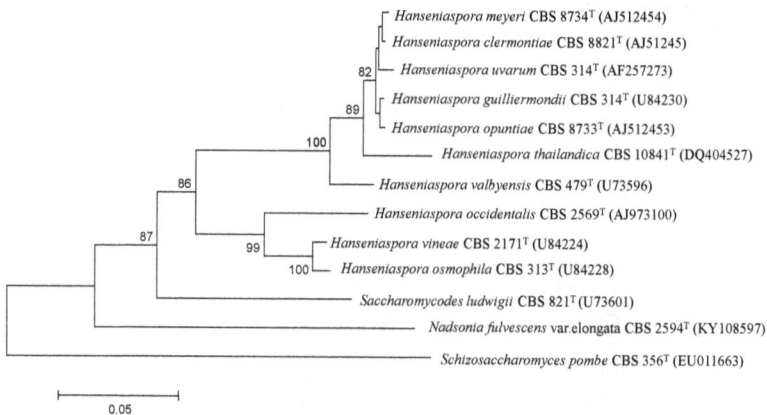

Figure 2. Phylogenetic relationships between type strains of *Hanseniaspora* species and other grape or wine-related apiculate yeasts. The dendrogram was constructed using partial *26S rRNA* gene sequences by the neighbor-joining method. The robustness of the branching is indicated by bootstrap values (%) calculated for 1000 subsets. The entries of the different genotypes include the accession numbers of the GenBank database sequences. *Schizosaccharomyces pombe* type strain was used as an outgroup.

2.2. Comparative Analysis of Hanseniaspora Genomes

Today, 10 strains belonging to six different *Hanseniaspora* species have been completely sequenced [40–43]. The data collected from these sequences (Table 1) evidence the close relation between *H. vineae* and *H. osmophila* that present a similar genome size and G + C percentage compared with the others. Moreover, the protein count is quite similar throughout the species from the genus, but the number of contigs and scaffolds reported vary widely among different species. As informed by karyotyping approaches, H/K species could present between seven and nine chromosomes [29,30]. In a recent study based on field inversion gel electrophoresis and the whole genome sequencing of type strain *H. uvarum* DSM2768, seven chromosomes were detected [44]. Notwithstanding, there are wide differences in genome size and chromosome number in karyotyping results from natural grape samples. Besides, *Hanseniaspora* genus belongs to the group of yeast that does not undergo whole-genome duplication, contrary to *Saccharomyces* [44]. These discrepancies were previously detected in the mitochondrial DNA of *H. uvarum*. It presented a reduced size compared with those from other yeasts and also a different organization of genes [45].

Table 1. *Hanseniaspora* genome assembly statistics.

Species Name (Number of Strains)	Protein Count	Number of Contigs	G + C	Scaffold Number	Assembly (Mb)	Reference
H. guillermondii (1)	4070	250	31%	208	9.04	Seixas et al., 2017 [39]
H. opuntiae (1)	4167	67	35%	18	8.53	Sternes et al., 2016 [38]
H. uvarum (4)	3552	44	32%	18	8.81	Sternes et al., 2016 [38]
H. valbyensis (1)	4772	1345	23%	647	11.46	Riley et al., 2016 [37]
H. vineae (2)	4733	277	37%	124	11.40	Giorello et al., 2014 [36]
H. osmophila (1)	4657	899	37%	17	11.37	Sternes et al., 2016 [38]

There are also some differences in the information about genes linked to interesting oenological traits. The highest number of alcohol dehydrogenases, like *ADH1, ADH2, ADH3, ADH4, ADH6,* and *ADH7*, from SC is found in the *H. vineae* genome. It presents eight genes for alcohol dehydrogenases, followed by *H. osmophila* with six. *H. uvarum, H. guilliermondii,* and *H. opuntiae* present just four. The highest number of copies could be related to the fermentation capacity, given the alcohol dehydrogenase activity is involved in the last step of the glycolytic pathway [46]. The fermentation ability is considered a hurdle in NS yeast relative to *Saccharomyces* strains, and thereby an improvement in fermentation performance is necessary to select a strain for wine inoculation. Limited information is available about the functional analysis of protein activities from *Hanseniaspora*. A key enzyme associated with the glycolytic pathway is pyruvate kinase. Langenberg et al. [44] recently demonstrated the correlation between pyruvate kinase activity and the enhanced fermentative ability of SC compared with *H. uvarum*. The authors explained this difference was due to a lowered specific activity rather than the structure of this enzyme. *H. vineae* and *H. osmophila* present higher sequence homology (Figure 3) in the predicted protein corresponding to the *CDC19* gene from SC than *H. uvarum* and other H/K species. Further biochemical studies will clarify the potential pyruvate kinase activity in H/K species compared to *H. uvarum* and SC.

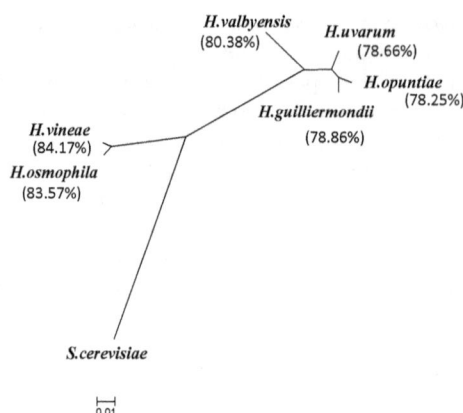

Figure 3. Relative homology of predicted protein sequences from the *CDC19* gene for pyruvate kinase activity in genome-sequenced *Hanseniaspora* strains (*H. vineae* T02/19AF; *H. osmophila* AWRI3579; *H. guilliermondii* UTAD222; *H. opuntiae* AWRI3578; *H. valbyensis* NRRL Y-1626; *H. uvarum* AWRI3580; *Saccharomyces cerevisiae* S288c). Data sequences have been collected from the NCBI protein database. Unrooted trees have been constructed using neighbor-joining analysis to calculate the percentage divergence. The percentage of identity with SC is expressed in brackets and calculated as the number of identical amino acids based on the total length.

The lack of nutrients, especially nitrogen, is a leading concern in wine fermentation that can cause stuck or sluggish fermentations [47,48]. Some genes linked to the regulation of nitrogen consumption have been identified in SC [49]. The general amino acid permease activity is attributed to *GAP1* in SC, and homologous sequences are present in a high copy number. For instance, 12 *GAP1* homologues were detected in the *H. guilliermondii* UTAD222 genome. Ammonium permeases are also involved in the regulation of nitrogen metabolism; *MEP2* homologues were found in all H/K species sequenced, and *MEP3* similar sequences were found just in *H. uvarum* and *H. osmophila*. The absence of *MEP3* in *H. vineae* might explain the inability of this species to use ammonium salts, as reported for agave juice fermentations [22].

Several enzymes that contribute to wine aroma have been extensively described in SC. One of them, *IAH1*, codifies for isoamyl acetate hydrolysing esterase, which adds to the production of desired volatile esters. The genomes of *H. osmophila*, *H. opuntiae*, *H. uvarum*, and *H. guilliermondii* present sequences that codify for a predicted protein highly similar to *IAH1*. Instead, *ATF2* and *EHT1* are both alcohol acetyltransferases. The activity of *ATF2* is affiliated with the formation of volatile esters during SC fermentation, and *EHT1* is linked to short-chain esterase activity [46]. Putative homologous alcohol acetyltransferases were predicted from DNA sequences only present in the genomes of *H. osmophila* and *H. vineae* [40].

The increase in sequences from whole genomes of *Hanseniaspora* strains available in databases is a good starting point to apply the biotechnological potential that these yeasts represent for oenology. Indeed, promising results were obtained in an attempt to genetically modify *H. guilliermondii* strains [50].

3. *Hanseniaspora/Kloeckera* Strains and Flavor Compounds

H/K yeasts may affect the wine fermentation directly, by producing flavors, and indirectly, by modulating the growth and metabolism of SC.

More recently, NS wine yeasts have received special attention by winemakers due to the search of different and desired oenological characteristics, compared to SC commercial strains. Diverse secondary metabolic pathways and higher enzymatic activities (esterases, β-glycosidases, lipases, and proteases), result in sensory complexity [1,51,52] that might contribute to an increased diversity of 'flavor phenotypes'. The 'flavor phenotype' is an interesting concept for yeast selection, considering that now more than 1300 volatile compounds can be determined in wine [2,53].

In recent years, the genus H/K has been the subject of considerable study and publications, due to its positive contribution to the sensory characteristics of wines. Specifically, the yeast *H. vineae* of this genus has been of great interest because it produces several key aromatic compounds and has a good fermentation capacity. A strain of *H. vineae* isolated from Uruguayan vineyards was selected because of its positive effect on wine fermentation and contribution to the aroma profile of the final wine [18]. *H. vineae* has been demonstrated to increase fruity aromas and produce a high amount of acetate esters, such as 2-phenylethyl acetate and ethyl acetate (Figure 4), both in laboratory assays and in wines elaborated by sequential fermentation with SC [18,54].

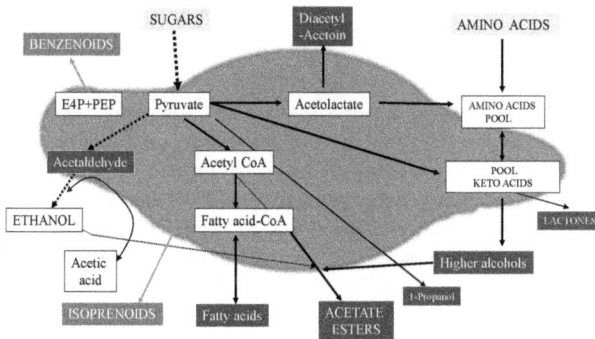

Figure 4. Fermentation flavor compounds produced by *Hanseniaspora*/*Kloeckera* yeasts during wine production (red and green boxes). Specific metabolic pathways that are highlighted in some species of this genus are shown in green (arrows and boxes). Pink boxes are the medium nutrients and doted arrows showed the main glycolysis pathway of primary fermentation.

Various groups of volatile compounds are produced during fermentation with H/K genus. For example, the use of a selected *H. uvarum* strain in mixed fermentation with commercial SC F5 increased the medium-chain fatty acid ethyl ester content in both synthetic media and grape must of Cabernet Gernischt grapes [55]. However, Medina et al. [18] did not find a significant increase in ethyl ester, using co-fermentation with *H. vineae* in Chardonnay grape must. In this work, decreases in the higher alcohols content—including 1,3-propanediol, 3-methyl 1-propanol, and tyrosol—were detected. Similar results were revealed in treatments with *H. uvarum*, finding a lower concentration of higher alcohols than the treatments inoculated with the SC isolates [56]. On the contrary, co-fermentation with *H. opuntiae* increased the amount of higher alcohols (phenylethanol and 3-methyl-butanol) and phenylacetaldehyde, in Cabernet Sauvignon grape must, intensifying the floral and sweet attributes of wine [57].

All H/K species increase the concentration of almost all the acetate esters. For example, all the acetate esters determined, except isoamyl acetate, were significantly affected by the inclusion of *H. osmophila* in a starter [58]. In this case, ethyl acetate and β-phenylethyl acetate concentrations in wine were increased when the proportion of *H. osmophila* in the culture increased. In wines fermented with the H/K:SC culture ratio of 90:10, the concentration of β-phenylethyl acetate was approximately 9-fold greater than that produced by SC pure culture [58]. In cold pre-maceration of Pinot noir grapes, inoculation with *H. uvarum* had the highest ethyl acetate level among the treatments evaluated, as well as high concentrations of the aforementioned branch-chained esters and, also, isoamyl acetate and isobutyl acetate [56]. In another report, the increased acetate ester levels were increased when *H. uvarum* was inoculated 48 h before SC, in different wine varieties, demonstrating that their enhancement could be induced by high population proportions of *H. uvarum* to SC. However, excessive *H. uvarum* yeasts in the inoculation slowed down the fermentation rate and produced a nail polish-like odour in Cabernet Sauvignon wines, by increasing the contents of acetate esters and volatile phenols [59]. Conversely, the wines produced from Negroamaro grapes by co-fermentation with *H. uvarum* showed an increment of acetate esters (ethyl acetate, isoamyl acetate and β-phenylethyl acetate) and fatty acids esters (ethyl hexanoate, ethyl octanoate and ethyl decanoate). In particular, an increase of isoamyl alcohol and β-phenyl alcohol was shown when compared to the wines produced by the SC starter [60].

Volatile compounds produced during fermentation of Macabeo grapes inoculated with *H. vineae* and separately with SC demonstrated significant differences in the acetates and higher alcohols. The *H. vineae* vinification produced low levels of higher alcohols and 5-fold greater concentration of the acetates [26]. Interesting, in this work, three compounds, 4-ethyl guaiacol, N-acetyltyramine and 1H-indole-3-ethanol acetate ester, were identified in wines with *H. vineae* but not in the wine fermented with SC [26].

3.1. De Novo Synthesis of Benzenoids and Isoprenoids

Benzyl alcohol, benzaldehyde, p-hydroxybenzaldehyde and p-hydroxybenzyl alcohol, compounds typically synthesised by plants, are synthesised de novo in the absence of grape-derived precursors by *H. vineae*. Levels of benzyl alcohol produced by 11 different *H. vineae* strains were 20–200 times higher than those measured in fermentations with SC strains. The absence of *PAL* in *H. vineae* suggests that benzenoids are necessarily dependent on de novo synthesis from chorismate [61,62]. It is worth noting that the increased use of diammonium phosphate, mainly applied in winemaking for increasing ester production or avoiding hydrogen sulphide formation, will decrease the production of phenylpropanoid compounds (Figure 5), compromising the final flavor complexity of the wine [61,62].

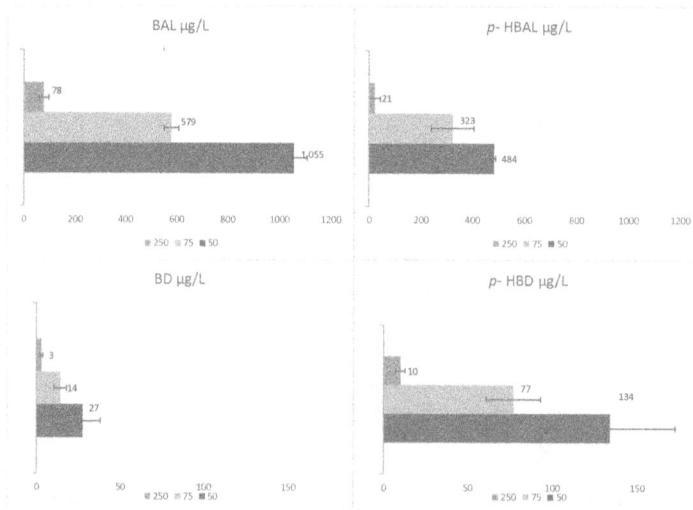

Figure 5. Formation of benzyl alcohol (BAL), benzaldehyde (BD), p-hydroxybenzyl alcohol (p-HBAL), and p-hydroxybenzaldehyde (p-HBD) by *Hanseniaspora vineae* 12/196 in the chemically-defined grape medium with three yeast assimilable nitrogen levels, where nitrogen levels of 75 and 250 mg/L were reached via the addition of diammonium phosphate. Fermentations were conducted at 20 °C; data are expressed in micrograms per liter.

Contrariwise, *H. vineae* produces high concentrations of the benzenoid and phenylpropanoid acetates. In the vinification of Macabeo grape must with *H. vineae*, 50 times more 2-phenylethyl acetate was generated than in vinifications with SC [26]. A similar trend was seen during de novo synthesis of monoterpenes by *H. uvarum*, where significant levels of citronellol were detected compared to SC strains. More recently, studies have shown the formation of terpenes and sesquiterpenes in vinifications with different *H. vineae* strains (Figure 6) exceeded the threshold values and reached higher concentrations than sole fermentation by SC [63].

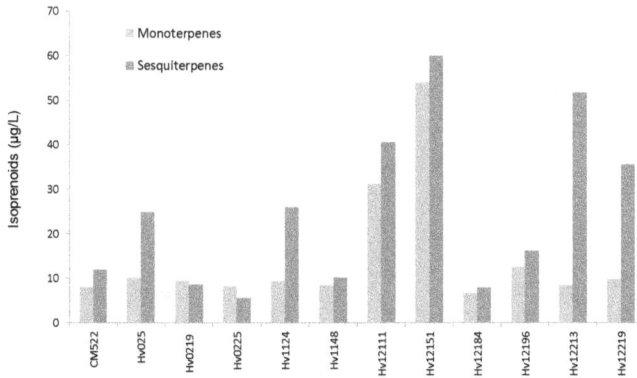

Figure 6. Production of isoprenoids (monoterpenes and sesquiterpenes) by various strains of *Hanseniaspora vineae* and the reference *Saccharomyces cerevisiae* CM522.

3.2. β-Glucosidase

Recently, Hu et al. [59] reported that β-glucosidase activity of *H. uvarum* yeast was 6.6-fold higher than that of the few naturally found SC strains. This characteristic explained why the participation of *H. uvarum* yeasts contributed to the increase of free terpene and C13-norisoprenoid contents with sensory impact [59]. However, high levels of β-glucosidase activity also increased the volatile phenol content, which might impart spicy odor traits to wines.

In a previous work, Mendes Ferreira et al. [64] studied the β-glucosidase activity using the *p*-nitrophenyl-β-D-glycoside (pNPG) as substrate in *H. uvarum* (formerly *K. apiculata*), *Pichia anomala*, and *Metschnikowia pulcherrima*, detecting the highest activity in *H. uvarum*. Furthermore, these authors demonstrated that *H. uvarum* was able to release some monoterpenols from an extract of Muscat grape juice, such as linalool, geraniol and in less quantity 3,7-dimethyl-1,7-octadien-3,6-diol and 3,7-dimethyl-1,5-octadien-3,7-diol, nerol, trans o-cimenol, α-terpineol, and citronellol [64].

The investigation of 31 H/K strains, including *H. guilliermondii*, *H. osmophila*, *H. uvarum*, and *H. vineae*, showed β-glucosidase and β-xylosidase activities (remarkable in one *H. uvarum* strain and two *H. vineae* strains) [65]. However, in this work, Muscat wine (13% v/v, initial alcohol) had only a moderate overall increase in terpene (1.1- to 1.3-fold) when treated with these strains. Specifically, these strains increase the levels of ho-trienol, β-phenylethanol, and 2,6-dimethyl-3,7-octadien-2,6-diol in the wine [65].

3.3. Effect of Hanseniaspora on the Volatile Compounds Produced during Tannat Red Grape Vinification

The vinification of Tannat grapes was conducted at three production scales: semi-pilot (20 kg), pilot (500 kg) and industrial (5000 kg) [66]. Figure 7 depicts the main flavor compound groups produced. The highest formation of acetates was detected in the vinifications with *H. vineae*, whereas, the maximum ethyl acetate concentration occurred in the vinification with *H. clermontiae*. Interestingly, the greatest concentration of norisoprenoid compounds was achieved by *H. vineae* vinification at industrial-scale compared to micro-fermentations.

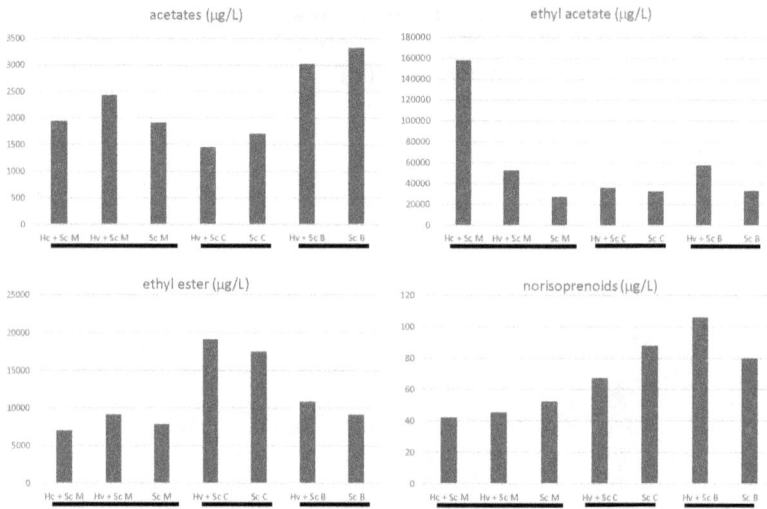

Figure 7. Concentration of principal groups of volatile compounds in vinifications at semi-pilot (M), pilot (C) and industrial (B) scale. Vinifications were inoculated with *Hanseniaspora vineae* (Hv), *Hanseniaspora clermontiae* (Hc) and *Saccharomyces cerevisiae* (Sc).

4. *Hanseniaspora/Kloeckera* Strains and Red Wine Color

The yeasts and grape maceration technology utilised during the vinification process affects pigment contents and the final red wine color [67–69]. Interactions between yeasts and anthocyanins during fermentation involve a range of mechanisms that might decrease or increase color. Yeast cell wall anthocyanin adsorption [70–72] and β-glucosidase activity, which releases the corresponding glycosylated anthocyanidin, exposing it to ready oxidation or conversion to colorless compounds [73], are well-known phenomena. Further research in the last decade has proved that some key compounds released during fermentation, such as pyruvic acid and acetaldehyde, are reactive precursors in the formation of new stable pigments. Vitisin A, vitisin B and ethyl-linked anthocyanin-flavanol pigments are examples of anthocyanin-derived compounds produced by SC strains [72,74–80]. Yeast strain selection strongly affects color intensity (CI) and the final concentrations of the anthocyanins [81,82] and other phenolic compounds [82].

Recently, studies have proven that some NS species might also be involved in wine color stabilisation [83,84]. As it was expected, some of these reactions could be attributed to the variable levels of acetaldehyde or pyruvate synthesis by different yeast species. For example, *Pichia* species generated significantly higher levels of acetaldehyde compared with *Saccharomyces* [85], and acetaldehyde increased linearly with increasing cell biomass concentration [86]. Except for a few reports on *Pichia*, *Schizosaccharomyces* and, more recently, some species of *Hanseniaspora* [83,87,88], limited information has been presented about the effect of NC strains on wine color composition. For the selection of new yeasts for the application with *Vitis vinifera* L. cv Tannat, a widely grown cultivar in Uruguay and one of the richest varieties in polyphenolic compounds [89,90], a program to select native NS yeasts and thereby increase the yeast diversity for fermentation without affecting wine color, was developed [91]. The Tannat grape juice model medium utilised allowed to screen the strains' capacity to synthesise anthocyanin-derived compounds while avoiding the interference of grape solids, such as the skin and seeds, was demonstrated previously for SC strains [72].

According to Medina [66], who evaluated 22 native H/K species for their effect on total anthocyanins (TA), CI, hue and total polyphenol index (TPI), the TPI values did not differ significantly between strains. The parameters with the greatest variation were CI with 32% (Figure 8), followed by TA with 30%, and then hue with 24%.

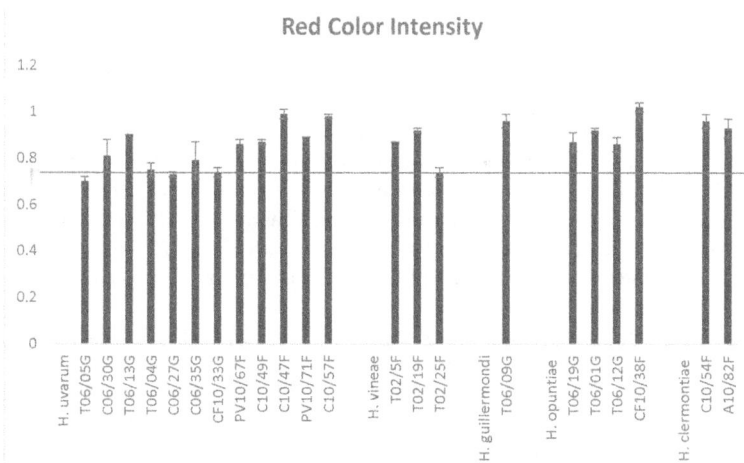

Figure 8. Mean normalised value and standard deviation of color intensity (sum of absorbance at 420, 520 and 620 nm), for 22 strains of five different *Hanseniaspora* species. The blue line indicates the average of the four lowest values obtained.

In the same study [26], consideration of the impact of CI and TA as the main color parameters in wines, the following strains were selected: *H. guilliermondii* (T06/09G), *H. opuntiae* (T06/01G), *H. vineae* (T02/5F) and *H. clermontiae* (A10/82F and C10/54F). Anthocyanin content and CI were evaluated against the best SC (882), previously selected for red grape fermentations [72]. All the previous selected H/K strains formed vitisin B, vitisin A, malvidin-3-O-glucoside-4-vinylphenol, malvidin-3-O-glucoside-4-vinylguaiacol (Figure 9, shows with letters A, B, C and D respectively).

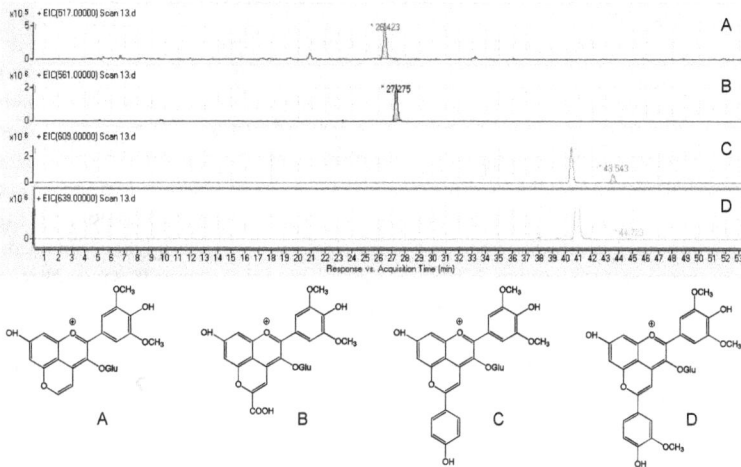

Figure 9. Identification of anthocyanin-derived pigments of Tannat grapes during fermentation by *Hanseniaspora/Kloeckera* yeasts that contribute to enhanced color stability.

Vitisin B formation has been reported previously only for SC yeasts [77,78,81,92]. Results of the anthocyanin-derived compounds formed during fermentation in the mentioned model grape medium indicated vitisin B could be linked to the increased acetaldehyde levels produced by SC when compared with NS yeasts [93]. All the NS strains selected showed vitisin B formation, despite some relatively low concentrations recorded relative to that formed by *Saccharomyces* yeast. For vitisin A, in contrast, there was a greater formation with NS strains than SC, possibly linked to the presence of pyruvic acid in the medium [78]. In corroboration with these findings, Morata et al. [83] noticed that in comparison to SC, *Schizosaccharomyces pombe* produced more pyruvic acid. Differences in the levels of pyruvic acid production might be explained by the particular "Crabtree effect" of each yeast species [94], which is defined as a system where respiration is repressed under high concentration of sugars. SC strains display a positive Crabtree effect and, consequently, this species presents a greater ethanol fermentation efficacy than many negative-Crabtree effect NS strains [94]. According to the literature, the production of vitisin A has been reported for SC [71,77,78], *Schizosaccharomyces pombe* [83,88] and, more recently, for some species of the H/K genus (*H. guilliermondii*, *H. opuntiae*, *H. vineae*, and *H. clermontiae*) [91,93].

Conversely, another anthocyanin-derived compound (malvidin-3-*O*-glucoside-4-vinylguaiacol) was detected during alcoholic fermentation with SC 882 [72], other SC strains [79–81,95] and *Pichia guilliermondii* [87]. The first report on the formation of this derived compound for the yeast genera *Hanseniaspora* and *Metschnikowia* was relatively more recent [91]. In that work, the authors argued that the high concentration of malvidin-3-glucoside-4-vinylguaicol found for all yeast treatments might also be a consequence of the differences in the grape variety and the concentrations of the respective hydroxycinnamic acids [96,97].

Formation of pigments derived from vinylphenol and vinylguaiacol could be explained by the hydroxycinnamate decarboxylase (HCDC) activity. The HCDC activity, specifically for supplying coumaric acid, has been mentioned for strains of the genera *Pichia*, *Torulaspora*, and *Zygosaccharomyces* [98,99]. A high HCDC activity of 90% for *P. guilliermondii* was recently noticed, which significantly influenced the formation of vinyl phenolic pyranoanthocyanins [87]. The work confirmed that during mixed or sequential fermentations carried out with NS or highly fermentative SC strains, with high HCDC activity, the content of stable pigments could be increased [87].

5. Applications of *Hanseniaspora/Kloeckera* Strains in Winemaking

Mixed-culture fermentation with *Saccharomyces* wine yeast is a controlled manner to apply NS strains, where the positive effects of NS yeasts and a complete dry fermentation is obtained. Even though SC produces most of the ethanol in wine, the NS yeasts present in the grape must, play a significant role in producing aroma compounds [16,100], contributing to diverse 'flavor phenotypes'.

As mentioned above, it is currently widely accepted that the secondary metabolites formed by properly selected NS yeasts, some of them already commercially available, during alcoholic fermentation positively affect the quality of wines [1,3,100–103]. The great variety of such yeasts allows designing different selected starter cultures (in conjunction with SC). Enhanced varietal and fermentative aromas, glycerol production, or specific enzymatic activities might be obtained, based on the ability of these yeasts to ferment different wine varieties [3]. As a result, winemakers can adapt wines to consumers searching for flavor diversity [37].

Yeasts of the genus H/K are frequently isolated during the first stages of the fermentation and are also found on the surface of the grapes, as well as in the soil, cellar, harvesting machinery, and during the processing of these fruit [104,105]. Based on current knowledge, H/K is one of the NS yeast genera with a major contribution to the sensory quality of wines. The H/K tend to be the dominant yeasts in the early stages of fermentation [26,37,106–108], perhaps attributed to their high population found in grapes or high tolerance to osmotic pressure (>200 g/L).

As the fermentation process progresses, the presence of H/K decreases, as a result of their low capacity to adapt to increasing levels of ethanol [109,110], although *H. uvarum* could be found until the end of fermentation, in some situations [55]. With the aid of culture-independent molecular techniques, it was possible to verify that some *H. vineae* strains are maintained until the end of fermentation, but their proportion decreases compared to SC [26]. This behavior can be seen in Figure 10, where the presence of *H. vineae* and other *Hanseniaspora* yeast is observed until the end of fermentation, at the semi-industrial scale of Merlot and Macabeo grapes [26].

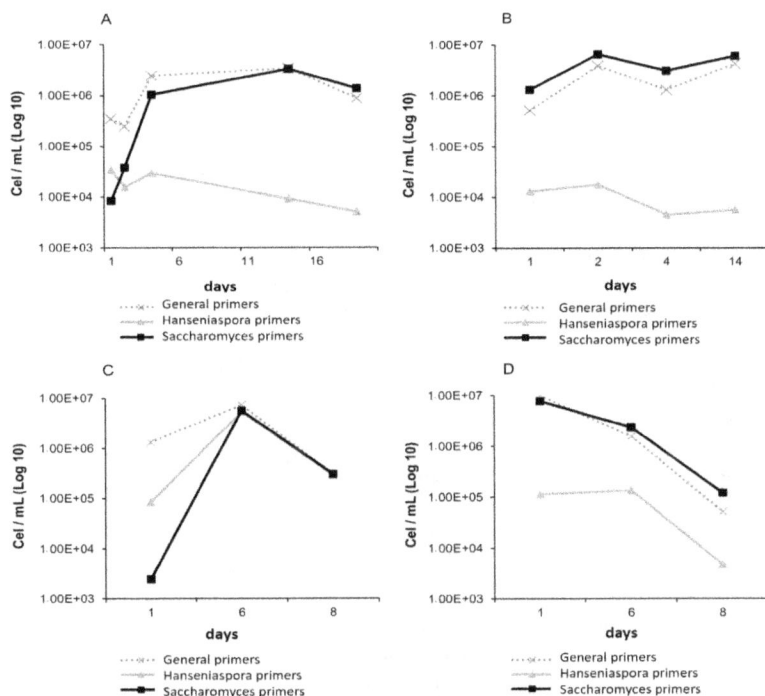

Figure 10. Monitoring of the yeast population by quantitative polymerase chain reaction, with general primers for yeasts (punctate), specific for *Hanseniaspora* spp. (grey) and specific for *Saccharomyces* (black), in Macabeo (**A,B**) and Merlot fermentations (**C,D**), in tanks inoculated with *H. vineae* (**A,C**) and tanks inoculated with *S. cerevisiae* (**B,D**).

In general, H/K shows a medium/low fermentative capacity (reaching up to 9% ethanol in some cases) [107]. Although the behavior of *Hanseniaspora* yeasts stands out from other NS yeasts [91,93] and, under certain conditions, give a better or no differences in performance against a *Saccharomyces* control [59,63], this is not a genus-dependent behavior but rather a strain-dependent characteristic. However, as seen in Figure 11, *H. vineae* is one of the main H/K fermenters, a character that corroborates the already-mentioned high homology of the pyruvate kinase gene with SC compared to the other H/K species (Figure 2). This result also justifies why is so difficult to isolate "*vineae*" species from grapes, yet readily detect them after two days of fermentation [63]. Likewise, the fermentation efficacy can be influenced by the inoculation procedures for mixed cultures. If the inoculation occurs sequentially, then the fermentation will be slowed down compared to simultaneous inoculations, due to cell retention of nutrients, and an additional nutrient addition will be necessary when the second inoculation is done [17]. Although SC has a higher capacity for fermentation than H/K strains, the lack of nutrients after 48 h will cause a sluggish fermentation process [8,59,93,111]. This tendency was also shown with different *H. clermontiae* strains [91,93]. The fermentative capacity and cell survival under mixed fermentation conditions can also be affected by the size of the inoculum. Good performances were obtained when *H. uvarum* was inoculated simultaneously and at twice the proportion of SC [60].

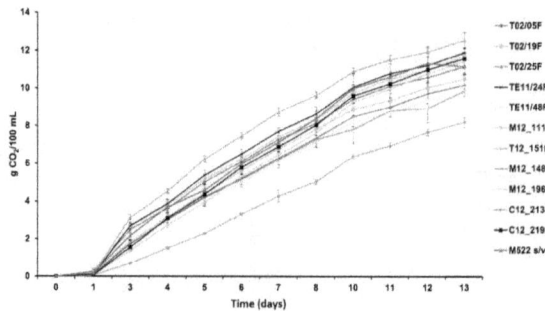

Figure 11. Fermentation kinetics of 11 strains of *Hanseniaspora vineae* at 20 °C, with a daily agitation. The yeast M522 *(Saccharomyces cerevisiae)* was used as a control (red). Data for CO_2 were obtained with cotton plug flask fermenters that include an average loss of 3 g of water vapor for every treatment.

H/K strains are considered important during vinification since they produce aromatic compounds of interest and modify the chemical composition of wines [9–11,18,112]. Two species stand out for producing high amounts of β-phenylethyl acetate, *H. guilliermondii* [113], and *H. vineae* [18,26,54,58,61,62,114]. However, this compound is not found at such high levels in the other species, as discussed above. This ester is associated with fruity, floral (rose) and honey sensory notes [100,115–117]. *H. uvarum* and *H. guilliermondii* have been reported as producing high levels of sulphur-containing aromatics [112]. At the same time, co-fermented wines with H/K strains presented more body and greater aromatic complexity in the mouth compared to SC solo fermentations [26], positively contributing to the final wine (Figure 12). Wines obtained at the semi-industrial scale from inoculums with *H. vineae* and then finished spontaneously with SC were preferred by a sensory panel than wines inoculated with a commercial SC, as a result of a higher floral descriptor, increased volume, increased structure and, ultimately, a better overall concept of the wine obtained. For body and mouth volume, no significant increase of glycerol or polysaccharides were recorded for *H. vineae* strains [63]. However, increased cell lysis was evident compared to SC commercial strains. Furthermore, the increase presence of C_{10} compounds found in wines fermented with *H. vineae*, suggest the existence of a faster autolysis rate compared to SC, as this flavor parameter was related to cell lysis by some authors for other yeast species [118,119]. The cell walls of *H. valbyensis* strains are reportedly about five times more sensitive to hydrolysis than those of SC, which is why they were used for a yeast glucan enzymatic tests [120]. Interestingly, Chardonnay barrel-fermented wines with mixed cultures of *H. vineae*/SC had significantly decreased biogenic amines and volatile acidity and increased glycerol and dry weight levels compared to pure SC fermentations [18]. The dry weight increase might also be associated with an increased cell lysis behavior of *H. vineae*. The authors of that study also showed cooperation between mixed cultures of *H. vineae*/SC with malolactic bacteria fermentation, by a significant stimulation compared to SC pure fermentation, finishing the process earlier by 45 days. More recently, these data were confirmed in red wine fermentations, also at an industrial scale [63], but further studies are needed to understand how the lactic bacteria were stimulated by this yeast.

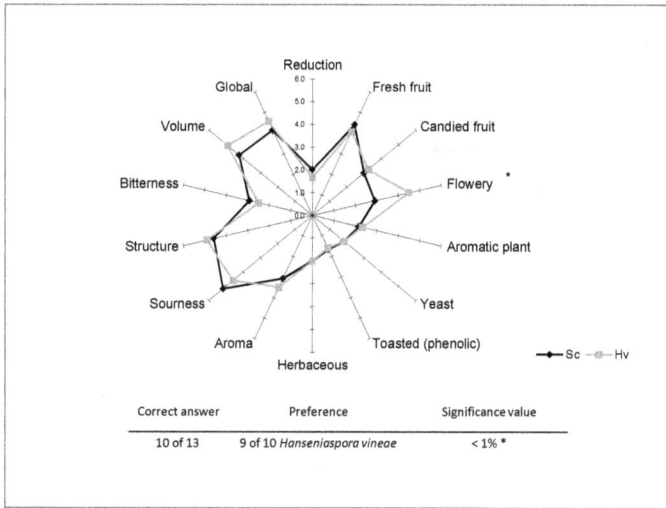

Figure 12. Results of the triangle (table) and descriptive (graphic) test of Macabeo wine fermented with *Hanseniaspora vineae* and *Saccharomyces cerevisiae*. Significant value * is indicated for flowery.

Most of these positive contributions by H/K yeasts can be explained by the presence of increased enzymatic activity compared to SC. The presence of active enzymes depends, in part, on the carbon and nitrogen sources present in the grape must. Small changes in the concentration of these nutrients can affect the nature, quantity, and diversity of the secreted enzymes [121]. The enzymes most commonly studied for their role during vinification are proteases, β-glucosidases, and pectinases since they intervene in sensory attributes, such as the color, aromas, and stability of wines [122]. Most NS yeasts have some enzymatic activity [123]. *Hanseniaspora* spp. are considered to be one of the primary producers of glycolytic and protease activities [63,105,124]. It was recently reported that within the NS yeast that contribute to the organoleptic quality of wines, *H. uvarum* had the highest enzymatic activity [1,3].

6. Conclusions

An insight into apiculate yeast biology showed that H/K is the principal genus found in mature grapes and these yeasts have interesting potential applications for wine fermentation. It is evident that selected strains of H/K yeasts might beneficially enhance the aroma and flavor attributes of wines and, more recently, this was proved for some H/K species. At the real winemaking scale, mixed cultures of *H. guilliermondii*/SC, *H. uvarum*/SC, and *H. vineae*/SC increase flavor diversity and thereby complexity. White and red grape varieties—such as Bobal, Macabeo, Chardonnay, Pinot Noir, Negroamaro, Tempranillo, Cabernet Sauvignon, Merlot, and Tannat—resulted in wines with improved sensorial profile.

Concurrently, it was confirmed that various *Hanseniaspora* species (*H. clermontiae*, *H. opuntiae*, *H. guilliermondii*, and *H. vineae*) contribute to the polyphenolic composition and color of the red wines. Thus, it was possible to demonstrate for the first time that the increased anthocyanin derived compounds generated from the mixed culture fermentation of these yeasts, enhanced the red wine color perception.

This set of mentioned characteristics (fermentative capacity, enzymatic activity, production of aromatic compounds, and ability to enhance the color of wines) makes the genus H/K a suitable stock for the selection of unconventional yeasts in commercial winemaking. Although H/K strains are still not easily available for their application, winemakers will have the opportunity to differentiate and increase regional characteristics to highlight their wines in a hugely competitive market.

Author Contributions: V.M., M.J.V., K.M., E.B. and F.C. contribute to the conception and search of data for this review, M.J.V., E.B. and F.C. analyze original data, M.J.V. and F.C. wrote the paper.

Funding: This research was funded by Agencia Nacional de Investigacion e Innovacion ANII of Uruguay.

Acknowledgments: We wish to thank ANII Postgraduate POS_NAC_2012_1_9099, Agencia Nacional de Investigación e Innovación (ANII) and to the postgraduate academic commission for the postgraduate completion scholarship granted to VM. We are also grateful for the financial support of ANII Project FMV_1_2011_1_6956, and the postdoctoral fellowships of MJV, ANII PD_NAC_2016_1_133945 and Clarín-COFUND from Principado de Asturias and European Union.

Conflicts of Interest: The authors declare no conflict of interest.

References

1. Jolly, N.P.; Varela, C.; Pretorius, I.S. Not your ordinary yeast: Non-*Saccharomyces* yeasts in wine production uncovered. *FEMS Yeast Res.* **2014**, *14*, 215–237. [CrossRef] [PubMed]
2. Carrau, F.; Gaggero, C.; Aguilar, P.S. Yeast diversity and native vigor for flavor phenotypes. *Trends Biotechnol.* **2015**, *33*, 148–154. [CrossRef] [PubMed]
3. Padilla, B.; Gil, J.V.; Manzanares, P. Past and Future of Non-*Saccharomyces* Yeasts: From Spoilage Microorganisms to Biotechnological Tools for Improving Wine Aroma Complexity. *Front. Microbiol.* **2016**, *7*, 411. [CrossRef] [PubMed]
4. Rosini, G.; Federici, F.; Martini, A. Yeast flora of grape berries during ripening. *Microb. Ecol.* **1982**, *8*, 83–89. [CrossRef] [PubMed]
5. Loureiro, V.; Ferreira, M.M.; Monteiro, S.; Ferreira, R.B. The Microbial Community of Grape Berry. In *The Biochemistry of the Grape Berry*; Gerós, H., Chaves, M., Delrot, S., Eds.; Bentham Science Publishers: Soest, The Netherlands, 2012; pp. 241–268.
6. Renouf, V.; Claisse, O.; Lonvaud-Funel, A. Inventory and monitoring of wine microbial consortia. *Appl. Microbiol. Biotechnol.* **2007**, *75*, 149–164. [CrossRef] [PubMed]
7. Varela, C.; Borneman, A.R. Yeasts found in vineyards and wineries. *Yeast* **2017**, *34*, 111–128. [CrossRef] [PubMed]
8. Di Maro, E.; Ercolini, D.; Coppola, S. Yeast dynamics during spontaneous wine fermentation of the Catalanesca grape. *Int. J. Food Microbiol.* **2007**, *117*, 201–210. [CrossRef] [PubMed]
9. Herraiz, T.; Reglero, G.; Herraiz, M.; Martin-Alvarez, P.J.; Cabezudo, M.D. The influence of the yeast and type of culture on the volatile composition of wines fermented without sulfur dioxide. *Am. J. Enol. Vitic.* **1990**, *41*, 313–318.
10. Mateo, J.; Jimenez, M.; Huerta, T.; Pastor, A. Contribution of different yeasts isolated from musts of Monastrell grapes to the aroma of wine. *Int. J. Food Microbiol.* **1991**, *14*, 153–160. [CrossRef]
11. Zironi, R.; Romano, P.; Suzzi, G.; Battistutta, F.; Comi, G. Volatile metabolites produced in wine by mixed and sequential cultures of *Hanseniaspora guilliermondii* or *Kloeckera apiculata* and *Saccharomyces cerevisiae*. *Biotechnol. Lett.* **1993**, *15*, 235–238. [CrossRef]
12. Gil, J.; Mateo, J.; Jimenez, M.; Pastor, A.; Huerta, T. Aroma compounds in wine as influenced by apiculate yeasts. *J. Food Sci.* **1996**, *61*, 1247–1266. [CrossRef]
13. Caridi, A.; Crucitti, P.; Ramondino, D. Winemaking of musts at high osmotic strength by thermotolerant yeasts. *Biotechnol. Lett.* **1999**, *21*, 617–620. [CrossRef]
14. Romano, P.; Suzzi, G.; Comi, G.; Zironi, R. Higher alcohol and acetic acid production by apiculate wine yeasts. *J. Appl. Bacteriol.* **1992**, *73*, 126–130. [CrossRef]
15. Ciani, M.; Maccarelli, F. Oenological properties of non-*Saccharomyces* yeasts associated with wine-making. *World J. Microbiol. Biotechnol.* **1998**, *14*, 199–203. [CrossRef]
16. Romano, P.; Suzzi, G.; Domizio, P.; Fatichenti, F. Secondary products formation as a tool for discriminating non-*Saccharomyces* wine strains. *Antonie van Leeuwenhoek Int. J. Gen. Mol. Microbiol.* **1997**, *71*, 239–242. [CrossRef]
17. Medina, K.; Boido, E.; Dellacassa, E.; Carrau, F. Growth of non-*Saccharomyces* yeasts affects nutrient availability for Saccharomyces cerevisiae during wine fermentation. *Int. J. Food Microbiol.* **2012**, *157*, 245–250. [CrossRef] [PubMed]

18. Medina, K.; Boido, E.; Fariña, L.; Gioia, O.; Gomez, M.E.; Barquet, M.; Gaggero, C.; Dellacassa, E.; Carrau, F. Increased flavour diversity of Chardonnay wines by spontaneous fermentation and co-fermentation with *Hanseniaspora vineae*. *Food Chem.* **2013**, *141*, 2513–2521. [CrossRef] [PubMed]

19. Arneborg, N.; Siegumfeldt, H.; Andersen, G.H.; Nissen, P.; Daria, V.R.; Rodrigo, P.J.; Glückstad, J. Interactive optical trapping shows that confinement is a determinant of growth in a mixed yeast culture. *FEMS Microbiol. Lett.* **2005**, *245*, 155–159. [CrossRef] [PubMed]

20. Bataillon, M.; Rico, A.; Salmon, J.-M.; Barre, P. Early Thiamin Assimilation by Yeasts under Enological Conditions: Impact on Alcoholic Fermentation Kinetics. *J. Ferment. Bioeng.* **1996**, *82*, 145–150. [CrossRef]

21. Wang, X.D.; Bohlscheid, J.C.; Edwards, C.G. Fermentative activity and production of volatile compounds by *Saccharomyces* grown in synthetic grape juice media deficient in assimilable nitrogen and/or pantothenic acid. *J. Appl. Microbiol.* **2003**, *94*, 349–359. [CrossRef] [PubMed]

22. Díaz-Montaño, D.M.; Favela-Torres, E.; Córdova, J. Improvement of growth, fermentative efficiency and ethanol tolerance of *Kloeckera africana* during the fermentation of agave tequilana juice by addition of yeast extract. *J. Sci. Food Agric.* **2010**, *90*, 321–328. [CrossRef] [PubMed]

23. Jolly, N.P.; Augustyn, O.P.H.; Pretorius, I.S. The Effect of Non-*Saccharomyces* Yeasts on Fermentation and Wine Quality. *S. Afr. J. Enol. Vitic.* **2003**, *24*, 55–62. [CrossRef]

24. Rojas, V.; Gil, J.V.; Piaga, F.; Manzanares, P. Studies on acetate ester production by non-*Saccharomyces* wine yeasts. *Int. J. Food Microbiol.* **2001**, *70*, 283–289. [CrossRef]

25. Moreira, N.; Pina, C.; Mendes, F.; Couto, J.A.; Hogg, T.; Vasconcelos, I. Volatile compounds contribution of *Hanseniaspora guilliermondii* and *Hanseniaspora uvarum* during red wine vinifications. *Food Control* **2011**, *22*, 662–667. [CrossRef]

26. Lleixa, J.; Martin, V.; Portillo, C.; Carrau, F.; Beltran, G.; Mas, A. Comparison of the performances of *Hanseniaspora vineae* and *Saccharomyces cerevisiae* during winemaking. *Front. Microbiol.* **2016**, *7*, 338. [CrossRef] [PubMed]

27. McLaughlin, R.; Wilson, C.; Droby, S.; Ben, R.; Chalutz, E. Biological control of postharvest diseases of grape, peach an apple with the yeast *Kloeckera apiculata* and *Candida guilliermondii*. *Plant Dis.* **1992**, *76*, 470–473. [CrossRef]

28. Rabosto, X.; Carrau, M.; Paz, A.; Boido, E.; Dellacassa, E.; Carrau, F.M. Grapes and vineyard soils as sources of microorganisms for biological control of *Botrytis cinerea*. *Am. J. Enol. Vitic.* **2006**, *57*, 332–338.

29. Esteve-Zarzoso, B.; Peris-Toran, M.J.; Ramón, D.; Querol, A. Navigating wall-sized displays with the gaze: A proposal for cultural heritage. *CEUR Workshop Proc.* **2001**, *80*, 85–92.

30. Cadez, N.; Raspor, P.; de Cock, A.W.; Boekhout, T.; Smith, M.T. Molecular identification and genetic diversity within species of the genera *Hanseniaspora* and *Kloeckera*. *FEMS Yeast Res.* **2002**, *1*, 279–289. [PubMed]

31. Mills, D.A.; Johannsen, E.A.; Cocolin, L. Yeast diversity and persistence in botrytis-affected wine fermentations. *Appl. Environ. Microbiol.* **2002**, *68*, 4884–4893. [CrossRef] [PubMed]

32. Andorrà, I.; Landi, S.; Mas, A.; Esteve-Zarzoso, B.; Guillamón, J.M. Effect of fermentation temperature on microbial population evolution using culture-independent and dependent techniques. *Food Res. Int.* **2010**, *43*, 773–779. [CrossRef]

33. Wang, C.; García-Fernández, D.; Mas, A.; Esteve-Zarzoso, B. Fungal diversity in grape must and wine fermentation assessed by massive sequencing, quantitative PCR and DGGE. *Front. Microbiol.* **2015**, *6*, 1156. [CrossRef] [PubMed]

34. Bokulich, N.A.; Ohta, M.; Richardson, P.M.; Mills, D.A. Monitoring Seasonal Changes in Winery-Resident Microbiota. *PLoS ONE* **2013**, *8*, 66437. [CrossRef] [PubMed]

35. Pinto, C.; Pinho, D.; Cardoso, R.; Custodio, V.; Fernandes, J.; Sousa, S.; Pinheiro, M.; Egas, C.; Gomes, A.C. Wine fermentation microbiome: A landscape from different Portuguese wine appellations. *Front. Microbiol.* **2015**, *6*, 905. [CrossRef] [PubMed]

36. Sabate, J.; Cano, J.; Esteve-Zarzoso, B.; Guillamón, J.M. Isolation and identification of yeasts associated with vineyard and winery by RFLP analysis of ribosomal genes and mitochondrial DNA. *Microbiol. Res.* **2002**, *157*, 267–274. [CrossRef] [PubMed]

37. Pretorius, I.S. Tailoring wine yeast for the new millennium: Novel approaches to the ancient art of winemaking. *Yeast* **2000**, *16*, 675–729. [CrossRef]

38. Yamada, Y.; Maeda, K.; Nagahama, T.; Banno, I. The phylogenetic relationships of the Q6-equiped species in the teleomorphic apiculate yeast genera *Hanseniaspora*, *Nadsonia*, and *Saccharomycodes* based on the partial sequences of 18S and 26S ribosomal ribonucleiic acids. *J. Gen. Appl. Microbiol.* **1992**, *38*, 585–596. [CrossRef]

39. Lleixà, J.; Manzano, M.; Mas, A.; del C. Portillo, M. *Saccharomyces* and non-*Saccharomyces* competition during microvinification under different sugar and nitrogen conditions. *Front. Microbiol.* **2016**, *7*, 1959. [CrossRef]

40. Giorello, F.M.; Berná, L.; Greif, G.; Camesasca, L.; Salzman, V.; Medina, K.; Robello, C.; Gaggero, C.; Aguilar, P.S.; Carrau, F. Genome sequence of the native apiculate wine yeast *Hanseniaspora vineae* T02/19AF. *Genome Announc.* **2014**, *2*, e00530-14. [CrossRef] [PubMed]

41. Riley, R.; Haridas, S.; Wolfe, K.H.; Lopes, M.R.; Hittinger, C.T.; Göker, M.; Salamov, A.A.; Wisecaver, J.H.; Long, T.M.; Calvey, C.H.; et al. Comparative genomics of biotechnologically important yeasts. *Proc. Natl. Acad. Sci. USA* **2016**, *113*, 9882–9887. [CrossRef] [PubMed]

42. Sternes, P.R.; Lee, D.; Kutyna, D.R.; Borneman, A.R. Genome Sequences of Three Species of *Hanseniaspora* Isolated from Spontaneous Wine Fermentations: TABLE 1. *Genome Announc.* **2016**, *4*, e01287-16. [CrossRef] [PubMed]

43. Seixas, I.; Barbosa, C.; Salazar, S.B.; Mendes-Faia, A.; Wang, Y.; Güldener, U.; Mendes-Ferreira, A.; Mira, N.P. Genome Sequence of the Nonconventional Wine Yeast *Hanseniaspora guilliermondii* UTAD222. *Genome Announc.* **2017**, *5*, e01515-16. [CrossRef] [PubMed]

44. Langenberg, A.; Bink, F.J.; Wolff, L.; Walter, S.; Grossmann, M.; Schmitz, H. Glycolytic Functions Are Conserved in the Genome of the Wine Yeast Hanseniaspora uvarum, and Pyruvate Kinase Limits Its Capacity for Alcoholic Fermentation. *Appl. Environ. Microbiol.* **2017**, *83*, e01580-17. [CrossRef] [PubMed]

45. Pramateftaki, P.V.; Kouvelis, V.N.; Lanaridis, P.; Typas, M. The mitochondrial genome of the wine yeast *Hanseniaspora uvarum*: A unique genome organization among yeast/fungal counterparts. *FEMS Yeast Res.* **2006**, *6*, 77–90. [CrossRef] [PubMed]

46. Molina, A.M.; Swiegers, J.H.; Varela, C.; Pretorius, I.S.; Agosin, E. Influence of wine fermentation temperature on the synthesis of yeast-derived volatile aroma compounds. *Appl. Microbiol. Biotechnol.* **2007**, *77*, 675–687. [CrossRef] [PubMed]

47. Bisson, L.F. Stuck and sluggish fermentations. *Am. J. Enol. Vitic.* **1999**, *50*, 107–119.

48. Taillandier, P.; Ramon Portugal, F.; Fuster, A.; Strehaiano, P. Effect of ammonium concentration on alcoholic fermentation kinetics by wine yeasts for high sugar content. *Food Microbiol.* **2007**, *24*, 95–100. [CrossRef] [PubMed]

49. Beltran, G.; Novo, M.; Rozès, N.; Mas, A.; Guillamón, J.M. Nitrogen catabolite repression in *Saccharomyces cerevisiae* during wine fermentations. *FEMS Yeast Res.* **2004**, *4*, 625–632. [CrossRef] [PubMed]

50. Nguong, D.L.S.; Jun, L.Y.; Yatim, N.I.; Nathan, S.; Murad, A.M.A.; Mahadi, N.M.; Bakar, F.D.A. Characterising yeast isolates from Malaysia towards the development of alternative heterologous protein expression systems. *Sains Malays* **2011**, *40*, 323–329.

51. Andorrà, I.; Berradre, M.; Mas, A.; Esteve-Zarzoso, B.; Guillamón, J.M. Effect of mixed culture fermentations on yeast populations and aroma profile. *LWT-Food Sci. Technol.* **2012**, *49*, 8–13. [CrossRef]

52. Tempère, S.; Marchal, A.; Barbe, J.C.; Bely, M.; Masneuf-Pomarede, I.; Marullo, P.; Albertin, W. The complexity of wine: Clarifying the role of microorganisms. *Appl. Microbiol. Biotechnol.* **2018**, *102*, 3995–4007. [CrossRef] [PubMed]

53. Cordente, A.G.; Curtin, C.D.; Varela, C.; Pretorius, I.S. Flavour-active wine yeasts. *Appl. Microbiol. Biotechnol.* **2012**, *96*, 601–618. [CrossRef] [PubMed]

54. Viana, F.; Belloch, C.; Vallés, S.; Manzanares, P. Monitoring a mixed starter of *Hanseniaspora vineae-Saccharomyces cerevisiae* in natural must: Impact on 2-phenylethyl acetate production. *Int. J. Food Microbiol.* **2011**, *151*, 235–240. [CrossRef] [PubMed]

55. Hu, K.; Jin, G.-J.; Mei, W.-C.; Li, T.; Tao, Y.-S. Increase of medium-chain fatty acid ethyl ester content in mixed *H. uvarum/S. cerevisiae* fermentation leads to wine fruity aroma enhancement. *Food Chem.* **2018**, *239*, 495–501. [CrossRef] [PubMed]

56. Hall, H.; Zhou, Q.; Qian, M.C.; Osborne, J.P. Impact of yeasts present during prefermentation cold maceration of pinot noir grapes on wine volatile aromas. *Am. J. Enol. Vitic.* **2017**, *68*, 81–90. [CrossRef]

57. Luan, Y.; Zhang, B.-Q.; Duan, C.-Q.; Yan, G.-L. Effects of different pre-fermentation cold maceration time on aroma compounds of *Saccharomyces cerevisiae* co-fermentation with *Hanseniaspora opuntiae* or *Pichia kudriavzevii*. *LWT-Food Sci. Technol.* **2018**, *92*, 177–186. [CrossRef]

58. Viana, F.; Gil, J.V.; Vallés, S.; Manzanares, P. Increasing the levels of 2-phenylethyl acetate in wine through the use of a mixed culture of *Hanseniaspora osmophila* and *Saccharomyces cerevisiae*. *Int. J. Food Microbiol.* **2009**, *135*, 68–74. [CrossRef] [PubMed]

59. Hu, K.; Jin, G.-J.; Xu, Y.-H.; Tao, Y.-S. Wine aroma response to different participation of selected *Hanseniaspora uvarum* in mixed fermentation with *Saccharomyces cerevisiae*. *Food Res. Int.* **2018**, *108*, 119–127. [CrossRef] [PubMed]

60. Tristezza, M.; Tufariello, M.; Capozzi, V.; Spano, G.; Mita, G.; Grieco, F. The oenological potential of *Hanseniaspora uvarum* in simultaneous and sequential co-fermentation with *Saccharomyces cerevisiae* for industrial wine production. *Front. Microbiol.* **2016**, *7*, 1–14. [CrossRef] [PubMed]

61. Martin, V.; Giorello, F.; Fariña, L.; Minteguiaga, M.; Salzman, V.; Boido, E.; Aguilar, P.S.; Gaggero, C.; Dellacassa, E.; Mas, A.; et al. De novo synthesis of benzenoid compounds by the yeast *Hanseniaspora vineae* increases the flavor diversity of wines. *J. Agric. Food Chem.* **2016**, *64*, 4574–4583. [CrossRef] [PubMed]

62. Martin, V.; Boido, E.; Giorello, F.; Mas, A.; Dellacassa, E.; Carrau, F. Effect of yeast assimilable nitrogen on the synthesis of phenolic aroma compounds by *Hanseniaspora vineae* strains. *Yeast* **2016**, *33*, 323–328. [CrossRef] [PubMed]

63. Martín, V. *Hanseniaspora vineae*: Caracterizacióon y su uso en la Vinificación. Ph.D. Thesis, Universidad de la Republica, Montevideo, Uruguay, 2016.

64. Mendes Ferreira, A.; Climaco, M.C.; Mendes Faia, A. The role of non-*Saccharomyces* species in releasing glycosidic bound fraction of grape aroma components—A preliminary study. *J. Appl. Microbiol.* **2001**, *91*, 67–71. [CrossRef] [PubMed]

65. López, S.; Mateo, J.J.; Maicas, S. Characterisation of *Hanseniaspora* isolates with potential aromaenhancing properties in muscat wines. *S. Afr. J. Enol. Vitic.* **2014**, *35*, 292–303.

66. Medina, K. Biodiversidad de Levaduras no-*Saccharomyces*: Efecto del Metabolismo Secundario en el Color y el Aroma de Vinos de *Calidad*. Ph.D. Thesis, Universidad de la Republica, Montevideo, Uruguay, 2014.

67. Boulton, R. The copigmentation of anthocyanins and its role in the color of red wine: A critical review. *Am. J. Enol. Vitic.* **2001**, *52*, 67–87.

68. Vivar-Quintana, A.M.; Santos-Buelga, C.; Rivas-Gonzalo, J.C. Anthocyanin-derived pigments and colour of red wines. *Anal. Chim. Acta* **2002**, *458*, 147–155. [CrossRef]

69. Schwarz, M.; Wabnitz, T.C.; Winterhalter, P. Pathway leading to the formation of anthocyanin-vinylphenol adducts and related pigments in red wines. *J. Agric. Food Chem.* **2003**, *51*, 3682–3687. [CrossRef] [PubMed]

70. Vasserot, Y.; Caillet, S.; Maujean, A. Study of anthocyanin adsorption by yeast lees. Effect of some physicochemical parameters. *Am. J. Enol. Vitic.* **1997**, *48*, 433–437.

71. Morata, A.; Gómez-Cordovés, M.C.; Suberviola, J.; Bartolomé, B.; Colomo, B.; Suárez, J.A. Adsorption of anthocyanins by yeast cell walls during the fermentation of red wines. *J. Agric. Food Chem.* **2003**, *51*, 4084–4088. [CrossRef] [PubMed]

72. Medina, K.; Boido, E.; Dellacassa, E.; Carrau, F. Yeast interactions with anthocyanins during red wine fermentation. *Am. J. Enol. Vitic.* **2005**, *56*, 104–109.

73. Manzanares, P.; Rojas, V.; Genovés, S.; Vallés, S. A preliminary search for anthocyanin-β-D-glucosidase activity in non-*Saccharomyces* wine yeasts. *Int. J. Food Sci. Technol.* **2000**, *35*, 95–103. [CrossRef]

74. Asenstorfer, R.E.; Markides, A.J.; Iland, P.G.; Jones, G.P. Formation of vitisin A during red wine fermentation and maturation. *Aust. J. Grape Wine Res.* **2003**, *9*, 40–46. [CrossRef]

75. Eglinton, J.; Griesser, M.; Henschke, P.; Kwiatkowski, M.; Parker, M.; Herderich, M. Yeast-Mediated Formation of Pigmented Polymers in Red Wine. In *Red Wine Color*; ACS Symposium Series: Washington, DC, USA, 2004.

76. Lee, D.F.; Swinny, E.E.; Jones, G.P. NMR identification of ethyl-linked anthocyanin-flavanol pigments formed in model wine ferments. *Tetrahedron Lett.* **2004**, *45*, 1671–1674. [CrossRef]

77. Morata, A.; Gómez-Cordovés, M.C.; Colomo, B.; Suárez, J.A. Pyruvic Acid and Acetaldehyde Production by Different Strains of *Saccharomyces cerevisiae*: Relationship with Vitisin A and B Formation in Red Wines. *J. Agric. Food Chem.* **2003**, *51*, 7402–7409. [CrossRef] [PubMed]

78. Morata, A.; Calderón, F.; González, M.C.; Gómez-Cordovés, M.C.; Suárez, J.A. Formation of the highly stable pyranoanthocyanins (vitisins A and B) in red wines by the addition of pyruvic acid and acetaldehyde. *Food Chem.* **2007**, *100*, 1144–1152. [CrossRef]

79. Morata, A.; González, C.; Suárez-Lepe, J.A. Formation of vinylphenolic pyranoanthocyanins by selected yeasts fermenting red grape musts supplemented with hydroxycinnamic acids. *Int. J. Food Microbiol.* **2007**, *116*, 144–152. [CrossRef] [PubMed]

80. Benito, S.; Palomero, F.; Morata, A.; Uthurry, C.; Suárez-Lepe, J.A. Minimization of ethylphenol precursors in red wines via the formation of pyranoanthocyanins by selected yeasts. *Int. J. Food Microbiol.* **2009**, *132*, 145–152. [CrossRef] [PubMed]

81. Morata, A.; Gómez-Cordovés, M.C.; Calderón, F.; Suárez, J.A. Effects of pH, temperature and SO$_2$ on the formation of pyranoanthocyanins during red wine fermentation with two species of *Saccharomyces*. *Int. J. Food Microbiol.* **2006**, *106*, 123–129. [CrossRef] [PubMed]

82. Monagas, M.; Gómez-Cordovés, C.; Bartolomé, B. Evaluation of different *Saccharomyces cerevisiae* strains for red winemaking. Influence on the anthocyanin, pyranoanthocyanin and non-anthocyanin phenolic content and colour characteristics of wines. *Food Chem.* **2007**, *104*, 814–823. [CrossRef]

83. Morata, A.; Benito, S.; Loira, I.; Palomero, F.; González, M.C.; Suárez-Lepe, J.A. Formation of pyranoanthocyanins by *Schizosaccharomyces pombe* during the fermentation of red must. *Int. J. Food Microbiol.* **2012**, *159*, 47–53. [CrossRef] [PubMed]

84. Loira, I.; Morata, A.; Comuzzo, P.; Callejo, M.J.; González, C.; Calderón, F.; Suárez-Lepe, J.A. Use of *Schizosaccharomyces pombe* and *Torulaspora delbrueckii* strains in mixed and sequential fermentations to improve red wine sensory quality. *Food Res. Int.* **2015**, *76*, 325–333. [CrossRef] [PubMed]

85. Clemente-Jimenez, J.M.; Mingorance-Cazorla, L.; Martínez-Rodríguez, S.; Las Heras-Vázquez, F.J.; Rodríguez-Vico, F. Influence of sequential yeast mixtures on wine fermentation. *Int. J. Food Microbiol.* **2005**, *3*, 301–308. [CrossRef] [PubMed]

86. Duff, S.J.; Murray, W.D. Production and application of methylotrophic yeast *Pichia pastoris*. *Biotechnol. Bioeng.* **1988**, *31*, 44–49. [CrossRef] [PubMed]

87. Benito, S.; Morata, A.; Palomero, F.; González, M.C.; Suárez-Lepe, J.A. Formation of vinylphenolic pyranoanthocyanins by *Saccharomyces cerevisiae* and *Pichia guillermondii* in red wines produced following different fermentation strategies. *Food Chem.* **2011**, *124*, 15–23. [CrossRef]

88. Benito, S.; Palomero, F.; Calderón, F.; Palmero, D.; Suárez-Lepe, J.A. Selection of appropriate *Schizosaccharomyces* strains for winemaking. *Food Microbiol.* **2014**, *42*, 218–224. [CrossRef] [PubMed]

89. Carrau, F.; Boido, E.; Gaggero, C.; Medina, K. *Vitis vinifera* Tannat, chemical characterization and functional properties. Ten years of research. *Transworld Res. Netw.* **2011**, *661*, 53–71.

90. Da Silva, C.; Zamperin, G.; Ferrarini, A.; Minio, A.; Dal Molin, A.; Venturini, L.; Buson, G.; Tononi, P.; Avanzato, C.; Zago, E.; et al. The High Polyphenol Content of Grapevine Cultivar Tannat Berries Is Conferred Primarily by Genes That Are Not Shared with the Reference Genome. *Plant Cell* **2013**, *25*, 4777–4788. [CrossRef] [PubMed]

91. Medina, K.; Boido, E.; Dellacassa, E.; Carrau, F. Effects of non-*Saccharomyces* yeasts on color, anthocyanin, and anthocyanin-derived pigments of Tannat grapes during fermentation. *Am. J. Enol. Vitic.* **2018**, *69*, 148–156. [CrossRef]

92. Suárez-Lepe, J.A.; Morata, A. New trends in yeast selection for winemaking. *Trends Food Sci. Technol.* **2012**, *23*, 39–50. [CrossRef]

93. Medina, K.; Boido, E.; Fariña, L.; Dellacassa, E.; Carrau, F. Non-*Saccharomyces* and *Saccharomyces* strains co-fermentation increases acetaldehyde accumulation: Effect on anthocyanin-derived pigments in Tannat red wines. *Yeast* **2016**, *33*, 339–343. [CrossRef] [PubMed]

94. Gonzalez, R.; Quirós, M.; Morales, P. Yeast respiration of sugars by non-*Saccharomyces* yeast species: A promising and barely explored approach to lowering alcohol content of wines. *Trends Food Sci. Technol.* **2013**, *29*, 55–61. [CrossRef]

95. Vanbeneden, N.; Gils, F.; Delvaux, F.; Delvaux, F.R. Formation of 4-vinyl and 4-ethyl derivatives from hydroxycinnamic acids: Occurrence of volatile phenolic flavour compounds in beer and distribution of Pad1-activity among brewing yeasts. *Food Chem.* **2008**, *107*, 221–230. [CrossRef]

96. Boursiquot, J. Contribution à l'étude des Esters Hidroxycinnamiques chez le Genre Vitis. Recherche D'application Taxonomique. Ph.D. Thesis, Ecole Nationale Supériore Agronomique de Montpellier, Montpellier, France, 1987.

97. Boido, E.; García-Marino, M.; Dellacassa, E.; Carrau, F.; Rivas-Gonzalo, J.C.; Escribano-Bailón, M.T. Characterisation and evolution of grape polyphenol profiles of *Vitis vinifera* L. cv. Tannat during ripening and vinification. *Aust. J. Grape Wine Res.* **2011**, *17*, 383–393. [CrossRef]

98. Dias, L.; Dias, S.; Sancho, T.; Stender, H.; Querol, A.; Malfeito-Ferreira, M.; Loureiro, V. Identification of yeasts isolated from wine-related environments and capable of producing 4-ethylphenol. *Food Microbiol.* **2003**, *20*, 567–574. [CrossRef]

99. Suezawa, Y.; Suzuki, M. Bioconversion of Ferulic Acid to 4-Vinylguaiacol and 4-Ethylguaiacol and of 4-Vinylguaiacol to 4-Ethylguaiacol by Halotolerant Yeasts Belonging to the Genus *Candida*. *Biosci. Biotechnol. Biochem.* **2007**, *71*, 1058–1062. [CrossRef] [PubMed]

100. Swiegers, J.H.; Bartowsky, E.J.; Henschke, P.A.; Pretorius, I.S. Yeast and bacterial modulaton of wine aroma and flavour. *Aust. J. Grape Wine Res.* **2005**, *11*, 139–173. [CrossRef]

101. Jolly, N.P.; Augustyn, O.P.H.; Pretorius, I.S. The Occurrence of Non-*Saccharomyces* cerevisiae Yeast Species Over Three Vintages in Four Vineyards and Grape Musts from Four Production Regions of the Western Cape, South Africa. *S. Afr. J. Enol. Vitic.* **2003**, *24*, 35–42.

102. Jolly, N.P.; Augustyn, O.P.H.; Pretorius, I.S. The Role and Use of Non-*Saccharomyces* Yeasts in Wine Production. *S. Afr. J. Enol. Vitic.* **2006**, *27*, 15–39. [CrossRef]

103. Varela, C. The impact of non-*Saccharomyces* yeasts in the production of alcoholic beverages. *Appl. Microbiol. Biotechnol.* **2016**. [CrossRef] [PubMed]

104. Strauss, M.L.A.; Jolly, N.P.; Lambrechts, M.G.; Rensburg, P. Van Screening for the production of extracellular hydrolytic enzymes by non-*Saccharomyces* wine yeasts. *J. Appl. Microbiol.* **2001**, *91*, 182–190. [CrossRef] [PubMed]

105. López, S.; Mateo, J.; Maicas, S. Screening of *Hanseniaspora* Strains for the Production of Enzymes with Potential Interest for Winemaking. *Fermentation* **2015**, *2*, 1. [CrossRef]

106. Padilla, B.; Zulian, L.; Ferreres, À.; Pastor, R.; Esteve-Zarzoso, B.; Beltran, G.; Mas, A. Sequential inoculation of native non-*Saccharomyces* and *Saccharomyces cerevisiae* strains for wine making. *Front. Microbiol.* **2017**, *8*, 1293. [CrossRef] [PubMed]

107. Zohre, D.E.; Erten, H. The influence of *Kloeckera apiculata* and *Candida pulcherrima* yeasts on wine fermentation. *Process Biochem.* **2002**, *38*, 319–324. [CrossRef]

108. Martini, A. Origin and Domestication of the Wine Yeast *Saccharomyces cerevisiae*. *J. Wine Res.* **1993**, *4*, 165–176. [CrossRef]

109. Díaz-Montaño, D.M.; de Jesús Ramírez Córdova, J. The Fermentative and Aromatic Ability of *Kloeckera* and *Hanseniaspora* Yeasts. In *Yeast Biotechnology: Diversity and Applications*; Satyanarayana, T., Kunze, G., Eds.; Springer: Dordrecht, The Netherlands, 2009; pp. 281–305.

110. Pina, C.; Santos, C.; Couto, J.A.; Hogg, T. Ethanol tolerance of five non-*Saccharomyces* wine yeasts in comparison with a strain of *Saccharomyces cerevisiae*-Influence of different culture conditions. *Food Microbiol.* **2004**, *21*, 439–447. [CrossRef]

111. Erten, H. Relations between elevated temperatures and fermentation behaviour of *Kloeckera apiculata* and *Saccharomyces cerevisiae* associated with winemaking in mixed cultures. *World J. Microbiol. Biotechnol.* **2002**, *18*, 373–378. [CrossRef]

112. Moreira, N.; Mendes, F.; Hogg, T.; Vasconcelos, I. Alcohols, esters and heavy sulphur compounds production by pure and mixed cultures of apiculate wine yeasts. *Int. J. Food Microbiol.* **2005**, *103*, 285–294. [CrossRef] [PubMed]

113. Rojas, V.; Gil, J.V.; Piñaga, F.; Manzanares, P. Acetate ester formation in wine by mixed cultures in laboratory fermentations. *Int. J. Food Microbiol.* **2003**, *86*, 181–188. [CrossRef]

114. Martínez-Rodríguez, A.; Polo, M.; Carrascosa, A. Structural and ultrastructural changes in yeast cells during autolysis in a model wine system and in sparkling wines. *Int. J. Food Microbiol.* **2001**, *71*, 45–51. [CrossRef]

115. Lambrechts, M.G.; Pretorius, I.S. Yeast and its importance to wine aroma. *S. Afr. J. Enol. Vitic.* **2000**, *21*, 97–129.

116. Swiegers, J.H.; Pretorius, I.S. Yeast modulation of wine flavor. *Adv. Appl. Microbiol.* **2005**, *57*, 131–175. [CrossRef] [PubMed]

117. Swiegers, J.H.; Saerens, S.M.G.; Pretorius, I.S. Novel yeast strains as tools for adjusting the flavor of fermented beverages to market specification. In *Biotechnology in Flavor Production*; Havkin-Frenkel, D., Dudai, N., Eds.; John Wiley and Sons, Ltd.: Hoboken, NJ, USA, 2016; pp. 62–135.

118. Ravaglia, S.; Delfini, C. Inhibitory effects of medium chain fatty acids on yeasts cells growing in synthetic nutrient medium and in the sparkling Moscato wine "Asti Spumante". *Wein-Wissenschaft* **1994**, *49*, 40–45.

119. Fleet, G.H. Wine yeasts for the future. *FEMS Yeast Res.* **2008**, *8*, 979–995. [CrossRef] [PubMed]

120. Jeffries, T.W.; Macmillan, J.D. Action patterns of (1→3)-β-d-glucanases from *Oerskovia xanthineolytica* on laminaran, lichenan, and yeast glucan. *Carbohydr. Res.* **1981**, *1*, 87–100. [CrossRef]

121. Buerth, C.; Heilmann, C.J.; Klis, F.M.; de Koster, C.G.; Ernst, J.F.; Tielker, D. Growth-dependent secretome of *Candida utilis. Microbiology* **2011**, *157*, 2493–2503. [CrossRef] [PubMed]

122. Mostert, T.T. Investigating the Secretome of Non *Saccharomyces* Yeast in Model Wine. Master's Thesis, Stellenbosch University, Stellenbosch, South Africa, 2013.

123. Ganga, M.A.; Martinez, C. Effect of wine yeast monoculture practice on the biodiversity of non-*Saccharomyces* yeasts. *J. Appl. Microbiol.* **2004**, *96*, 76–83. [CrossRef] [PubMed]

124. Pérez, G.; Fariña, L.; Barquet, M.; Boido, E.; Gaggero, C.; Dellacassa, E.; Carrau, F. A quick screening method to identify β-glucosidase activity in native wine yeast strains: Application of Esculin Glycerol Agar (EGA) medium. *World J. Microbiol. Biotechnol.* **2011**, *27*, 47–55. [CrossRef]

fermentation

MDPI

Review

Applications of *Metschnikowia pulcherrima* in Wine Biotechnology

Antonio Morata, Iris Loira *, Carlos Escott, Juan Manuel del Fresno, María Antonia Bañuelos and José Antonio Suárez-Lepe

Departamento de Química y Tecnología de Alimentos, Universidad Politécnica de Madrid, Av. Puerta de Hierro, nº 2, 28040 Madrid, Spain
* Correspondence: iris.loira@upm.es

Received: 11 June 2019; Accepted: 5 July 2019; Published: 9 July 2019

Abstract: *Metschnikowia pulcherrima* (*Mp*) is a ubiquitous yeast that frequently appears in spontaneous fermentations. The current interest in *Mp* is supported by the expression of many extracellular activities, some of which enhance the release of varietal aromatic compounds. The low fermentative power of *Mp* makes necessary the sequential or mixed use with *Saccharomyces cerevisiae* (*Sc*) to completely ferment grape musts. *Mp* has a respiratory metabolism that can help to lower ethanol content when used under aerobic conditions. Also, *Mp* shows good compatibility with *Sc* in producing a low-to-moderate global volatile acidity and, with suitable strains, a reduced level of H_2S. The excretion of pulcherrimin gives *Mp* some competitive advantages over other non-*Saccharomyces* yeasts as well as providing some antifungal properties.

Keywords: *Metschnikowia pulcherrima*; oenological uses; enzymes; stable pigments; pulcherrimin

1. Ecology and Physiology

Metschnikowia pulcherrima (*Mp*) is a globous/elliptical yeast that cannot be distinguished from *Saccharomyces cerevisiae* (*Sc*) by microscopy (Figure 1). Sometimes, it can be observed a single large, highly refractive oil droplet inside the cell. *Mp* is a teleomorph yeast belonging to an ascomycetous genus [1]. Its anamorph form is called *Candida pulcherrima*. *Mp* is a ubiquitous yeast that has been found in grapes, fruits (fresh and spoiled), flowers, nectars and tree sap fluxes. Several insects can work as vectors for this yeast. *Mp* strains can be identified through the use of selective and differential substrates; *Mp* strains showed both positive β-glucosidase enzyme activity and proteolytic activity [2]. *Mp* grows properly in either YPD or L-lysine media, and it can also can use arbutin as a carbon source in agar plates, indicating the expression of β-glucosidase activity (Figure 2) [3]. Recently, its nitrogen requirement was evaluated and slower consumption rates of ammonium were observed in *Mp* in comparison to other yeast genera [4]. This slow nitrogen uptake is indicative of its low fermentative ability [5].

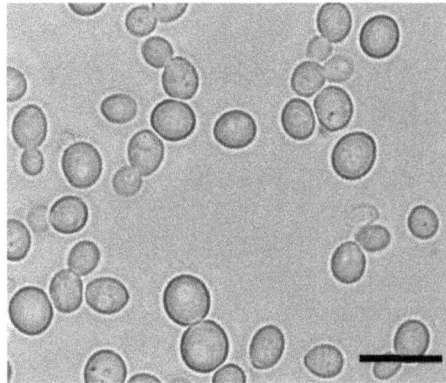

Figure 1. Cell morphology and shape of *Metschnikowia pulcherrima*. Graphical scale 10 μm.

The β-glucosidase activity related to *Mp* has been associated with different intracellular β-glucosidases, with the identification of three different bands observed when using fluorogenic substrates via an electrophoretic technique [6]. Of these three bands, the major band has similar physicochemical properties to those found in other studied yeasts, with high activity in ethanol and glucose concentrations often found in wines but low stability below pH 4. *Mp* is unable to develop in YPD at 37 °C and shows very weak or no growth in nitrate agar (Figure 2). It is able to use glucose, sucrose, fructose, galactose and maltose as carbon sources but shows weak or inexistent development in lactose [7]. It can grow properly under low temperature (15–20 °C) and pH conditions (3–6) [8]. Under environmental stress conditions such as a shortage of nitrogen, its recognition in optical microscopy is easy thanks to the appearance of a fat globule inside the cell at the beginning of the sporulation process [8]. In its sporulated form, the asci of *Metschnikowia* are long and clavate, containing one to two acicular to filiform spores [1].

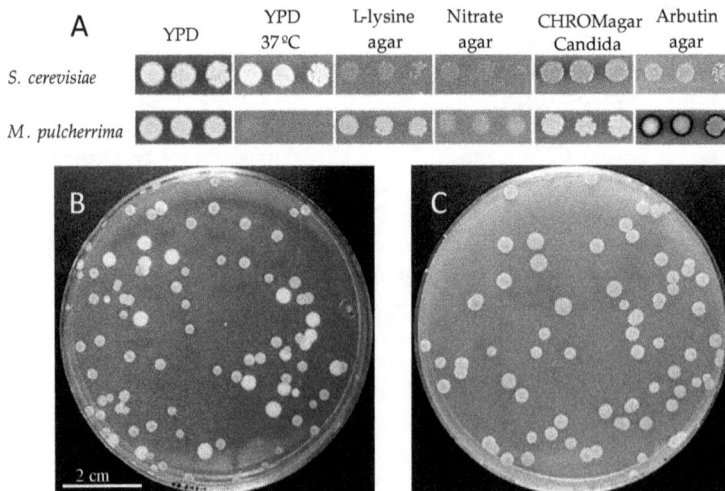

Figure 2. (**A**) Development and colony appearance in several growth media and different culture conditions (temperature). (**B**) *Metschnikowia pulcherrima* (*Mp*) orange colonies, some of them surrounded with white halos and *Saccharomyces cerevisiae* (*Sc*) white/creamy colonies in YPD media. (**C**) *Mp* and *Sc* in CHROMagar® media. Sc: bigger colonies with light pink color, *Mp*: smaller orange colonies, some of them with white halos.

The fermentative power of *Mp* is low, with many strains easily reaching 4% *v/v* in ethanol [3], although previous studies have observed the production of ethanol up to 6–7% *v/v* [9]. This feature, together with the fact that the presence of *Mp* in freshly pressed must is about 19–39% of the yeast ecology [9], makes it necessary to use *Mp* together with other yeast with a high fermentative power such as *Sc* or *Schizosaccharomyces pombe* to fully ferment grape sugars [10]. Its volatile acidity is also quite moderate, ranging from 0.3 to 0.4 g/L expressed as acetic acid [3]. Moreover, some strains are able to decrease the formation of H_2S during fermentation [11].

The fermentative performance of *Mp* is lower than that observed for other non-*Saccharomyces* species. The CO_2 production during fermentation yielded lower amounts for *Mp* than for *Sc* with 4.5 g per 100 mL vs. 12.9 g per 100 mL, respectively [12]. *Mp* has an intermediate acetoin production during alcoholic fermentation with respect to other species, such as *S. cerevisiae* and *B. bruxellensis* with low acetoin production and *C. stellata* and *K. apiculata* with the highest production of acetoin. The metabolic pathway for the production of this secondary metabolite from fermentation is shown in Figure 3. In addition, the amount of 2,3-butanediol produced by *Mp* is usually lower than that produced by *Sc*.

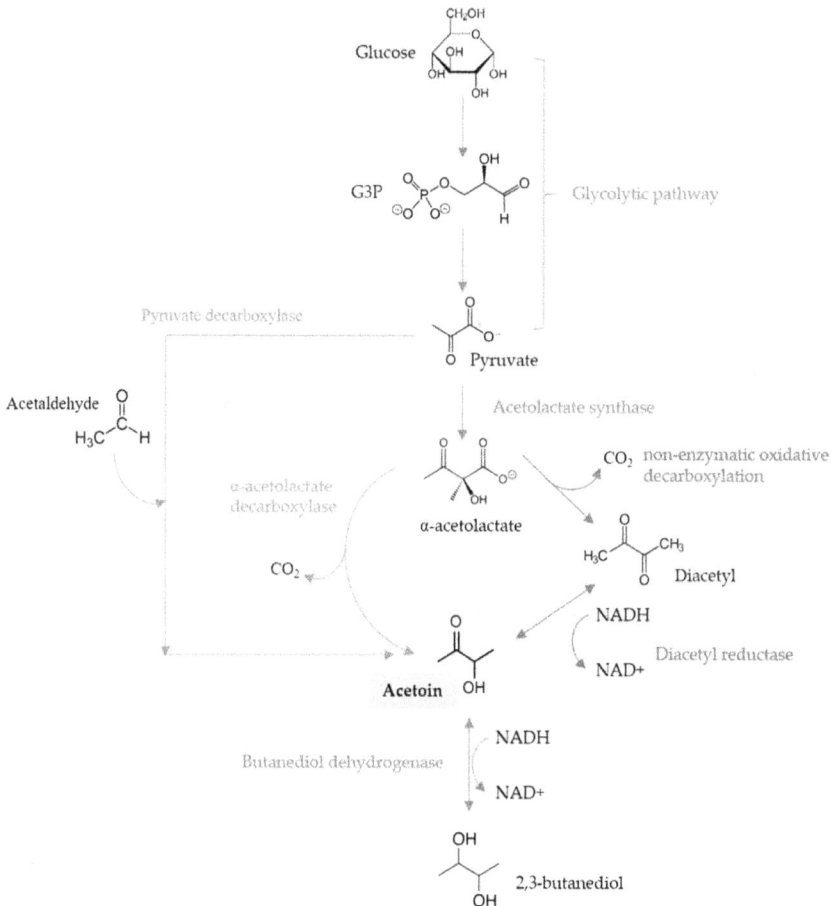

Figure 3. Metabolic route for the biosynthesis of acetoin by yeasts (adapted from Romano and Suzzi, [13]).

In mixed cultures with *S. cerevisiae*, viability was found to decrease rapidly after a few days of fermentation because of the low resistance to the ethanol produced by *S. cerevisiae* [14,15]. The use of emerging physical technologies that are able to strongly reduce the wild yeast content in grapes [16] can facilitate the prevalence of *Mp* during a longer period until the sequential inoculation of Sc, thus also increasing its effect on the sensory profile of the wines.

The sensibility of *Mp* to SO_2 is lower than that observed in Sc, *Saccharomycodes ludwigii* or *S. pombe*, but *Mp* shows a medium resistance compared with other non-*Saccharomyces* species [7]. A certain sensibility to some antimicrobials such as carvacrol and thymol has also been observed [17]. Regarding the use of dimethyl dicarbonate (DMDC), the growth of *Mp* strains during the fermentation of grape must is delayed, but not inhibited, after the addition of 400 mg/L DMDC [18]. The total inhibition of the microbial population can be achieved with 500 mg/L of DMDC. *Sc* can survive the addition of 200 mg/L DMDC, whereas the growth of other species of the genus *Saccharomyces* is inhibited with 150 mg/L DMDC.

2. Antimicrobial Bio-Tool

Mp can be used as a biological control agent thanks to its ability to produce natural antimicrobial compounds, namely pulcherrimin, an insoluble red pigment with antifungal activity. This peculiar antimicrobial activity is produced by the depletion of iron in the medium through the precipitation of iron(III) ions caused by the interaction with pulcherriminic acid, a precursor of pulcherrimin secreted by *Mp*. In this way, the environment becomes inhospitable to other microorganisms that require iron for their development. Pulcherrimin has shown effective inhibitory activity against several yeasts: *Candida tropicalis* and *Candida albicans*, as well as the *Brettanomyces/Dekkera*, *Hanseniaspora* and *Pichia* genera; and fungi: *Botrytis cinerea*, as well as *Penicillium*, *Alternaria* and *Monilia* spp. [19–24]. However, *S. cerevisiae* seems not to be affected by this antimicrobial activity [21,22]. Therefore, the use of *Mp* as a selected starter in sequential or mixed biotechnologies with *Sc* could be of great interest in modern enology.

Mp, as well as other yeast species such as *Wickerhamomyces anomala* (formerly *Pichia anomala*) and *Torulaspora delbrueckii* (*Td*), has a broad killer spectrum against some spoilage yeasts [25,26], of which *C. glabrata* had the highest sensitivity against the toxins from this species [27]. *Mp* has also been described as biofungicide capable of effectively reducing the incidence of *Botrytis* development in postharvest fruits [28]. Its antagonistic mechanism is mainly based on its competition for nutrients [29].

3. Aroma Compounds

The single use of *Mp* has led to excessive production of ethyl acetate with negative sensory repercussions [30]. However, the mixed use of *Mp* with *Saccharomyces uvarum* reduces the production of ethyl acetate, simultaneously favoring the formation of 2-phenyl ethanol and 2-phenylethyl acetate [30]. The use of co-inoculations of this type (mixed fermentations with *Mp*/*Sc*) has produced high contents of acetate esters and β-damascenone with lower levels of C_6 alcohols in ice wines made from the Vidal blanc grape variety [31]. An improvement in the aromatic complexity of the wines can be obtained by the use of *Mp* as a co-starter with *Sc* [3,32], mainly due to its high production of esters derived from its intense extracellular enzymatic activity [10,33]. Similarly, sequential fermentations with *Mp* showed a higher production of higher alcohols, with particularly high concentrations of isobutanol and phenylethanol [4].

4. Enzymatic Activities

Activities of the following enzymes have been described in *Mp*: pectinase, protease, glucanase, lichenase, β-glucosidase, cellulase, xylanase, amylase, sulphite reductase, lipase and β-lyase [11,33–35]. This is because *Mp* one of the non-*Saccharomyces* yeast species able to express more extracellular hydrolytic enzymes. Its high proteolytic activity makes it a very interesting fermentation partner for *Sc*, since the amino acids released (including those from autolysis) can serve as a source of nutrients for

Sc [36]. In addition, its intense glucosidase activity [2], higher under aerobic conditions [37], promotes the release of varietal aromas from the grape by hydrolyzing bound monoterpenes. However, it is important to always remember that the intensity of the enzymatic activity depends not only on the species, but also on the strain [32].

Concerning aroma enhancement, the expression of β-D-glucosidase favors the release of free terpenes and this activity has been evaluated using the substrates 4-methylumbelliferyl-β-D-glucoside (MUG) and *p*-nitrophenyl-β-D-glucoside (*p*NPG), showing a good intensity with medium-to-low degradation of color by the effect on anthocyanin glucosides [38]. The commercial *Mp* L1781 (Flavia™ MP346, Lallemand) expresses α-arabinofuranosidase; this activity helps to release precursors of volatile terpenes [39,40] (Figure 4) and thiols [32,41], which help to enhance fruity smells in some varieties. This strain has shown an enzymatic specific activity of 0.22 U/mg when used as a dry yeast or fresh culture [41]. This has been measured by the hydrolysis of 11 μmol de *p*-nitrophenyl-α-L-arabinofuranosidase (*p*NPA) per minute [42].

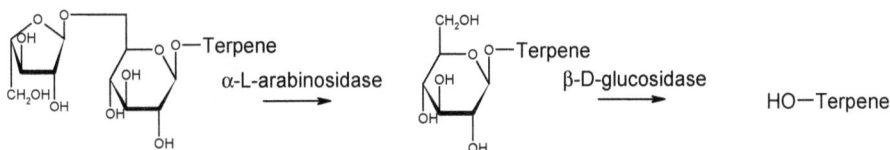

Figure 4. Effect of sequential α-arabinofuranosidase and β-D-glucosidase activities on the transformation of bonded terpenes into free forms, enhancing the aromatic profile.

Intracellular β-glucosidase of *Mp* has been purified by ion-exchange chromatography on amino agarose gel [6] and subsequently characterized. The optimum catalytic activity was observed at 50 °C and pH 4.5. The enzyme shows hydrolytic activity on β-(1→4) and β-(1→2) glycosidic bonds. The stability in alcoholic media (12% *v/v*) is good but it is affected by low pH.

5. Aerobic Metabolism/Alcohol Degree Reduction

The sequential use of *Mp* and *Sc* has proved to be somewhat effective in lowering the ethanol content of wine [11,43–46]. This is connected with the aerobic respiratory metabolisms of *Mp* that, in suitable aeration conditions, can aerobically metabolize more than 40% of sugars, thus significantly reducing the ethanol yield. An example of this application can be seen in the study developed by Contreras et al. (2014), where an average reduction in the alcoholic strength of 1.6% *v/v* was achieved when *Mp* was used in sequential fermentation with *Sc* (inoculated on the fourth day) in the production of red wine of the Syrah variety from a must with 240 g/L of sugars (potential alcoholic strength of 14% *v/v*). Therefore, the use of certain non-*Saccharomyces* yeast species, such as *Mp*, has been suggested as a biotechnological strategy aimed at producing wines with lower levels of ethanol [47]. In this last study, a kind of "collaboration" was seen between populations of *Mp* and *S. uvarum*, that is, a synergistic effect, achieving a lower ethanol production than in pure fermentations with each yeast. Recently, Mestre Furlani et al. (2017) evaluated the metabolic behavior of different non-*Saccharomyces* native yeasts to reduce the ethanol content during winemaking. They report that two out of the three strains of *Mp* isolated from grapes have a sugar to ethanol conversion ratio greater than >19 g/L/% *v/v* [48]. This confirms the usefulness of *Mp* to obtain wines with lower ethanol content.

6. Improvement of Wine Color Stability

Some non-*Saccharomyces* adsorb lower contents of anthocyanins during fermentation than *Sc* [49]. In *Sc*, the adsorption can range between 1 and 6% in total content of anthocyanins [50], but can reach up to 30% for some specific anthocyanins [51]. Adsorption is influenced by the composition and structure of the yeast cell wall. *Mp* shows a low adsorption of anthocyanins in cell walls when compared with other yeasts such as *Sc*, *Td* or *Lachancea thermotolerans* (*Lt*) in grape skin agar (Figure 5), according to the methodology described by Caridi et al. [52].

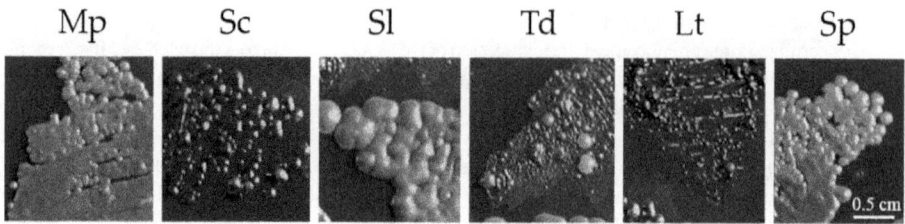

Figure 5. Adsorption of grape anthocyanins in yeast cell walls (*Saccharomyces* and non-*Saccharomyces*) during growth in a specific plating medium containing pigments. *Metschnikowia pulcherrima* (*Mp*), *Saccharomyces cerevisiae* (*Sc*), *Saccharomycodes ludwigii* (*Sl*), *Torulaspora delbrueckii* (*Td*), *Lachancea thermotolerans* (*Lt*), *Schizosaccharomyces pombe* (*Sp*).

The effect of *Mp* in the formation of stable pigments (pyranoanthocyanins and polymers) during fermentation has been studied in sequential fermentations with *Sc* and *S. pombe* [10].

7. Conclusions

The versatility of *Metschnikowia pulcherrima* lies in its ability to ferment in combination with other yeast species as well as modulate the synthesis of secondary metabolites of fermentation to improve the sensory profile of the wine. It is characterized by a medium fermentation power and a high enzymatic capacity to release aromatic precursors from the grape. In addition, this yeast has potential as a biocontrol agent in order to limit competition with other yeasts in the fermentation medium.

The abovementioned applications and features of *Metschnikowia pulcherrima* may be of great interest in order to address one of the major concerns in today's winemaking industry, such as excessive alcoholic strengths and the increasing prevalence in the market of flat wines from a sensory point of view. *Mp* could help solve these issues. The only important thing is to select the proper combination, as well as the right time and ratio of inoculation, between *Mp* and another yeast species capable of completing the alcoholic fermentation.

Author Contributions: A.M., C.E. and I.L.: literature review, writing, and editing; A.M.: images design; J.M.d.F.: literature review and critical reading; M.A.B.: critical reading; J.A.S.-L.: critical reading.

Funding: This research received no external funding.

Conflicts of Interest: The authors declare that there are no conflicts of interest.

Compliance with Ethics Requirements: This article does not contain any studies with human or animal subjects.

References

1. Kurtzman, C.P.; Fell, J.W. *The Yeasts: A Taxonomic Study*, 4th ed.; Elsevier Science Publishers: Amsterdam, The Netherlands, 1998; ISBN 9780444813121.
2. Fernández, M.; Ubeda, J.F.; Briones, A.I. Typing of non-*Saccharomyces* yeasts with enzymatic activities of interest in wine-making. *Int. J. Food Microbiol.* **2000**, *59*, 29–36. [CrossRef]
3. Comitini, F.; Gobbi, M.; Domizio, P.; Romani, C.; Lencioni, L.; Mannazzu, I.; Ciani, M. Selected non-*Saccharomyces* wine yeasts in controlled multistarter fermentations with *Saccharomyces cerevisiae*. *Food Microbiol.* **2011**, *28*, 873–882. [CrossRef] [PubMed]
4. Prior, K.J.; Bauer, F.F.; Divol, B. The utilisation of nitrogenous compounds by commercial non-*Saccharomyces* yeasts associated with wine. *Food Microbiol.* **2019**, *79*, 75–84. [CrossRef] [PubMed]
5. Ribéreau-Gayon, P. *Handbook of Enology, Volume 1: The Microbiology of Wine and Vinifications*; John Wiley & Sons: Hoboken, NJ, USA, 2006; Volume 1, ISBN 9780470010341.

6. González-Pombo, P.; Pérez, G.; Carrau, F.; Guisán, J.M.; Batista-Viera, F.; Brena, B.M. One-step purification and characterization of an intracellular β-glucosidase from *Metschnikowia pulcherrima*. *Biotechnol. Lett.* **2008**, *30*, 1469–1475. [CrossRef] [PubMed]

7. Loira, I.; Morata, A.; Bañuelos, M.A.; Suárez-Lepe, J.A. Isolation, selection and identification techniques for non-*Saccharomyces* yeasts of oenological interest. In *Biotechnological Progress and Beverage Consumption*; Volume 19: The Science of Beverages; Grumezescu, A., Holban, A.-M., Eds.; Elsevier Academic Press: Cambridge, MA, USA, Chapter 15, in press.

8. Santamauro, F.; Whiffin, F.M.; Scott, R.J.; Chuck, C.J. Low-cost lipid production by an oleaginous yeast cultured in non-sterile conditions using model waste resources. *Biotechnol. Biofuels* **2014**, *7*, 34. [CrossRef] [PubMed]

9. Combina, M.; Elía, A.; Mercado, L.; Catania, C.; Ganga, A.; Martinez, C. Dynamics of indigenous yeast populations during spontaneous fermentation of wines from Mendoza, Argentina. *Int. J. Food Microbiol.* **2005**, *99*, 237–243. [CrossRef] [PubMed]

10. Escott, C.; Del Fresno, J.M.; Loira, I.; Morata, A.; Tesfaye, W.; del Carmen González, M.; Suárez-Lepe, J.A. Formation of polymeric pigments in red wines through sequential fermentation of flavanol-enriched musts with non-*Saccharomyces* yeasts. *Food Chem.* **2018**, *239*, 975–983. [CrossRef]

11. Barbosa, C.; Lage, P.; Esteves, M.; Chambel, L.; Mendes-Faia, A.; Mendes-Ferreira, A.; Barbosa, C.; Lage, P.; Esteves, M.; Chambel, L.; et al. Molecular and phenotypic characterization of *Metschnikowia pulcherrima* strains from Douro wine region. *Fermentation* **2018**, *4*, 8. [CrossRef]

12. Romano, P.; Granchi, L.; Caruso, M.; Borra, G.; Palla, G.; Fiore, C.; Ganucci, D.; Caligiani, A.; Brandolini, V. The species-specific ratios of 2,3-butanediol and acetoin isomers as a tool to evaluate wine yeast performance. *Int. J. Food Microbiol.* **2003**, *86*, 163–168. [CrossRef]

13. Romano, P.; Suzzi, G. Origin and production of acetoin during wine yeast fermentation. *Appl. Environ. Microbiol.* **1996**, *62*, 309.

14. Sadoudi, M.; Tourdot-Maréchal, R.; Rousseaux, S.; Steyer, D.; Gallardo-Chacón, J.J.; Ballester, J.; Vichi, S.; Guérin-Schneider, R.; Caixach, J.; Alexandre, H. Yeast-yeast interactions revealed by aromatic profile analysis of Sauvignon Blanc wine fermented by single or co-culture of non-*Saccharomyces* and *Saccharomyces* yeasts. *Food Microbiol.* **2012**, *32*, 243–253. [CrossRef] [PubMed]

15. Wang, C.; Mas, A.; Esteve-Zarzoso, B. The interaction between *Saccharomyces cerevisiae* and non-*Saccharomyces* yeast during alcoholic fermentation is species and strain specific. *Front. Microbiol.* **2016**, *7*, 502. [CrossRef]

16. Morata, A.; Loira, I.; Vejarano, R.; González, C.; Callejo, M.J.; Suárez-Lepe, J.A. Emerging preservation technologies in grapes for winemaking. *Trends Food Sci. Technol.* **2017**, *67*, 36–43. [CrossRef]

17. Escott, C.; Loira, I.; Morata, A.; Bañuelos, M.A.; Antonio Suárez-Lepe, J. Wine Spoilage Yeasts: Control Strategy. In *Yeast—Industrial Applications*; Morata, A., Loira, I., Eds.; InTech: London, UK, 2017; p. 98.

18. Delfini, C.; Gaia, P.; Schellino, R.; Strano, M.; Pagliara, A.; Ambrò, S. Fermentability of grape must after inhibition with Dimethyl Dicarbonate (DMDC). *J. Agric. Food Chem.* **2002**, *50*, 5605–5611. [CrossRef] [PubMed]

19. Csutak, O.; Vassu, T.; Sarbu, I.; Stoica, I.; Cornea, P. Antagonistic activity of three newly isolated yeast strains from the surface of fruits. *Food Technol. Biotechnol.* **2013**, *51*, 70–77.

20. Janisiewicz, W.J.; Tworkoski, T.J.; Kurtzman, C.P. Biocontrol potential of *Metchnikowia pulcherrima* strains against blue mold of apple. *Phytopathology* **2001**, *91*, 1098–1108. [CrossRef]

21. Kántor, A.; Hutková, J.; Petrová, J.; Hleba, L.; Kačániová, M. Antimicrobial activity of pulcherrimin pigment produced by *Metschnikowia pulcherrima* against various yeast species. *J. Microbiol. Biotechnol. Food Sci.* **2015**, *5*, 282–285. [CrossRef]

22. Oro, L.; Ciani, M.; Comitini, F. Antimicrobial activity of *Metschnikowia pulcherrima* on wine yeasts. *J. Appl. Microbiol.* **2014**, *116*, 1209–1217. [CrossRef] [PubMed]

23. Saravanakumar, D.; Ciavorella, A.; Spadaro, D.; Garibaldi, A.; Gullino, M.L. *Metschnikowia pulcherrima* strain MACH1 outcompetes *Botrytis cinerea*, *Alternaria alternata* and *Penicillium expansum* in apples through iron depletion. *Postharvest Biol. Technol.* **2008**, *49*, 121–128. [CrossRef]

24. Sipiczki, M. *Metschnikowia* strains isolated from botrytized grapes antagonize fungal and bacterial growth by iron depletion. *Appl. Environ. Microbiol.* **2006**, *72*, 6716–6724. [CrossRef]

25. Lopes, C.A.; Sáez, J.S.; Sangorrín, M.P. Differential response of *Pichia guilliermondii* spoilage isolates to biological and physico-chemical factors prevailing in Patagonian wine fermentations. *Can. J. Microbiol.* **2009**, *55*, 801–809. [CrossRef]

26. Sangorrín, M.P.; Lopes, C.A.; Jofré, V.; Querol, A.; Caballero, A.C. Spoilage yeasts from Patagonian cellars: Characterization and potential biocontrol based on killer interactions. *World J. Microbiol. Biotechnol.* **2008**, *24*, 945–953. [CrossRef]

27. Lopes, C.A.; Sangorrín, M.P. Optimization of killer assays for yeast selection protocols. *Rev. Argent. Microbiol.* **2010**, *42*, 298–306.

28. Spadaro, D.; Ciavorella, A.; Dianpeng, Z.; Garibaldi, A.; Gullino, M.L. Effect of culture media and pH on the biomass production and biocontrol efficacy of a *Metschnikowia pulcherrima* strain to be used as a biofungicide for postharvest disease control. *Can. J. Microbiol.* **2010**, *56*, 128–137. [CrossRef]

29. Piano, S.; Neyrotti, V.; Migheli, Q.; Gullino, M.L. Biocontrol capability of *Metschnikowia pulcherrima* against *Botrytis* postharvest rot of apple. *Postharvest Biol. Technol.* **1997**, *11*, 131–140. [CrossRef]

30. Varela, C.; Sengler, F.; Solomon, M.; Curtin, C. Volatile flavour profile of reduced alcohol wines fermented with the non-conventional yeast species *Metschnikowia pulcherrima* and *Saccharomyces uvarum*. *Food Chem.* **2016**, *209*, 57–64. [CrossRef]

31. Zhang, B.-Q.; Shen, J.-Y.; Duan, C.-Q.; Yan, G.-L. Use of indigenous *Hanseniaspora vineae* and *Metschnikowia pulcherrima* co-fermentation with *Saccharomyces cerevisiae* to improve the aroma diversity of Vidal Blanc icewine. *Front. Microbiol.* **2018**, *9*, 2303. [CrossRef]

32. Zott, K.; Thibon, C.; Bely, M.; Lonvaud-Funel, A.; Dubourdieu, D.; Masneuf-Pomarede, I. The grape must non-*Saccharomyces* microbial community: Impact on volatile thiol release. *Int. J. Food Microbiol.* **2011**, *151*, 210–215. [CrossRef]

33. Jolly, N.P.; Augustyn, O.P.H.; Pretorius, I.S. The Role and Use of Non-Saccharomyces Yeasts in Wine Production. Available online: https://pdfs.semanticscholar.org/2c05/c11843cafd3d1ed97dcbefce97964537386a.pdf (accessed on 7 June 2019).

34. Ganga, M.A.; Martínez, C. Effect of wine yeast monoculture practice on the biodiversity of non-*Saccharomyces* yeasts. *J. Appl. Microbiol.* **2004**, *96*, 76–83. [CrossRef]

35. Reid, V.J.; Theron, L.W.; Du Toit, M.; Divol, B. Identification and partial characterization of extracellular aspartic protease genes from *Metschnikowia pulcherrima* IWBT Y1123 and *Candida apicola* IWBT Y1384. *Appl. Environ. Microbiol.* **2012**, *78*, 6838–6849. [CrossRef]

36. Romano, P.; Capece, A.; Jespersen, L. Taxonomic and ecological diversity of food and beverage yeasts. In *Yeasts in Food and Beverages*; Springer: Berlin/Heidelberg, Germany, 2006; pp. 13–53.

37. Mendes Ferreira, A.; Clímaco, M.C.; Mendes Faia, A. The role of non-*Saccharomyces* species in releasing glycosidic bound fraction of grape aroma components-a preliminary study. *J. Appl. Microbiol.* **2001**, *91*, 67–71. [CrossRef]

38. Manzanares, P.; Rojas, V.; Genovés, S.; Vallés, S. A preliminary search for anthocyanin-β-D-glucosidase activity in non-*Saccharomyces* wine yeasts. *Int. J. Food Sci. Technol.* **2000**, *35*, 95–103. [CrossRef]

39. Gunata, Y.Z.; Bayonove, C.L.; Baumes, R.L.; Cordonnier, R.E. The aroma of grapes I. Extraction and determination of free and glycosidically bound fractions of some grape aroma components. *J. Chromatogr. A* **1985**, *331*, 83–90. [CrossRef]

40. Gunata, Z.; Bitteur, S.; Brillouet, J.-M.; Bayonove, C.; Cordonnier, R. Sequential enzymic hydrolysis of potentially aromatic glycosides from grape. *Carbohydr. Res.* **1988**, *184*, 139–149. [CrossRef]

41. Ganga, M.A.; Carriles, P.; Raynal, C.; Heras, J.M.; Ortiz-Julien, A.; Dumont, A. Vincular la *Metschnikowia pulcherrima* y la *Saccharomyces cerevisiae* Para una Máxima Revelación del Aroma en Vinos Blancos. Available online: https://www.lallemandwine.com/wp-content/uploads/2014/10/Flavia-Lee-el-documento.pdf (accessed on 7 June 2019).

42. Gunata, Z.; Brillouet, J.M.; Voirin, S.; Baumes, R.; Cordonnier, R. Purification and some properties of an alpha-L-arabinofuranosidase from *Aspergillus niger*. Action on grape monoterpenyl arabinofuranosylglucosides. *J. Agric. Food Chem.* **1990**, *38*, 772–776. [CrossRef]

43. Quirós, M.; Rojas, V.; Gonzalez, R.; Morales, P. Selection of non-*Saccharomyces* yeast strains for reducing alcohol levels in wine by sugar respiration. *Int. J. Food Microbiol.* **2014**, *181*, 85–91. [CrossRef]

44. Contreras, A.; Hidalgo, C.; Henschke, P.A.; Chambers, P.J.; Curtin, C.; Varela, C. Evaluation of non-*Saccharomyces* yeasts for the reduction of alcohol content in wine. *Appl. Environ. Microbiol.* **2014**, *80*, 1670–1678. [CrossRef]

45. Morales, P.; Rojas, V.; Quirós, M.; Gonzalez, R. The impact of oxygen on the final alcohol content of wine fermented by a mixed starter culture. *Appl. Microbiol. Biotechnol.* **2015**, *99*, 3993–4003. [CrossRef]

46. Varela, J.; Varela, C. Microbiological strategies to produce beer and wine with reduced ethanol concentration. *Curr. Opin. Biotechnol.* **2019**, *56*, 88–96. [CrossRef]

47. Contreras, A.; Curtin, C.; Varela, C. Yeast population dynamics reveal a potential 'collaboration' between *Metschnikowia pulcherrima* and *Saccharomyces uvarum* for the production of reduced alcohol wines during Shiraz fermentation. *Appl. Microbiol. Biotechnol.* **2015**, *99*, 1885–1895. [CrossRef]

48. Mestre Furlani, M.V.; Maturano, Y.P.; Combina, M.; Mercado, L.A.; Toro, M.E.; Vazquez, F. Selection of non-*Saccharomyces* yeasts to be used in grape musts with high alcoholic potential: A strategy to obtain wines with reduced ethanol content. *FEMS Yeast Res.* **2017**, *17*. [CrossRef]

49. Morata, A.; Loira, I.; Suárez Lepe, J.A. Influence of yeasts in wine colour. In *Grape and Wine Biotechnology*; Morata, A., Loira, I., Eds.; InTech: London, UK, 2016; pp. 298–300.

50. Morata, A.; Gómez-Cordovés, M.C.; Suberviola, J.; Bartolomé, B.; Colomo, B.; Suárez, J.A. Adsorption of anthocyanins by yeast cell walls during the fermentation of red wines. *J. Agric. Food Chem.* **2003**, *51*, 4084–4088. [CrossRef]

51. Morata, A.; Gómez-Cordovés, M.C.; Colomo, B.; Suárez, J.A. Cell wall anthocyanin adsorption by different *Saccharomyces* strains during the fermentation of *Vitis vinifera* L. cv Graciano grapes. *Eur. Food Res. Technol.* **2005**, *220*, 341–346. [CrossRef]

52. Caridi, A.; Sidari, R.; Kraková, L.; Kuchta, T.; Pangallo, D. Assessment of color adsorption by yeast using Grape Skin Agar and impact on red wine color. *OENO One* **2015**, *49*, 195–203. [CrossRef]

fermentation

Review
Advances in the Study of *Candida stellata*

Margarita García [1,*], Braulio Esteve-Zarzoso [2], Juan Mariano Cabellos [1] and Teresa Arroyo [1]

[1]   Department of Food and Agricultural Science, IMIDRA, 28800 Alcalá de Henares, Spain;
      juan.cabellos@madrid.org (J.M.C.); teresa.arroyo@madrid.org (T.A.)
[2]   Department of Chemistry and Biotechnology, Rovira i Virgili University, 43007 Tarragona, Spain;
      braulio.esteve@urv.cat
*   Correspondence: margarita_garcia_garcia@madrid.org

Received: 31 July 2018; Accepted: 1 September 2018; Published: 4 September 2018

Abstract: *Candida stellata* is an imperfect yeast of the genus *Candida* that belongs to the order *Saccharomycetales*, while phylum *Ascomycota*. *C. stellata* was isolated originally from a must overripe in Germany but is widespread in natural and artificial habitats. *C. stellata* is a yeast with a taxonomic history characterized by numerous changes; it is either a heterogeneous species or easily confused with other yeast species that colonize the same substrates. The strain DBVPG 3827, frequently used to investigate the oenological properties of *C. stellata*, was recently renamed as *Starmerella bombicola*, which can be easily confused with *C. zemplinina* or related species like *C. lactis-condensi*. Strains of *C. stellata* have been used in the processing of foods and feeds for thousands of years. This species, which is commonly isolated from grape must, has been found to be competitive and persistent in fermentation in both white and red wine in various wine regions of the world and tolerates a concentration of at least 9% (v/v) ethanol. Although these yeasts can produce spoilage, several studies have been conducted to characterize *C. stellata* for their ability to produce desirable metabolites for wine flavor, such as acetate esters, or for the presence of enzymatic activities that enhance wine aroma, such as β-glucosidase. This microorganism could also possess many interesting technological properties that could be applied in food processing. Exo and endoglucosidases and polygalactosidase of *C. stellata* are important in the degradation of β-glucans produced by *Botrytis cinerea*. In traditional balsamic vinegar production, *C. stellata* shapes the aromatic profile of traditional vinegar, producing ethanol from fructose and high concentrations of glycerol, succinic acid, ethyl acetate, and acetoin. Chemical characterization of exocellular polysaccharides produced by non-*Saccharomyces* yeasts revealed them to essentially be mannoproteins with high mannose contents, ranging from 73–74% for *Starmerella bombicola*. Numerous studies have clearly proven that these macromolecules make multiple positive contributions to wine quality. Recent studies on *C. stellata* strains in wines made by co-fermentation with *Saccharomyces cerevisiae* have found that the aroma attributes of the individual strains were apparent when the inoculation protocol permitted the growth and activity of both yeasts. The exploitation of the diversity of biochemical and sensory properties of non-*Saccharomyces* yeast could be of interest for obtaining new products.

Keywords: *Candida stellata*; ecology; taxonomy; metabolism; processing foods; co-fermentation

1. Characteristics of the Genus *Candida*

The genus *Candida* belongs to the order *Saccharomycetales* of the phylum *Ascomycota* and is defined as incerta sedis (of uncertain placement). *Candida* is phylogenetically heterogeneous and included 314 species and the type species *C. vulgaris* (syn. *C. tropicalis*) [1].

Candida are widespread distributed in natural and artificial habitats, being damp and wet with a high content of organic material, including organic acids and ethanol, a broad range of temperatures, and high salt and sugar osmolarity. Some species have been implicated in the conversion of foods and feeds for thousands of years. Their high biochemical potency makes *Candida* useful for commercial and biotechnological processes.

The diversity of the genus is reflected by an amplitude of unique species with respect to colony texture, microscopic morphology, and fermentation and assimilation profiles. The members of this genus may ferment a lot of sugars, assimilate the nitrate, and form pellicles and films on the surface of liquid media. Extracellular starch-like compounds are not produced. Some species assimilate the inositol and normally the urease is not produced, and gelatin may be liquefied. The reaction with blue of diazonium blue B is negative. The sugars (xylose, rhamnose, and fucose) are not found in cell hydrolysates. The dominant ubiquinones are Q9, Q7, Q8, and Q6. Additionally, the inositol assimilation might be positive or negative; in the case of the inositol-positive response, most strains develop pseudomycelia [2].

2. Ecological and Physiological Properties of Genus *Candida*

Candida covers numerous habitats that determine a wide range of physiological properties. Most of *Candida* is mesophilic, growing well at temperatures of 25–30 °C, with extremes of below 0 °C and up to 50 °C. The genus *Candida* does not have photosynthetic capacity or fix nitrogen and normally cannot grow anaerobically. *Candida* yeasts are employed to obtain a wide variety of biotechnologically interesting compounds like higher alcohols, organic acids, esters, diacetyl, aldehydes, ketones, acids, long chain dicarboxylic acids, xylitol, and glycerol. Other products are nicotinic acid, biotin, and D-β-hydroxyisobutyric acid. Another property exhibited by some strains of *Candida* is the ability to synthesize sophorosides [3] when they are growing on substrates like n-alkanes, alkenes, fatty acids, esters, or triglycerides. Also, the genus *Candida* is able to liberate extracellular enzymes, such as pectinases, β-glucosidases, proteases, invertases, amylases, and lipases, that are of high commercial interest [4]. *Candida* dominates in a vast variety of nutrient-rich habitats. These habitats are associated with plants, rotting vegetation, and insects that feed on plants. Insects (*Drosophila*, bees and bumblebees, etc.) act as vectors, and yeasts are an important food source for both the larval and adult stages of numerous insects [5]. Some species of *Candida* such as *C. famata*, *C. guilliermondii*, *C. tropicalis*, *C. parapsilosis*, and others may be isolated from natural and polluted water or sediments. Other species like *C. glabrata* and *C. parapsilosis* are often isolated from seafood; *Candida inconspicua* and *C. parapsilosis* from fish; and *C. stellata*, *C. sake*, and *C. parapsilosis* from oysters. *C. krusei* and *C. valida* grow better on polluted sediments. The presence of the *C. krusei* complex may be an index of sewage pollution. *C. boidinii* is associated with tanning solutions containing sugars, nitrogenous compounds, and mineral salts (pH 4.0–5.9) [2].

The presence of non-*Saccharomyces* yeasts in wine fermentation process has been widely documented [6,7]. During the early stages of fermentation, a lot of species can grow simultaneously in the grape must; the species *C. stellata* has been described in this stage and can survive even with a high level of ethanol in the medium [8,9], and it is supposed to play an important role in the contribution of aroma properties of certain wines [10]. In recent years, recent taxonomic studies revealed that *C. stellata* can be mistaken for the closely related species *C. zemplinina* [11,12]. This confusion around the taxonomic position of the strains may explain some of the controversial descriptions of the oenological properties of *C. stellata* [13].

3. Methods of Isolation and Identification of Genus *Candida*

In general, the identification and enumeration of microorganisms present in wine involve enrichment techniques [14,15]. These methods are considered indirect, because they do not reflect the number of original cells in the sample, but they are also considered to be progeny, because they are enriched in a selective growth media for cultivating yeast and bacteria from wine. The characterization of *Candida* at the species level is laborious, since they are widely disseminated, highly variable, change their physiology with varying conditions, and normally are associated with other yeasts, bacteria, and molds. Nonselective media most commonly used for yeast separation, cultivation, and enumeration are composed of glucose such as carbon source, and these media may be employed at the beginning. Examples include dextrose agar (pH 6.9), dextrose broth (pH 7.2), Sabouraud medium, dextrose tryptone agar, rice agar, malt extract medium, or plate count agar. The use of lactic, tartaric, or citric acid (10%, final pH 3.5) for acidification of the media, as well as the incorporation of antibiotics (up to 100 mg/L), such as cycloheximide, streptomycin, chloramphenicol, and gentamycin, enhances their selectivity in order to inhibit the development of acid lactic bacteria and other yeasts. Biphenyl, propionic acid, and dichloran control overgrowth of filamentous fungi. The culture temperatures are also an important factor; those between 25 °C and 30–32 °C should be chosen. Incubation times are fixed in the range of 3–5 days and must be increased for osmotolerant and osmophilic yeasts to 5–10 days and 14–28 days, respectively. Many specific commercial media are available for isolating and enumerating the genus *Candida* in different food products, including the brewing industry and wine industry [12].

Although selecting wine yeast strains have been addressed for decades, the unequivocal characterization has been possible with the knowledge of molecular techniques. Pramateftaki et al. [16] applied the PCR amplification and restriction pattern analysis of the ITS1-5.8S-ITS2 regions of the nuclear ribosomal gene complex for species characterization of isolated yeasts based in the techniques developed by different authors [17–19].

Most PCR-DGGE studies have been employed to discriminate both yeasts and bacteria in wine. Cocolin et al. [20] were the first to apply PCR-DGGE method in wine fermentation, developing primers for the D1/D2 domain of the large-subunit rDNA amplification of the yeast species. That work demonstrated that the population shifts of different wine-related yeasts could be easily followed using PCR-DGGE [21]. This study also confirmed the persistence of *Candida* sp. throughout wine fermentation, detecting populations until 104 days later. Supplementary studies on commercial sweet wine fermentation showed that non-*Saccharomyces* yeasts could be found in late stages of the fermentation process by PCR-DGGE and even a long time after could be cultured on specific media [7]. This fact was particularly evident for the *Candida* sp. population, *C. zemplinina* [11]. DGGE signatures from both RNA and DNA templates directly extracted from wine revealed *C. zemplinina* signatures remained throughout the fermentation, even when direct plating manifested clearly a relative low number of cells. Applying RNA dot blot analysis with *C. zemplinina*-specific probes shows that the size of that population could be relatively high ($>10^6$ cells per mL) at the end of the fermentation, while only 100–1000 CFU per mL could be detected by plating. These results provided some of the first evidence of the presence of metabolically active but nonculturable yeasts in wine fermentation.

Endpoint PCR assays have been developed and applied for several wine yeast and bacteria. López et al. [22] used a multiplex PCR approach amplifying different segments of the yeast *S. cerevisiae* COX1 gene to enumerate different starter strains. Cocolin et al. [23] also developed 26S rRNA gene PCR primers for specific amplification of *Hanseniaspora uvarum* and *C. zemplinina*. In that experiment, the authors founded a persistence of both RNA and DNA signatures for *H. uvarum* and *C. zemplinina* in sulfited wine, even though no growth of either strain was witnessed on plating media. After 20 days of SO_2 addition and without grow on plates, the detection of *H. uvarum* and *C. zemplinina* RNA signatures in wine provides a useful example of how PCR results must be considered with caution, since both live and dead cells may be detected.

The more recent QPCR system is being widely applied in wine fermentation. This technique is used to the exponential amplification of target DNA sequences together with a fluorescent molecule (SYBR Green dye is commonly used by wine-related species) [24]. The application of QPCR to specific bacteria or non-*Saccharomyces* yeasts in wine fermentation allows for their enumeration in combination with high populations of *Saccharomyces*. Organisms such as *Candida* sp. can be detected and quantified in as little as one to two hours, which is a considerable improvement on the five to 10 days necessary to develop the conventional analysis by plates [25,26].

4. Characteristics of *Candida stellata*

The non-*Saccharomyces C. stellata* is an *Ascomycete*, anamorph yeast belonging to the genus *Candida*, with a taxonomic history subject to numerous changes; it is either a heterogeneous species or is easily confused with other yeast species present in the same substrates.

C. stellata, a habitual member of the early yeast strains in both white and red wines in certain wine regions of the world [7,9,27–38], is able to remain active throughout most of the alcoholic fermentation and much longer than most other non-*Saccharomyces* yeasts [29,32,35,39–41]. The habitual presence of *C. stellata* in the samples confirmed that this yeast is frequently associated with overripe and botrytized grape berries and musts proceeding from botrytized grapes [42].

Among the genus/species linked to *Candida stellata*, the most notable are *Saccharomyces stellatus*, *Torulopsis stellata*, *Cryptococcus stellatus*, *Cryptococcus bacillaris*, *Saccharomyces bacillaris*, *Torulopsis bacillaris*, and *Brettanomyces italicus*. The cells are spherical to ovoid; they are usually found as single cells but may be arranged in a star-like configuration of cells; no hyphae or pseudohyphae are formed. Growing in YPD, colonies are grayish-white to brownish, glossy soft, and smooth. In malt agar, there are large, round cream, or white colonies. *C. stellata* does not form spores. A whitish cheese-like film can appear in liquid medium. This non-*Saccharomyces* yeast ferments glucose, sucrose, and raffinose (sometimes it does this slowly). On the other hand, it can assimilate sucrose and raffinose but not nitrate. *C. stellata* uses lysine as sole N source. Its growth requires vitamins such as biotin, pantothenate, inositol, and thiamin. With regard to medium conditions, its growth is variable at 37 °C but is sensitive to heat, while it is able to grow at lower temperatures and higher pH values. Moreover, it is not sensitive to ethanol and under aerobic conditions; by contrast, it is sensitive to cycloheximide, sorbate, DMDC, low pH, and acids.

5. Taxonomic Reclassification of *Candida stellata*

Initially, two types of *Candida* were isolated from a must elaborated in Germany from overripe grape berries and raisins with a high sugar concentration (ca. 60%). One type had elongated cells and was denominated *Saccharomyces bacillaris*. The other type was nominated *Saccharomyces stellatus*, because in liquid media the star-like chains presented cells with spherical shape. Both species were later included in genus *Torulopsis* due to the lack of spores' generation. In another study on taxonomic research carried out in Italy, a third type of species was isolated from grapes and named as *Brettanomyces italicus*. Lastly, these taxa were unified in a single species named *C. stellata*, and the strain originally described as *S. stellatus* was considered as type strain (CBS 157) (Figure 1).

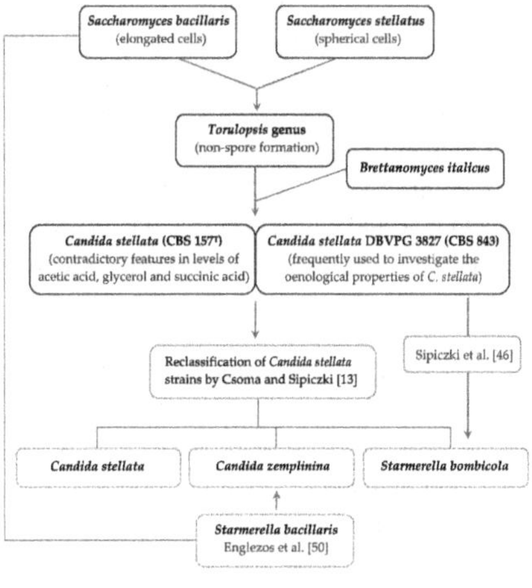

Figure 1. Taxonomic reclassification of *Candida stellata* [13,43,44].

Traditionally, *C. stellata* is associated with overripe and botrytized grape berries [27,29,34,38]. *Candida* is present almost until the final stages of the alcoholic fermentation [29,32,39,41], suggesting that *C. stellata* might significantly take part in the ecology of fermentation and the wine quality. Nevertheless, the role of *C. stellata* in wine attributes seems to be controversial owing to the contradictory enological features attributed to this yeast by several research groups. Some authors report on the high production of acetic acid [45], glycerol [10,46], and succinic acid [47]; conversely, other works found low acetic acid levels and low glycerol production [35]. These controversial results of *Candida* show that *C. stellata* is either a heterogeneous species or is easily confused with other yeast species of *Candida*, which are present in the same substrates. Sipiczki [11] found a new osmotolerant and psychrotolerant species studying four yeast strains isolated from fermenting botrytized grape musts in the Tokaj wine region of Hungary, corresponding to *C. zemplinina*. Traditional taxonomic test shows small differences between these isolates and *C. stellata* strains CBS 157^T and DBVPG 3827 (Dipartimento di Biologia Vegetale, Perugia, Italia) (CBS 843).

The species *C. zemplinina* was discovered among wine yeasts that showed a taxonomic profile characteristic of *C. stellata* [11]. Both species grow in similar environments (must with high sugar concentration) but may form mixed populations in the colonized substrates. Lastly, the strain DBVPG 3827, frequently used to investigate the oenological properties of *C. stellata*, has been reclassified as *Starmerella bombicola* [43] (Figure 1). Considering the recent identification of these new species *C. zemplinina* and *S. bombicola*, they may be confused with *C. stellata* when conventional taxonomic tests and routine PCR-restriction fragment length polymorphism (RFLP) analysis are used for identification [11,43]. In view of these results, Csoma and Sipiczki [13] report that the name of the species *C. stellata* has been used for group yeasts not conspecific with the type of strain of *C. stellata*. Most strains originally identified as *C. stellata* and examined by the authors turned out to belong to species that were not known yet at the time of their isolation, such as *C. zemplinina*, *C. lactis-condensi*, *C. davenportii*, or *S. bombicola*. Csoma and Spiczki [13] studied 41 strains deposited in six culture collections originally identified as *C. stellata* (Figure 1). The ITS1-5.8S rRNA-ITS2 sequence region was studied in all strains by PCR-RFLP. The enzymes MBoI, DraI, and HaeIII were used separately during the digestion of amplified fragments. Digestion with MboI is known to generate specific patterns for each of *C. stellata*, *C. zemplinina*, and *S. bombicola* [12]. As result of the digestion of the fragments of amplification, all strains gave three or four patterns. Thirty-nine out of the 41 strains examined showed combinations of patterns different from those of the type strain of *C. stellata*; this result highlights the fact that only two of investigated strains might belong to *C. stellata*. The digestion of the amplified region with MboI and DraI distinguished *C. stellata* from *C. zemplinina* and the CfoI, HaeIII, and HinfI restriction patterns separated *C. stellata* from *S. bombicola* [43]. Later, the D1/D2 domains of the LSU rRNA gene of all strains studied were amplified and sequenced. The Blast search with the sequences identified high degrees of similarity (98–100%) with the sequences of the type strains of 11 species. Based on these sequences, most strains originally isolated from grapes or wine fermentation belonged to *C. zemplinina* or *S. bombicola* (DBVPG strains). At the same time, the results of the taxonomic physiological test were contrasted with the molecular results and all *C. zemplinina* strains growing in the presence of 1% acetic acid, which inhibited the growth of *C. stellata*. The wine yeasts deposited in DBVPG as *C. stellata* strains turned out to be strains of *S. bombicola*, which were identified species [48], unknown at the time of their deposition.

As a result, it can be concluded that most wine strains preserved in CBS or described in recent publications as *C. stellata* proved to belong to *C. zemplinina* [13]. *C. stellata* was not found among yeasts newly isolated from noble rotted grapes and botrytized wines either, although overripe grapes and fermenting grape musts with high sugar concentrations are environmental conditions in which strains identified as *C. stellata* were frequently detected [27,28,37,38]. *C. stellata* is far less present in grapes and natural wine fermentation than hitherto thought. Regarding botrytized wines, the higher appearance of *C. zemplinina* is ligated to the capacity to resist higher acetic acid concentrations. It is known that *C. zemplinina* can grow in presence of 1% of acetic acid, which is inhibitory to *C. stellata*. The grapes infected by *Botrytis cinerea* present a high number of acetic acid bacteria, with can grow in grape must with production of gluconic and acetic acids [49].

To probe the hypothesis about the broad presence of *C. zemplinina*, the *C. stellata* LSU rRNA gene sequences published by others authors were reviewed [13]. The results showed that D1/D2 domain sequences of the *C. stellata* strains isolated from French cider by Coton et al. [50] and from Spanish wine by Hierro et al. [38] and the corresponding sequence of *Candida* sp. isolated from Californian sweet botrytized wine by Mills et al. [7] are coincident with those of the type strain of *C. zemplinina*. Moreover, *C. zemplinina*, but not *C. stellata*, was found in fermented red wine from Portugal grape variety Castelao. Besides, *C. zemplinina* was also identified in other Portuguese wine (accession number AY394855) and in Greek botrytized wines (accession number DQ872872).

The results obtained from electrophoretic karyotypes suppose another means for the differentiation of these species. Although both species had three chromosomes and showed length polymorphism, their chromosomes differed in size. *C. stellata* had a somewhat larger genome, and each chromosome differed in size from the comparative used strains, CBS 157^T and CBS 843. *C. stellata* appears to be prone to undergo chromosomal rearrangements. In contrast, the *C. zemplinina* strains did not show chromosomal polymorphism [13].

C. zemplinina also proved to be much more acidogenic; this aspect may significantly affect the quality of the wine. *C. stellata* grew much more slowly at all conditions tested. This observation is in accordance with earlier reports that described *C. stellata* as a slow-growing yeast [29].

On the other hand, different studies about the original type *Saccharomyces bacillaris* described together with *Saccharomyces stellata* from overripe grapes and concentrated musts concluded that this species is not synonymous with *C. stellata*. Different profiles were observed for the type strain of *C. stellata* (CBS 157) for both the isoenzyme and rDNA restriction analysis, and only 91% similarity was found between the D1/D2 sequence of this strain and *S. bacillaris*. In view of the results, *S. bacillaris* has been recently reinstated as *Starmerella bacillaris* comb. nov., with *C. zemplinina* as an obligate synonym [44] (Figure 1). This reorganization is in line with the latest edition of the "International Code of Nomenclature for Algae, Fungi and Plants" [51], which eliminated the rule that was in force for a long time that anamorphic yeasts with ascomycetous affiliation had to be classified to *Candida*; this species was recently moved to the genus *Starmerella*, leaving the name *C. zemplinina* as obligate synonym [52]. Following this reclassification of most of the yeasts, *S. bacillaris*, previously identified as *C. stellata*, became *Starmerella bacillaris* [44].

6. Characteristics of *Candida zemplinina* sp. nov. Sipiczki

C. zemplinina was discovered studying wine yeasts with a taxonomic profile characteristic of *C. stellata* [11]. Both species grown in similar environments (overripe grapes and grape must with high sugar concentration) presumably may form mixed populations in the colonized substrates [13].

C. zemplinina owes its name to the Zemplin mountain range, whose south and south-east facing slopes form the Tokaj wine region. The type strain is 10-372^T (=CBS 9494^T = NCAIM Y016667^T), which was isolated from white wine in Zemplin, Hungary [11]. Growing on morphologic agar, the cells are ellipsoid to elongated (2.2–3.0 × 3.0–5.2 µm) alone and in pairs after 3 days incubation at 25 °C. Their budding is multilateral. In contrast, after 7 days incubation at 25 °C on the same culture media, colonies are low convex with smooth to finely lobed margins, and their texture is butyrose. Neither hyphae nor pseudohyphae are generated. Ascospores formation is not seen after 25 days incubation at 25 °C on the agar culture media for corn-meal, potato dextrose, or Gorodkowa. *C. zemplinina* ferments sugars, glucose, sucrose, and raffinose but does not ferment galactose, maltose, and lactose. On the other hand, it can assimilate glucose, sucrose, L-sorbose (slowly), raffinose, and lysine but does not assimilate the following compounds: galactose, D-glucosamine, D-ribose, D-xylose, L-arabinose, D-arabinose, L-rhamnose, maltose, trehalose, methyl α-D-glucoside, cellobiose, salicin, melibiose, lactose, melezitose, inulin, starch, glycerol, erythritol, ribitol, D-glucitol, D-manitol, galactitol, inositol, D-glucono-1,5-lactone, succinate, citrate, methanol, ethanol, potassium nitrate, cadaverine, *N*-acetyl-D-glucosamine, and lysine. Vitamins are essential to its growth. Finally, this yeast species is able to grow in the presence of 60% (*w/v*) glucose. Additionally, no growth is observed in the presence of 10 μg/mL cycloheximide or at 37 °C.

C. zemplinina cannot be considered a wine-specific yeast. *C. zemplinina* has been detected in yeast populations associated with Ghanaian cocoa fermentation and two CBS strains identified as *C. zemplinina* originate from soil (CBS 2799) and *Drosophila* sp. (CBS 4729) [53,54]. The association of *C. zemplinina* with *Drosophila* confirms that fruit flies can be important vectors of yeasts from winery to ripening grapes in the vineyard. The related species as *C. bombi*, *C. lactis-condensis*, and *S. bombicola* are also associated with insects [55,56].

The genome of *C. zemplinina* is similar in size to the genome of *C. stellata* and the genomes of the other related species, *C. bombi*, *C. lactis-condensi*, and *Starmerella bombicola*, but appears to differ from it in stability. The *C. zemplinina* chromosomes show less variability than those of *C. stellata*. This stability indicates that chromosome rearrangements may not be as important in this species as in *S. cerevisiae* [57] for adaptation conditions during wine fermentation.

Kurtzman and Robnett [58] observed that strains showing greater than 1% difference in the D1/D2 domain of the 26S rRNA are usually different species. The D1/D2 domains of the 26S rDNA of four isolates of *C. zemplinina* and the control strain *C. stellata* CBS 157[T] were amplified and sequenced to confirm the taxonomic separation of *C. zemplinina* from *C. stellata*. The amplified fragments of the *C. zemplinina* strains had identical nucleotide sequences, which differed from the homologous sequence of *C. stellata* 157[T] at 39 positions (8.1% sequence difference).

C. zemplinina stand out against *C. stellata* for being osmotolerant and psychrotolerant and thus could be better adapted to grow under high sugar concentrations and at low temperatures. These physiological attributes can be especially favorable for propagating botrytized grape musts, which normally contain high sugar content and are fermented at low temperatures, as in the case of Tokaj wines generally below 15 °C.

7. Characteristics of *Starmerella bombicola*

Starmerella bombicola is the type species of genus *Starmerella* (Rosa and Lachance, 1998) [48]. The strain studied, CBS 6009 (type strain), was isolated from honey of bumble bee (*Bombus* sp.).

S. bombicola is the anamorph of *C. bombicola* and the synonym of *Torulaspora bombicola* and *C. bombicola* [59]. On YM agar after 3 days at 25 °C, the cells are ovoidal to elongated, 1–2 × 2–4 μm, and occur singly and in pairs. The colonies are small, convex, and white and have an entire margin. In glucose-yeast extract broth, a ring forms after 1 month. In Dalmau plate culture on corn meal agar, pseudohyphae and true hyphae are not formed. Positive formation of ascospores after 1 day on YCBAS (yeast carbon base, Difco, with. 0.01 % ammonium sulphate) agar mixed compatible mating types fuse in pairs. After 3 days, the conjugated asci contained a single spherical ascospore with a convoluted wall and a membranous basal ledge. The ascospores are released terminally and tend to agglutinate. This species presents positive fermentation of glucose and sucrose, and variable to raffinose. The fermentation of galactose, maltose, lactose, and trehalose is negative. It can grow on agar media of glucose, ethanol, glycerol, and mannitol, and provides a positive answer to the additional growth test of glucono-δ-lactone, cadaverine, 50% glucose, amino acid-free, and 30 °C CoQ 9.

This yeast had been previously assigned to other species now known to be members of the *Starmerella* clade. This species shows the ability to excrete extracellular hydroxyl fatty acid sophorosides. *S. bombicola* is associated with bees and flowers, with the bees as the principal vector. Sophorolipid biosynthesis by *S. bombicola* may be industrially useful for the production of biodegradation detergents [60].

To confirm the taxonomic affiliation of species of *Candida* deposited in DBVPG, their growth under various conditions was studied. *C. zemplinina* and *C. stellata* differed from *Candida* strains deposited in DBVPG with regard to temperature profile, osmotolerance, and greater sensitivity to ethanol compared with these two *Candida* species. Comparing its electrophoretic karyotype, *Candida*-type strain differed in the banding pattern; although both had three chromosomal bands, their chromosomes differed in size, and the genome of *C. zemplinina* was smaller than the genome of *C. stellata*. The karyotype of DBVPG 3827 was indistinguishable from that the karyotype of *S. bombicola* CBS 6009[T], which only had two bands, one of which corresponded in size to one of the *Candida* chromosomal bands.

Table 1 shows the principal aspects that can help to distinguish between *Candida stellata*, *Candida zemplinina*, and *Starmerella bombicola* to achieve a correct identification of these species.

Table 1. Main differential characteristics of species *Candida stellata* (CBS 157), *Candida zemplinina*, and *Starmerella bombicola*.

	C. stellata (CBS 157)	*C. zemplinina*	*S. bombicola*
Growth in high sugar concentration	+	+	+
Growth in botrytized grape berries	−	++	+
Growth in presence of 1% of acetic acid	−	+	v[1]
Formation of ascospores	−	−	+
Banding pattern (electrophoretic karyotype)	3	3	2
Chromosomal polymorphism	yes	no	no
% D1/D2 sequence in difference with *C. stellata* (CBS 157)		8.1	nd[2]
MboI and DraI digestion distingue between the species	yes	yes	
CfoI, HaeIII, and HinfI digestion distingue between the species	yes		yes

[1] v, variable; [2] nd, not determined.

8. Characteristics of *Starmerella bacillaris* (synonym *C. zemplinina*)

Starmerella bacillaris (synonym *Candida zemplinina*) [52] is a non-*Saccharomyces* yeast, isolated for the first time in Napa Valley (Napa, CA, USA) in 2002, under the name EJ1 [7]. This yeast is characterized by ellipsoid to elongate cells upon growth in yeast malt agar. It ferments glucose, sucrose, and raffinose, but not galactose, maltose, or lactose. It assimilates very few carbon and nitrogen sources, namely, glucose and L-lysine, and it experiences no growth in the presence of high glucose concentration. Additionally, it presents high fructophily, average volatile acidity and alcoholic degree production, and high glycerol production [52].

Starm. bacillaris can be distinguished from the closely related species *C. stellata* by EST, G6PD, ACP, LDH, and ADH isoenzymes profiles; restriction profiles of a region of 26S rDNA digested with endonucleases HinfI, MseI, CfoI, and HaeIII clearly distinguish both species. The nucleotides sequence of D1/D2 region of 26S rDNA of *Starm. bacillaris* shows an 8% difference at 39 positions [52] thereby justifying the separation of the two species.

From the point of view of its enological application, this strain was able to ferment exclusively the fructose from Chardonnay must without affecting the concentration of glucose. *Starm. bacillaris* has since been reported to have a potentially important role in the winemaking industry, due to its extremely fructophilic character and the poor ethanol yield from sugar consumed [61,62].

Starm. bacillaris presents other interesting characteristics, such as growth at high concentrations of sugars and low temperatures [11,63] and production of low levels of acetic acid and acetaldehyde and significant amounts of glycerol from consumed sugars [64].

9. Metabolic Features and By-Products from *Candida stellata* Activity

Candida spp. are found as food-associated and beverage-associated yeasts. In particular, *C. stellata* has been typically isolated during must fermentation process in different wine regions worldwide, where this yeast species is normally associated with the fermentation of botrytized wines and other wines produced from overripe grapes in cooked musts and in traditional balsamic vinegars [2,3,28,38,65]. Thus, there are several studies that allow one to better understand the metabolic characteristics of *Candida* spp. with interesting applications in food and fermented beverages industries, with special focus on wine elaboration.

9.1. Fructophilic Character

The strong fructophilic character of *Candida* is one of distinctive features of this yeast genera. Several studies have described sugar depletion (glucose and fructose) during grape juice fermentation [10,64,66–68]. All *C. stellata* strains studied in these works showed a significant lower fermentation rate for glucose than the rate measured for the fructose. In the work with *C. stellata* CBS 2649 strain [10], the extreme fructophilic nature of this strain has been reported, since glucose was not consumed until the fructose was completely depleted. Similar behavior was observed by Mills et al. [7] when studying a *Candida* sp. isolate (EJ1) in Chardonnay wine elaboration. However, it is still unknown how preferential consumption of fructose can be beneficial, since vigorous growth on glucose has been observed when this sugar is the only energy and carbon source available in fructophilic yeasts [63,69]. As previously observed in *Zygosaccharomyces bailii* [70] and *Z. rouxii* [71], Gonçalves et al. [72] have noted the presence of the transporter Ffz1 as a prerequisite for fructophily in *S. bombicola*. This Ffz1 is a specific fructose transporter codified by FFZ1 gene [73]. The reason for the preferential use of fructose by *C. stellata* may be result of a wider remodeling of central carbon metabolism, together with an adaptation to high-sugar environment [72,74].

9.2. Alternative Carbon Metabolism: Glycerol Production

Glycerol can be used as food additive produced from fats and oils, from chemical synthesis, or by microbial fermentation [75]. Glycerol biosynthesis is an important side-reaction of glycolysis pathway produced by reduction of dihydroxyacetone phosphate (DHAP) to glycerol-3-phosphate (G3P) and by dephosphorylation of G3P to glycerol (Figure 2). The first step of this conversion is catalyzed by the enzyme NAD-dependent glycerol-3-phosphate dehydrogenase (Gpd) and, subsequently, the glycerol is formed by glycerol-3-phosphatase (Gpp). The enzyme Gpd is encoded as two isoforms by the GPD1 and GPD2 genes [76]. Yeast growth under hyperosmotic stress situation leads to the expression of GPD1 through the so-called HOG (High Osmolarity Glycerol) signaling pathway [77–79]. On the other hand, GPD2 is believed to help maintain the cell's intracellular redox balance.

As mentioned previously, *C. stellata* species exhibits unusual metabolism of sugar; it is usually considered a facultatively fermentative yeast characterized by a very low fermentation rate and high production of secondary metabolites as glycerol, acetaldehyde, acetoin, and succinic acid [67,80]. In regard to glycerol formation, this behavior of *C. stellata* is probably owing to low alcohol dehydrogenase activity (4-fold lesser than *S. cerevisiae*) and high glycerol-3-phosphate dehydrogenase activity (40-fold higher than *S. cerevisiae*); thus, this higher Gpd activity causes a strong deviation towards glycerol production [67] (Figure 2, in red).

Glucose

Hxt1

MEDIUM

CYTOPLASM

Glucose-6-Phosphate

Glyceraldehyde-6-Phosphate ⟷ Dihydroxyacetone Phosphate

Tpi1p

Gpd1p
Gpd2p — NAD+

1,3-Diphosphoglycerate

Glycerol-3-Phosphate

Gpp1p
Gpp2p — P$_i$

CO$_2$ NAD+

Pyruvate — Ethanol

Glycerol

Pdc1p Pdc5p

Fps1

Glycerol

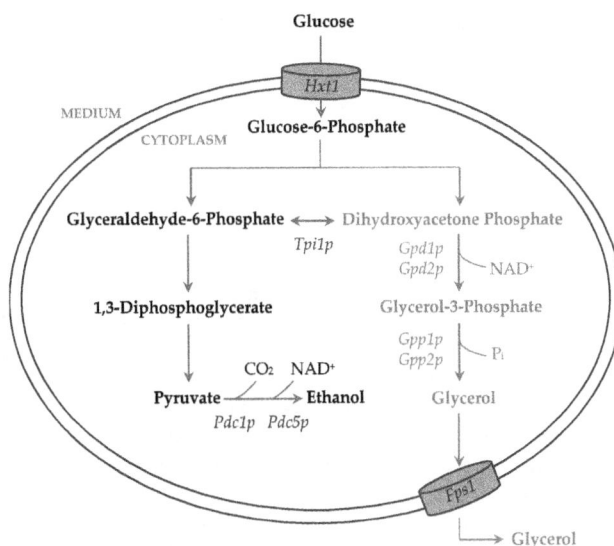

Figure 2. Glycerol biosynthesis in yeasts. Glycolysis and the reduction of intermediate DHAP to G3P, followed by oxidation of NADH to NAD+ leads to glycerol formation (adapted from Scanes et al. [81]).

In oenology, the glycerol content is appreciated, because it imparts some sensory attributes to the wine. It is an important alcohol with a slightly sweet taste and viscous nature that contributes to the smoothness, consistency, and overall body in wine [82,83]. Typically, glycerol concentration is higher in red than in white wines ranging from 1 to 15 g/L. The threshold taste level of glycerol is observed to 5.2 g/L in wine, whereas a change of viscosity is only perceived at 25 g/L of glycerol [84]. Also, it is known that its production is raised by the presence of sulfur dioxide, higher incubation temperature, and high-sugar concentration, but it is significantly influenced by yeast strain and species [85]. In particular, *C. stellata* has typically been described as glycerol producer in wine elaboration [46,68,86,87]. Glycerol concentrations between 9 and 14 g/L have been reported in wines elaborated with *C. stellata*, in contrast with lower amounts produced by *S. cerevisiae* monoculture [46,64,87]. However, glycerol and ethanol content are inversely related; as consequence, the tendency of *C. stellata* to form glycerol seems to be the reason for its low growth and fermentation rate [67,81]. Other authors found an ethanol yield produced by *C. stellata* comparable with that of *S. uvarum/bayanus* strains, although both produced significantly lower ethanol than *S. cerevisiae* [64]. By contrast, Gobbi et al. [88] reported one *C. stellata* strain with an ethanol yield (9.09 g/100 mL) and fermentative power (19 g CO$_2$ evolved) without significant differences from *S. cerevisiae* (9.05 g/100 mL and 19.2 g CO$_2$ evolved, respectively).

9.3. Biotechnological Application of Extracellular Enzymes Secreted by Candida stellata

Enzymes are the bio-catalysts that play an important role in metabolism and biochemical reactions [89]. Microorganisms are the primary source of enzymes that have a more active and stable nature than those of plants and animals [90]. Specifically, yeast strains with enzymatic activity could be a potential source of commercial enzymes and an important factor with which to improve the food and beverages processing. The *Saccharomyces* genus is not considered as a good producer of exogenous enzymes. Instead, several non-*Saccharomyces* yeast species exhibit natural enzymatic activities [91]. The enzymes of interest produced by these yeasts include esterases, lipases, glycosidases, proteases, and cellulases usually related to hydrolysis of structural components [4,92].

Specifically, *Candida* spp. have been described as extracellular enzymes producer. The enzymatic capacities of this non-*Saccharomyces* genus have been widely researched in oenology, as they can improve the process of winemaking and enhance wine quality [93,94]. However, it is well known that the secretion of enzymes with technological interest is not characteristic of a particular genus or species but depends specifically on yeast strain analyzed [33,92]. In the following paragraphs, a brief overview will be given of enzymes used in oenology with a special focus on those produced by *C. stellata*.

9.3.1. Pectinases

Pectic substances are the major component of the D plant cell wall and comprise a network in which cellulose microfibrils are linked [95]. The high viscosity of pectin prevents juice extraction, clarification, and filtration when it is dissolved after berry crushing. Furthermore, pectin impedes the phenolic and aroma compounds' diffusion into the must during wine fermentation [94]. Thus, pectinases such as polygalacturonase, pectin lyase, pectin methyl esterase, and polygalactosidase have the capacity to reduce the molecular size of pectin polymers by cleaving neutral side chain residues, facilitating the pressing and filtration processes of wines and ciders [96,97]. In addition to their use in winemaking, these enzymes are also utilized in oil extraction [98], coffee and cocoa curing [99], the extraction and clarification of fruit juices, and the retting of textile fibers [100].

Several authors have reported the production of polygalacturonase and pectin methyl esterase by *Candida* in wine [4,101,102]. In a study realized by Cordero-Bueso et al. [103], *C. stellata* CLI 920 strain, which was isolated during spontaneous fermentation in Malvar (*Vitis vinifera* cv. L.) must, produced the highest quantity of pectinases (polygalacturonases) in comparison with other non-*Saccharomyces*. This pectinase activity of *C. stellata* CLI 920 could be correlated with the higher galacturonic acid content observed into the oligosaccharides fraction of the wine produced with this strain alone [104]. Also, polygalactosidases enzymes produced by *C. stellata* together with exo and endoglucosidases are important in the degradation of the β-glucans by *Botrytis cinerea* [2].

9.3.2. Proteases

Protein haze supposes the most common physical instability in white wine and fruit juices. Proteases activity hydrolyzes the proteins into smaller stable molecules promoting clarification and stabilization of beverages and helping to prevent stuck and sluggish fermentations due to low level of assimilable nitrogen in the must [101,105]. Yeast producers of proteases can be a good substitute with which to bentonite for removal undesirable wine proteins [106]. In the study of Strauss et al. [4], 38% of *C. stellata* yeast strains presented protease activity. Also, other works have recorded protease activity in several strains of *Candida* species [101,107].

9.3.3. Cellulases and Hemicellulases

Hemicelluloses are a group of polysaccharides strongly bound to cellulose in plant cell walls. In winemaking, cellulases (glucanases) and hemicellulases (xylanases) enzymes have an impact on organoleptic properties of wine by promoting extraction of pigments and volatile compounds from grape skins, thus improving the filtration and clarification processes and reducing the time of maceration [4,108]. Only a few yeast strains have been known as major producers of these enzymes, but *Candida* species have been reported as able to produce cellulases and hemicellulases [4,102,109,110].

9.3.4. Glycosidases

The organoleptic characteristics of beverages (taste and aroma) can be enhanced by glycosidases that hydrolyse odourless and non-volatile glycosidic precursors of the fruits [111]. Glycosidase activities comprise β-D-glucosidase, β-D-xylosidase, β-D-apiosidase, α-L-rhamnosidase, and α-L-arabinofuranosidase. The bound aroma complex includes glucosides and diglycosides, and compounds such as terpenols, terpene diols, benzene derivatives, aliphatic alcohols, phenols, and C-13 norisoprenoids; additionally, the enzymatic hydrolysis of these sugar-conjugated precursors released very aromatic volatile monoterpenes (aglycons) through two-step reaction [112]. Numerous works have been based on glycosidase activities in yeasts of an oenological origin; in particular, some of them have observed β-glucosidase activity in *C. stellata* strains possibly related to the fruity and floral aroma found in the wines elaborated with these strains [4,87,103,113,114]. Hock et al. [115] had already documented the terpenes production (β-myrcene, limonene, linalool, α-terpineol, and farnesol) of *C. stellata*. Another study using one *C. stellata* strain isolated from Denomination of Origin (D.O.) "Vinos de Madrid" showed the highest concentration of β-phenylethyl alcohol (roses) in wine compared to other *Saccharomyces* and non-*Saccharomyces* strains analyzed [87]; the flowery and fruity aroma of pure culture with this *C. stellata* strain could be related to β-glucosidase activity previously documented by Cordero-Bueso et al. [103]. Similar results were obtained by other authors [105,116]; they concluded that the use of *C. stellata*, alone or combined with *S. cerevisiae*, enhanced the final quality and complexity of wines.

9.3.5. Invertases

Invertase enzyme, also known as β-D-fructofuranosidase, is commonly used in industries with numerous applications as production of lactic acid [117], fermentation of sugarcane to ethanol [118], and production of fructose syrup. Furthermore, it is employed in pharmaceutical industry, child nutrition, and fortified wines [119]. These enzymes hydrolyse the glycosidic linkage from sucrose in its respective monomers, glucose and fructose, to form "inverted sugar syrup" with special characteristics: 40% sweeter than sucrose, stable at high temperatures, more soluble than sucrose and higher point of boiling and lower of freezing [119]. Yeast production of these enzymes is typically studied in *S. cerevisiae* [120]. Recently, Gargel et al. [121] have observed that one *C. stellata* strain (N5) isolated from Brazilian grapes is a potential invertase producer. They propose this new invertase as a promising catalytic agent for use in biotechnological processes in the food industry and alcoholic fermentations.

9.4. Production of Sophorolipids Biosurfactants by Candida

The worldwide production of surfactants is about 10 million tons per year, divided between domestic and laundry detergents and different industrial applications. Currently, the surfactants are usually petroleum-derived, although the aim is to produce these compounds from renewable substances. Sophorolipids (SLs), which are composed of sophorose (a dimeric sugar) linked to a long-chain hydroxy fatty acid, are good candidates as surfactant product from renewable sources. These molecules are produced in high concentrations by phylogenetically diverse group of yeasts [122], and their biosynthesis is clearly influenced by aeration, initial glucose concentration, and pH values [123,124]. SLs present two different forms: a closed lactone and an open acidic form. Each form has different properties: Lactonic SLs have antimicrobial activity and are better in surface tension reduction, while acidic SLs have better foaming attributes [122].

The yeast *S. bombicola* has been widely studied as a major producer of SLs together with *Candida apicola* within *Starmerella* clade [122]. The highest *C. bombicola* (ATCC 22214) SL yield of 400 g/L was obtained when corn oil and honey served as the carbon sources [125]; also, Cavalero and Cooper [126] showed that the same strain synthetized SLs with antibacterial activity mainly against Gram-positive bacteria. In a study with 19 species of *Starmerella* yeast clade [123], *C. stellata* NRRL Y-1446 strain from Rovello bianco grape variety was one of 19 species with a significant production of SLs with 11.9 g/L predominantly as di-*O*-acetyl free-acid form, plus lesser amounts of mono-*O*-acetyl and non-acetyl SLs. Parekh et al. [124] obtained similar SLs concentration (18.2 g/L) using *S. bombicola* NRRL Y-17069 and determining the optimal fermentation method to generate these surfactant compounds. Recently, a novel lactone esterase enzyme from *S. bombicola*, which catalyzes the intramolecular lactonization of acidic SLs in an aqueous environment, is being investigated to become an ecological tool in industry applications [127].

10. Co-Fermentations between *Candida stellata* and *Saccharomyces cerevisiae*: A Way against Standardized Wines

The use of co-fermentation strategies between non-*Saccharomyces* and *S. cerevisiae* yeast species in a controlled manner can be a useful tool for wine production. Several aspects support this consideration, such as

1. Effect on some analytical compounds as increased glycerol concentration, enhanced total acidity, and reduced acetic acid concentration of wine.
2. Enhancement of desirable aromatic compounds (esters, volatile thiols).
3. Reduction of final ethanol content of the wine.
4. Improvement of complexity and overall quality of wine.
5. Larger release of polysaccharides (mannoproteins).

In the last few years, the use of *C. stellata* yeast in multi-starter fermentations with *S. cerevisiae* has been widely investigated for its ability to increase the glycerol content in wines and their special fructophilic character [66], its capacity to contribute to greater aroma complexity of the wine [128], and its capacity to minimize the risk of fermentation problems [68]. Ciani and Ferraro [66] carried out mixed and sequential fermentations with *C. stellata* and *S. cerevisiae*; the final wines were rich in glycerol and succinic acid, and with less alcohol and acetic acid in comparison with the mono-inoculated *S. cerevisiae* control. Milanovic et al. [68] concluded that *S. bombicola* influenced the alcohol production ability of *S. cerevisiae* under mixed inoculation, since pyruvate decarboxylase (Pdc1) activity in mixed fermentation was lower than pure culture of *S. cerevisiae*, while alcohol dehydrogenase (Adh1) activity showed opposite behavior.

The wines made through *C. stellata*/*S. cerevisiae* co-fermentations usually present higher aroma complexity and overall quality. In a study using Malvar white grape [87], an autochthonous grape variety from Madrid (Spain), different inoculation strategies were applied with *C. stellata* CLI 920 (Cs) and *S. cerevisiae* CLI 889 (Sc). Mixed and sequential were significantly different with regard to their volatile composition and the control of *S. cerevisiae*. These wines were characterized by increased esters concentration and β-phenylethyl alcohol (Figure 3). After sensory analysis, the sequential inoculation was well appreciated by tasters for its pleasant fruity (green apple, grapefruit) and floral aroma and its freshness and full-bodied on the palate. These results were corroborated by pilot scale fermentations [26].

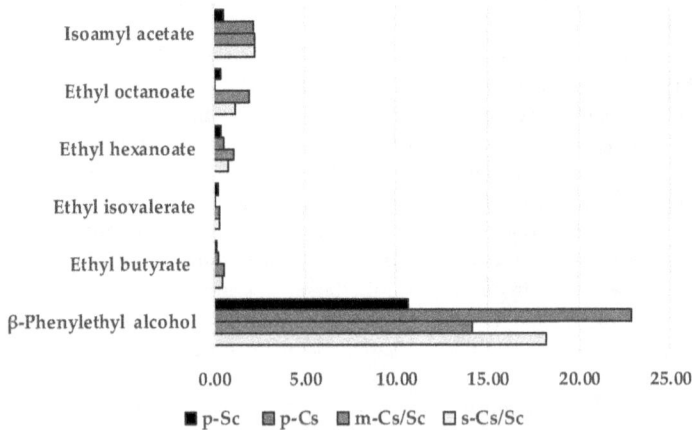

Figure 3. Relevant volatile compounds (mg/L) of pure (p), mixed (m), and sequential (s) fermentations made with *C. stellata* CLI 920 (Cs) and *S. cerevisiae* CLI 889 (Sc) native strains (adapted from García et al. [87]).

In agreement with above, Soden et al. [10] described the aroma of banana, flowers, and lime in wines conducted by sequential inoculation in comparison with the control of *S. cerevisiae*. Other works have also shown the fruity and flowery aroma in cocultures between *C. stellata* and *S. cerevisiae*, which is the result of greater concentration of desirable aromatic compounds including some higher alcohols; β-phenylethyl alcohol and ethyl esters correlated well with its medium-chain fatty acids [26,64,105,116,129].

In recent years, multiples studies have focused on polysaccharides content in wines, giving special attention to the mannoproteins. These molecules are one of the major polysaccharide groups in wines from yeast cell walls [130], and they are secreted into wine during alcoholic fermentation and yeast autolysis during ageing on lees [131]. Mannoproteins composition consists mainly of mannose (80 to 90%) and small amounts of glucose, associated with 10–20% of protein. Numerous investigations have clearly confirmed that these macromolecules are related to technological and sensorial properties in wines, such as prevention of protein haze in white wines [132], protection against crystallization of tartrate salts [133], interaction with aroma compounds [134], improvement of foam stability and flocculation in sparkling wines [135], reduction of astringency and increased body and mouthfeel [136], and increase of the growth of malolactic bacteria [137]. Moreover, it has been noted that the utilization of *Saccharomyces*/non-*Saccharomyces* co-fermentations results in increased release of polysaccharides into the wine, since the high capacity of non-*Saccharomyces* wine yeasts to release polysaccharides (including mannoproteins) has been verified [104,138–140]. Giovani et al. [139] characterized the monosaccharide composition of mannoproteins produced by *S. bombicola* 3827; they noted that the polysaccharides produced by *S. bombicola* were essentially mannoproteins with 73–74% of mannose residues.

In the previously mentioned study [87] (Figure 3), the polysaccharides' content and structure were studied in Malvar wines elaborated with *C. stellata* CLI 920 and *S. cerevisiae* CLI 889. The greater content of arabinose, galactose, and mannose in the total colloids means that mannoproteins from yeast cell walls and Polysaccharides Rich in Arabinose and Galactose (PRAGs) were the main macromolecules in Malvar wines regardless of the inoculation strategy used (Figure 4a). The high content of galactose observed, especially in *C. stellata* pure culture (p-Cs), could also be explained by the presence of this monosaccharide-like galactomannan in yeast cell walls, as in *Schizosaccharomyces pombe*. However, a phylogenetic study with 33 species of *Candida* carried out by Suzuki et al. [141] determined that the cell wall of *C. stellata* lacked galactose.

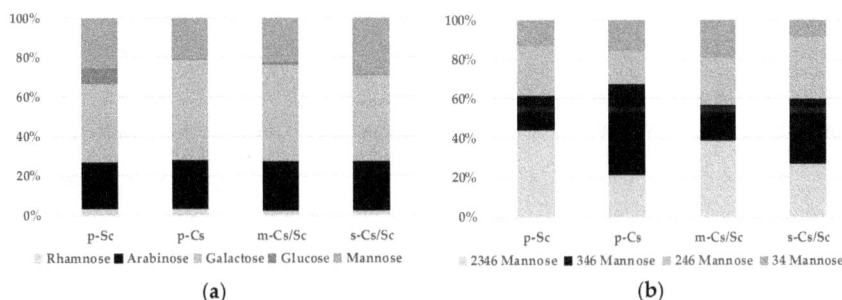

Figure 4. Study of polysaccharides content and structure in Malvar wines elaborated under different inoculation strategies with *C. stellata* and *S. cerevisiae* native strains *: (a) Glycosyl residue composition of polysaccharides from Malvar white wines and (b) Glycosil-linkage composition of mannose residue isolated from Malvar white wines. * Abbreviations associated with type of fermentation and yeast strains are explained in Figure 3.

Mannose residues are larger in *C. stellata*/*S. cerevisiae* sequential fermentation (s-Cs/Sc) than control (Figure 4a); s-Cs/Sc could be the best combination for mannoproteins release into the wine using these yeast strains. Other studies also showed that mixed inoculations with *C. zemplinina*/*S. cerevisiae* supposed an increase of polysaccharides mainly mannoproteins in the final wines [142,143]. Regarding the structure of mannose residues from mannoproteins (Figure 4b), these results are consistent with the *Candida* mannoproteins structure described by Ballou [144]. The structure of mannoproteins consists of a 6-linked backbone, substituted on the 2-position with 2- and 3- linked mannose. This 3-linked mannose (2,4,6-tri-*O*-mannose) proportion is substantially lower in p-Cs than in the control, which agrees with the results previously reported [144]. The high proportion of 3,4,6-tri-*O*-methyl mannose (2-linked mannose) in p-Cs can be observed in comparison with the control (p-Sc); therefore, the *C. stellata* mannoproteins released into the wine present a greater branched structure than those released by the control (Figure 4b). Also, sequential fermentation contained mannoproteins structurally similar to those in the monoculture with *C. stellata*. This could be explained by the important contribution of *C. stellata* strain to wine composition before the inoculation of *S. cerevisiae* strain.

11. Conclusions

At present, a preliminary genetic study needs to be used before the application of *Candida stellata* in food and beverage processing. This research should help to distinguish it from other closely related species within *Starmerella* clade.

Author Contributions: M.G.: revision of articles, writing, and editing; B.E.-Z.: revision and critical reading; J.M.C.: revision and critical reading; and T.A.: revision of articles, writing, and editing.

Funding: This research received no external funding.

Conflicts of Interest: The authors declare no conflict of interest.

References

1. Lachance, M.A.; Boekhout, T.; Scorzetti, G.; Fell, J.W.; Kurtzman, C.P. Candida Berkhout (1923). In *The Yeasts*; Kurtzman, C.P., Fell, J.W., Boekhout, T., Eds.; Elsevier, B.V.: Amsterdam, The Netherlands, 2011; pp. 987–1278. ISBN 9780444521491.
2. Hommel, R.K. *Candida*: Introduction. In *Encyclopedia of Food Microbiology: Second Edition*; Batt, C.A., Tortorello, M.L., Eds.; Elsevier: London, UK, 2014; pp. 367–373. ISBN 9780123847331.
3. Van Bogaert, I.N.A.; Zhang, J.; Soetaert, W. Microbial synthesis of sophorolipids. *Process Biochem.* **2011**, *46*, 821–833. [CrossRef]
4. Strauss, M.L.A.; Jolly, N.P.; Lambrechts, M.G.; van Rensburg, P. Screening for the production of extracellular hydrolytic enzymes by non-*Saccharomyces* wine yeasts. *J. Appl. Microbiol.* **2001**, *91*, 182–190. [CrossRef] [PubMed]
5. Günther, C.S.; Goddard, M.R. Do yeasts and *Drosophila* interact just by chance? *Fungal Ecol.* **2018**. [CrossRef]
6. Pretorius, I.S.; van der Westhuizen, T.J.; Augustyn, O.P.H. Yeast biodiversity in vineyards and wineries and its importance to the South African wine industry. *S. Afr. J. Enol. Vitic.* **1999**, *20*, 61–75. [CrossRef]
7. Mills, D.A.; Johannsen, E.A.; Cocolin, L. Yeast diversity and persistence in botrytis-affected wine fermentations. *Appl. Environ. Microbiol.* **2002**, *68*, 4884–4893. [CrossRef] [PubMed]
8. Mora, J.; Mulet, A. Effects of some treatments of grape juice on the population and growth of yeast species during fermentation. *Am. J. Enol. Vitic.* **1991**, *42*, 133–136.
9. Antunovics, Z.; Csoma, H.; Sipiczki, M. Molecular and genetic analysis of the yeast flora of botrytized Tokaj wines. *Bull. O.I.V.* **2003**, *76*, 380–397.
10. Soden, A.; Francis, I.L.; Oakey, H.; Henschke, P.A. Effects of co-fermentation with *Candida stellata* and *Saccharomyces cerevisiae* on the aroma and composition of Chardonnay wine. *Aust. J. Grape Wine Res.* **2000**, *6*, 21–30. [CrossRef]
11. Sipiczki, M. *Candida zemplinina* sp. nov., an osmotolerant and psychrotolerant yeast that ferments sweet botrytized wines. *Int. J. Syst. Evol. Microbiol.* **2003**, *53*, 2079–2083. [CrossRef] [PubMed]
12. Sipiczki, M. Species identification and comparative molecular and physiological analysis of *Candida zemplinina* and *Candida stellata*. *J. Basic Microbiol.* **2004**, *44*, 471–479. [CrossRef] [PubMed]

13. Csoma, H.; Sipiczki, M. Taxonomic reclassification of *Candida stellata* strains reveals frequent occurrence of *Candida zemplinina* in wine fermentation. *FEMS Yeast Res.* **2008**, *8*, 328–336. [CrossRef] [PubMed]

14. Boulton, R.B.; Singleton, V.L.; Bisson, L.F.; Kunkee, R.E. *Principles and Practices of Winemaking*; Chapman & Hall: New York, NY, USA, 1996.

15. Fuselsang, K.C. *Wine Microbiology*; Chapman & Hall: New York, NY, USA, 1997.

16. Pramateftaki, P.V.; Lanaridis, P.; Typas, M.A. Molecular identification of wine yeasts at species or strain level: A case study with strains from two vine-growing areas of Greece. *J. Appl. Microbiol.* **2000**, *89*, 236–248. [CrossRef] [PubMed]

17. Valente, P.; Gouveia, F.C.; De Lemos, G.A.; Pimentel, D.; Van Elsas, J.D.; Mendonça-Hagler, L.C.; Hagler, A.N. PCR amplification of the rDNA internal transcribed spacer region for differentiation of *Saccharomyces* cultures. *FEMS Microbiol. Lett.* **1996**, *137*, 253–256. [CrossRef] [PubMed]

18. Guillamón, M.; Sabat, J.; Barrio, E.; Cano, J.; Querol, A. Rapid identification of wine yeast species based on RFLP analysis of the ribosomal internal transcribed spacer (ITS) region. *Arch. Microbiol.* **1998**, *169*, 387–392. [CrossRef] [PubMed]

19. Esteve-Zarzoso, B.; Belloch, C.; Uruburu, F.; Querol, A. Identification of yeasts by RFLP analysis of the 5.8S rRNA gene and the two ribosomal internal transcribed spacers. *Int. J. Syst. Bacteriol.* **1999**, *49*, 329–337. [CrossRef] [PubMed]

20. Cocolin, L.; Bisson, L.F.; Mills, D.A. Direct profiling of the yeast dynamics in wine fermentations. *FEMS Microbiol. Lett.* **2000**, *189*, 81–87. [CrossRef] [PubMed]

21. Cocolin, L.; Heisey, A.; Mills, D.A. Direct identification of the indigenous yeasts in commercial wine fermentations. *Am. J. Enol. Vitic.* **2001**, *52*, 49–53.

22. López, V.; Fernández-Espinar, M.T.; Barrio, E.; Ramón, D.; Querol, A. A new PCR-based method for monitoring inoculated wine fermentations. *Int. J. Food Microbiol.* **2003**, *81*, 63–71. [CrossRef]

23. Cocolin, L.; Mills, D.A. Wine yeast inhibition by sulfur dioxide: A comparison of culture-dependent and independent methods. *Am. J. Enol. Vitic.* **2003**, *54*, 125–130.

24. Vitzthum, F.; Bernhagen, J. SYBR Green I: An ultrasensitive fluorescent dye for double-stranded DNA quantification in solution and other applications. *Recent Res. Devel. Anal. Biochem.* **2002**, *2*, 65–93.

25. Phister, T.G.; Mills, D. A Real-time PCR assay for detection and enumeration of *Dekkera bruxellensis* in wine. *Appl. Environ. Microbiol.* **2003**, *69*, 7430–7434. [CrossRef] [PubMed]

26. García, M.; Esteve-Zarzoso, B.; Crespo, J.; Cabellos, J.M.; Arroyo, T. Yeast monitoring of wine mixed or sequential fermentations made by native strains from D.O. "Vinos de Madrid" using real-time quantitative PCR. *Front. Microbiol.* **2017**, *8*. [CrossRef] [PubMed]

27. Minarik, E.; Hanikova, A. Die hefeflora konzentrierer traubenmoste und deren einfluss auf die stabilitat der weine. *Wein-Wissen* **1982**, *3*, 187–192.

28. Rosini, G.; Federici, F.; Martini, A. Yeast flora of grape berries during ripening. *Microb. Ecol.* **1982**, *8*, 83–89. [CrossRef] [PubMed]

29. Fleet, G.H. Evolution of yeasts and lactic acid bacteria during fermentation and storage of Bordeaux wines. *Appl. Environ. Microbiol.* **1984**, *48*, 1034–1038. [PubMed]

30. Pardo, I.; Garcia, M.I.; Zuniga, M.; Uruburu, F. Dynamics of microbial populations during fermentation of wine from the Utiel-Requena region of Spain. *Appl. Environ. Microbiol.* **1989**, *50*, 539–541.

31. Holloway, P.; van Twest, R.A.; Subden, R.E.; Lachance, M.A. A strain of *Candida stellata* of special interest to oenologists. *Food Res. Int.* **1992**, *25*, 147–149. [CrossRef]

32. Constanti, M.; Poblet, M.; Arola, L.; Mas, A.; Guillamón, J.M. Analysis of yeast populations during alcoholic fermentation in a newly established winery. *Am. J. Enol. Vitic.* **1997**, *48*, 339–344.

33. Fernández, M.T.; Úbeda, J.F.; Briones, A.I. Comparative study of non-*Saccharomyces* microflora of musts in fermentation, by physiological and molecular methods. *FEMS Microbiol. Lett.* **1999**, *173*, 223–229. [CrossRef]

34. Torija, M.J.; Rozès, N.; Poblet, M.; Guillamón, J.M.; Mas, A. Yeast population dynamics in spontaneous fermentations: Comparison between two different wine-producing areas over a period of three years. *Antonie Van Leeuwenhoek* **2001**, *79*, 345–352. [CrossRef] [PubMed]

35. Clemente-Jiménez, J.M.; Mingorance-Cazorla, L.; Martínez-Rodríguez, S.; Las Heras-Vázquez, F.J.; Rodríguez-Vico, F. Molecular characterization and oenological properties of wine yeasts isolated during spontaneous fermentation of six varieties of grape must. *Food Microbiol.* **2004**, *21*, 149–155. [CrossRef]

36. Combina, M.; Elía, A.; Mercado, L.; Catania, C.; Ganga, A.; Martínez, C. Dynamics of indigenous yeast populations during spontaneous fermentation of wines from Mendoza, Argentina. *Int. J. Food Microbiol.* **2005**, *99*, 237–243. [CrossRef] [PubMed]

37. Divol, B.; Lonvaud-Funel, A. Evidence for viable but nonculturable yeasts in botrytis-affected wine. *J. Appl. Microbiol.* **2005**, *99*, 85–93. [CrossRef] [PubMed]

38. Hierro, N.; González, Á.; Mas, A.; Guillamón, J.M. Diversity and evolution of non-*Saccharomyces* yeast populations during wine fermentation: Effect of grape ripeness and cold maceration. *FEMS Yeast Res.* **2006**, *6*, 102–111. [CrossRef] [PubMed]

39. Mora, J.; Barbas, J.I.; Mulet, A. Growth of yeast species during the fermentation of musts inoculated with *Kluyveromyces thermotolerans* and *Saccharomyces cerevisiae*. *Am. J. Enol. Vitic.* **1990**, *41*, 156–159.

40. Povhe Jemec, K.; Raspor, P. Initial *Saccharomyces cerevisiae* concentration in single or composite cultures dictates bioprocess kinetics. *Food Microbiol.* **2005**, *22*, 293–300. [CrossRef]

41. Xufre, A.; Albergaria, H.; Inácio, J.; Spencer-Martins, I.; Gírio, F. Application of fluorescence in situ hybridisation (FISH) to the analysis of yeast population dynamics in winery and laboratory grape must fermentations. *Int. J. Food Microbiol.* **2006**, *108*, 376–384. [CrossRef] [PubMed]

42. Jackson, R.S. *Wine Science-Principles, Practice, Perception*, 2nd ed.; Academic Press: San Diego, CA, USA, 2000.

43. Šipiczki, M.; Ciani, M.; Csoma, H. Taxonomic reclassification of *Candida stellata* DBVPG 3827. *Folia Microbiol. (Praha)* **2005**, *50*, 494–498. [CrossRef] [PubMed]

44. Englezos, V.; Giacosa, S.; Rantsiou, K.; Rolle, L.; Cocolin, L. *Starmerella bacillaris* in winemaking: Opportunities and risks. *Curr. Opin. Food Sci.* **2017**, *17*, 30–35. [CrossRef]

45. Soles, R.M.; Ough, C.S.; Kunkee, R.E. Ester concentration differences in wine fermented by various species and strains of yeasts. *Am. J. Enol. Vitic.* **1982**, *33*, 94–98.

46. Ciani, M.; Ferraro, L. Enhanced glycerol content in wines made with immobilized *Candida stellata* cells. *Appl. Environ. Microbiol.* **1996**, *62*, 128–132. [PubMed]

47. Ciani, M.; Maccarelli, F. Oenological properties of non-*Saccharomyces* yeasts associated with wine-making. *World J. Microbiol. Biotechnol.* **1998**, *14*, 199–203. [CrossRef]

48. Rosa, C.A.; Lachance, M.A. The yeast genus *Starmerella* gen. nov. and *Starmerella bombicola* sp. nov., the teleomorph of *Candida bombicola* (Spencer, Gorin & Tullock) Meyer & Yarrow. *Int. J. Syst. Bacteriol.* **1998**, *48*, 1413–1417. [CrossRef] [PubMed]

49. Barbe, J.C.; De Revel, G.; Joyeux, A.; Bertrand, A.; Lonvaud-Funel, A. Role of botrytized grape micro-organisms in SO$_2$ binding phenomena. *J. Appl. Microbiol.* **2001**, *90*, 34–42. [CrossRef] [PubMed]

50. Coton, E.; Coton, M.; Levert, D.; Casaregola, S.; Sohier, D. Yeast ecology in French cider and black olive natural fermentations. *Int. J. Food Microbiol.* **2006**, *108*, 130–135. [CrossRef] [PubMed]

51. McNeill, J.; Barrie, F.; Buck, W.; Demoulin, V.; Greuter, W.; Hawkworth, D.L.; Herendeen, P.; Knapp, S.; Marhold, K.; Prado, J.; et al. *International Code of Nomenclature for Algae, Fungi and Plants*; Regnum Vegetabile: Melbourne, Australia, 2012.

52. Duarte, F.L.; Pimentel, N.H.; Teixeira, A.; Fonseca, A. *Saccharomyces bacillaris* is not a synonym of *Candida stellata*: Reinstatement as *Starmerella bacillaris* comb. nov. *Antonie van Leeuwenhoek* **2012**, *102*, 653–658. [CrossRef] [PubMed]

53. Nielsen, D.S.; Hønholt, S.; Tano-Debrah, K.; Jespersen, L. Yeast populations associated with Ghanaian cocoa fermentations analysed using denaturing gradient gel electrophoresis (DGGE). *Yeast* **2005**, *22*, 271–284. [CrossRef] [PubMed]

54. Nielsen, D.S.; Teniola, O.D.; Ban-Koffi, L.; Owusu, M.; Andersson, T.S.; Holzapfel, W.H. The microbiology of Ghanaian cocoa fermentations analysed using culture-dependent and culture-independent methods. *Int. J. Food Microbiol.* **2007**, *114*, 168–186. [CrossRef] [PubMed]

55. Lachance, M.A.; Starmer, W.T.; Rosa, C.A.; Bowles, J.M.; Barker, J.S.F.; Janzen, D.H. Biogeography of the yeasts of ephemeral flowers and their insects. *FEMS Yeast Res.* **2001**, *1*, 1–8. [CrossRef] [PubMed]

56. Loureiro, V.; Malfeito-Ferreira, M. Spoilage yeasts in the wine industry. *Int. J. Food Microbiol.* **2003**, *86*, 23–50. [CrossRef]

57. Puig, S.; Querol, A.; Barrio, E.; Pérez-Ortín, J.E. Mitotic recombination and genetic changes in Saccharomyces cerevisiae during wine fermentation. *Appl. Environ. Microbiol.* **2000**, *66*, 8–13. [CrossRef]

58. Kurtzman, C.P.; Robnett, C.J. Identification and phylogeny of ascomycetous yeasts from analysis of nuclear large subunit (26S) ribosomal DNA partial sequences. *Antonie Van Leeuwenhoek* **1998**, *73*, 331–371. [CrossRef] [PubMed]

59. Yarrow, D.; Meyer, S.A. Proposal for amendment of the diagnosis of the genus *Candida* Berkhout nom. cons. *Int. J. Syst. Bacteriol.* **1978**, *28*, 611–615. [CrossRef]

60. de Koster, C.G.; Heerma, W.; Pepermans, H.A.M.; Groenewegen, A.; Peters, H.; Haverkamp, J. Tandem mass spectrometry and nuclear magnetic resonance spectroscopy studies of *Candida bombicola* sophorolipid and product formed on hydrolysis by cutinase. *Anal. Biochem.* **1995**, *230*, 135–148. [CrossRef] [PubMed]

61. Pfliegler, W.P.; Horváth, E.; Kállai, Z.; Sipiczki, M. Diversity of *Candida zemplinina* isolates inferred from RAPD, micro/minisatellite and physiological analysis. *Microbiol. Res.* **2014**, *169*, 402–410. [CrossRef] [PubMed]

62. Englezos, V.; Rantsiou, K.; Torchio, F.; Rolle, L.; Gerbi, V.; Cocolin, L. Exploitation of the non-*Saccharomyces* yeast *Starmerella bacillaris* (synonym *Candida zemplinina*) in wine fermentation: Physiological and molecular characterizations. *Int. J. Food Microbiol.* **2015**, *199*, 33–40. [CrossRef] [PubMed]

63. Tofalo, R.; Schirone, M.; Torriani, S.; Rantsiou, K.; Cocolin, L.; Perpetuini, G.; Suzzi, G. Diversity of *Candida zemplinina* strains from grapes and Italian wines. *Food Microbiol.* **2012**, *29*, 18–26. [CrossRef] [PubMed]

64. Magyar, I.; Tóth, T. Comparative evaluation of some oenological properties in wine strains of *Candida stellata*, *Candida zemplinina*, *Saccharomyces uvarum* and *Saccharomyces cerevisiae*. *Food Microbiol.* **2011**, *28*, 94–100. [CrossRef] [PubMed]

65. Solieri, L.; Landi, S.; De Vero, L.; Giudici, P. Molecular assessment of indigenous yeast population from traditional balsamic vinegar. *J. Appl. Microbiol.* **2006**, *101*, 63–71. [CrossRef] [PubMed]

66. Ciani, M.; Ferraro, L. Combined use of immobilized *Candida stellata* cells and *Saccharomyces cerevisiae* to improve the quality of wines. *J. Appl. Microbiol.* **1998**, *85*, 247–254. [CrossRef] [PubMed]

67. Ciani, M.; Ferraro, L.; Fatichenti, F. Influence of glycerol production on the aerobic and anaerobic growth of the wine yeast *Candida stellata*. *Enzyme Microb. Technol.* **2000**, *27*, 698–703. [CrossRef]

68. Milanovic, V.; Ciani, M.; Oro, L.; Comitini, F. *Starmerella bombicola* influences the metabolism of *Saccharomyces cerevisiae* at pyruvate decarboxylase and alcohol dehydrogenase level during mixed wine fermentation. *Microb. Cell Fact.* **2012**, *3*, 11–18. [CrossRef] [PubMed]

69. Alves-Araújo, C.; Pacheco, A.; Almeida, M.J.; Spencer-Martins, I.; Leão, C.; Sousa, M.J. Sugar utilization patterns and respiro-fermentative metabolism in the baker's yeast *Torulaspora delbrueckii*. *Microbiology* **2007**, *153*, 898–904. [CrossRef] [PubMed]

70. Pina, C.; Gonçalves, P.; Prista, C.; Loureiro-Dias, M.C. Ffz1, a new transporter specific for fructose from *Zygosaccharomyces bailii*. *Microbiology* **2004**, *150*, 2429–2433. [CrossRef] [PubMed]

71. Leandro, M.J.; Cabral, S.; Prista, C.; Loureiro-Dias, M.C.; Sychrová, H. The high-capacity specific fructose facilitator ZrFfz1 is essential for the fructophilic behavior of *Zygosaccharomyces rouxii* CBS 732T. *Eukaryot. Cell* **2014**, *13*, 1371–1379. [CrossRef] [PubMed]

72. Gonçalves, C.; Wisecaver, J.H.; Kominek, J.; Salema-Oom, M.; Leandro, M.J.; Shen, X.-X.; Opulente, D.; Zhou, X.; Peris, D.; Kurtzman, C.P.; et al. Evidence for loss and adaptive reacquisition of alcoholic fermentation in an early-derived fructophilic yeast lineage. *Elife* **2018**, *7*, e33034. [CrossRef] [PubMed]

73. Gonçalves, C.; Coelho, M.A.; Salema-Oom, M.; Gonçalves, P. Stepwise functional evolution in a fungal sugar transporter family. *Mol. Biol. Evol.* **2016**, *33*, 352–366. [CrossRef] [PubMed]

74. Flores, C.L.; Rodriguez, C.; Petit, T.; Gancedo, C. Carbohydrate and energy-yielding metabolism in non-conventional yeasts. *FEMS Microbiol. Rev.* **2000**, *24*, 507–529. [CrossRef]

75. Mortensen, A.; Aguilar, F.; Crebelli, R.; Di Domenico, A.; Dusemund, B.; Frutos, M.J.; Galtier, P.; Gott, D.; Gundert-Remy, U.; Leblanc, J.; et al. Re-evaluation of glycerol (E 422) as a food additive. *EFSA J.* **2017**, *15*. [CrossRef]

76. Eriksson, P.; Andre, L.; Ansell, R.; Blomberg, A.; Alder, L. Cloning and characterization of GPD2, a second gene encoding sn-glycerol 3-phosphate dehydrogenase (NAD$^+$) in *Saccharomyces cerevisiae*, and its comparison with GPD1. *Mol. Microbiol.* **1995**, *17*, 95–107. [CrossRef] [PubMed]

77. Hohmann, S. Osmotic stress signalling and osmoadaptation in yeasts. *Microbiol. Mol. Biol. Rev.* **2002**, *66*, 300–372. [CrossRef] [PubMed]

78. Dihazi, H.; Kessler, R.; Eschrich, K. High osmolarity glycerol (HOG) pathway-induced phosphorylation and activation of 6-phosphofructo-2-kinase are essential for glycerol accumulation and yeast cell proliferation under hyperosmotic stress. *J. Biol. Chem.* **2004**, *279*, 23961–23968. [CrossRef] [PubMed]

79. Rodríguez-Peña, J.M.; García, R.; Nombela, C.; Arroyo, J. The high-osmolarity glycerol (HOG) and cell wall integrity (CWI) signalling pathways interplay: A yeast dialogue between MAPK routes. *Yeast* **2010**, *27*, 495–502. [CrossRef] [PubMed]

80. Ciani, M. Wine vinegar production using base wines made with different yeast species. *J. Sci. Food Agric.* **1998**, *78*, 290–294. [CrossRef]

81. Scanes, K.T.; Hohmann, S.; Prior, B.A. Glycerol production by the yeast *Saccharomyces cerevisiae* and its relevance to wine: A review. *S. Afr. J. Enol. Vitic.* **1998**, *19*, 17–24. [CrossRef]

82. Prior, B.A.; Toh, T.H.; Jolly, N.; Baccari, C.; Mortimer, R.K. Impact of yeast breeding for elevated glycerol production on fermentative activity and metabolite formation in Chardonnay wine. *S. Afr. J. Enol. Vitic.* **2000**, *21*, 92–99. [CrossRef]

83. Pretorius, I.S. Tailoring wine yeast for the new millennium: Novel approaches to the ancient art of winemaking. *Yeast* **2000**, *16*, 675–729. [CrossRef]

84. Noble, A.C.; Bursick, G.F. The contribution of glycerol to perceived viscosity and sweetness in white wine. *Am. J. Enol. Vitic.* **1984**, *35*, 110–112.

85. Fleet, G.H. Wine. In *Food Microbiology: Fundamentals and Frontiers*; Doyle, M.P., Beuchat, L.R., Eds.; ASM Press: Washington, DC, USA, 2007; pp. 863–890.

86. Ferraro, L.; Fatichenti, F.; Ciani, M. Pilot scale vinification process using immobilized *Candida* stellata cells and *Saccharomyces cerevisiae*. *Process Biochem.* **2000**, *35*, 1125–1129. [CrossRef]

87. García, M.; Arroyo, T.; Crespo, J.; Cabellos, J.M.; Esteve-Zarzoso, B. Use of native non-*Saccharomyces* strain: A. new strategy in D.O. "Vinos de Madrid" (Spain) wines elaboration. *Eur. J. Food Sci. Technol.* **2017**, *5*, 1–31.

88. Gobbi, M.; De Vero, L.; Solieri, L.; Comitini, F.; Oro, L.; Giudici, P.; Ciani, M. Fermentative aptitude of non-*Saccharomyces* wine yeast for reduction in the ethanol content in wine. *Eur. Food Res. Technol.* **2014**, *239*, 41–48. [CrossRef]

89. Nigam, P.S. Microbial enzymes with special characteristics for biotechnological applications. *Biomolecules* **2013**, *3*, 597–611. [CrossRef] [PubMed]

90. Anbu, P.; Gopinath, S.C.B.; Chaulagain, B.P.; Lakshmipriya, T. Microbial enzymes and their applications in industries and medicine 2016. *Biomed. Res. Int.* **2017**. [CrossRef] [PubMed]

91. Pando Bedriñana, R.; Lastra Queipo, A.; Suárez Valles, B. Screening of enzymatic activities in non-*Saccharomyces* cider yeasts. *J. Food Biochem.* **2012**, *36*, 683–689. [CrossRef]

92. Maturano, Y.P.; Rodríguez, L.A.; Toro, M.E.; Nally, M.C.; Vallejo, M.; Castellanos de Figueroa, L.I.; Combina, M.; Vazquez, F. Multi-enzyme production by pure and mixed cultures of *Saccharomyces* and non-*Saccharomyces* yeasts during wine fermentation. *Int. J. Food Microbiol.* **2012**, *155*, 43–50. [CrossRef] [PubMed]

93. Esteve-Zarzoso, B.; Manzanares, P.; Ramón, D.; Querol, A. The role of non-*Saccharomyces* yeasts in industrial winemaking. *Int. Microbiol.* **1998**, *1*, 143–148. [CrossRef] [PubMed]

94. Claus, H.; Mojsov, K. Enzymes for wine fermentation: Current and perspective applications. *Fermentation* **2018**, *4*, 52. [CrossRef]

95. Carpita, N.C.; Gibeaut, D.M. Structural models of primary cell walls in flowering plants: Consistency of molecular structure with the physical properties of the walls during growth. *Plant J.* **1993**, *3*, 1–30. [CrossRef] [PubMed]

96. Canal-Llaubères, R.-M. Enzymes in winemaking. In *Wine Microbiology and Biotechnology*; Fleet, G.H., Ed.; Harwood Academic Publishers: Chur, Switzerland, 1993; pp. 477–506.

97. Hadfield, K.A.; Bennett, A.B. Polygalacturonases: Many genes in search of a function. *Plant Physiol.* **1998**, *117*, 337–343. [CrossRef] [PubMed]

98. Grassin, C.; Fauquembergue, P. Application of pectinases in beverages. In *Pectin and pectinases*; Visser, J., Voragen, A.G.J., Eds.; Elsevier: Amsterdam, The Netherlands, 1996; pp. 453–462.

99. Boccas, F.; Roussos, S.; Gutiérrez, M.; Serrano, L.; Viniegra, G.G. Production of pectinase from coffee pulp in solid-state fermentation system-selection of wild fungal isolate of high potency by a simple 3-step screening technique. *J. Food Sci. Technol.* **1994**, *31*, 22–26.

100. Evans, J.D.; Akin, D.E.; Foulk, J.A. Flax-retting by polygalacturonase-containing enzyme mixtures and effects on fiber properties. *J. Biotechnol.* **2002**, *97*, 223–231. [CrossRef]

101. Fernández, M.; Úbeda, J.F.; Briones, A.I. Typing of non-*Saccharomyces* yeasts with enzymatic activities of interest in wine-making. *Int. J. Food Microbiol.* **2000**, *59*, 29–36. [CrossRef]

102. Merín, M.G.; Martín, M.C.; Rantsiou, K.; Cocolin, L.; De Ambrosini, V.I.M. Characterization of pectinase activity for enology from yeasts occurring in Argentine Bonarda grape. *Braz. J. Microbiol.* **2015**, *46*, 815–823. [CrossRef] [PubMed]

103. Cordero-Bueso, G.; Esteve-Zarzoso, B.; Cabellos, J.M.; Gil-Díaz, M.; Arroyo, T. Biotechnological potential of non-*Saccharomyces* yeasts isolated during spontaneous fermentations of Malvar (*Vitis vinifera* cv. L.). *Eur. Food Res. Technol.* **2013**, *236*, 193–207. [CrossRef]

104. García, M.; Apolinar-Valiente, R.; Williams, P.; Esteve-Zarzoso, B.; Arroyo, T.; Crespo, J.; Doco, T. Polysaccharides and oligosaccharides produced on Malvar wines elaborated with *Torulaspora delbrueckii* CLI 918 and *Saccharomyces cerevisiae* CLI 889 native yeasts from D.O. "Vinos de Madrid". *J. Agric. Food Chem.* **2017**, *65*, 6656–6664. [CrossRef]

105. Andorrà, I.; Berradre, M.; Rozès, N.; Mas, A.; Guillamón, J.M.; Esteve-Zarzoso, B. Effect of pure and mixed cultures of the main wine yeast species on grape must fermentations. *Eur. Food Res. Technol.* **2010**, *231*, 215–224. [CrossRef]

106. Theron, L.W.; Divol, B. Microbial aspartic proteases: Current and potential applications in industry. *Appl. Microbiol. Biotechnol.* **2014**, *98*, 8853–8868. [CrossRef] [PubMed]

107. Dizy, M.; Bisson, L.F. Proteolytic activity of yeast strains during grape juice fermentation. *Am. J. Enol. Vitic.* **2000**, *51*, 155–167.

108. Romero-Cascales, I.; Fernández-Fernández, J.I.; Ros-García, J.M.; López-Roca, J.M.; Gómez-Plaza, E. Characterisation of the main enzymatic activities present in six commercial macerating enzymes and their effects on extracting colour during winemaking of Monastrell grapes. *Int. J. Food Sci. Technol.* **2008**, *43*, 1295–1305. [CrossRef]

109. Capozzi, V.; Garofalo, C.; Chiriatti, M.A.; Grieco, F.; Spano, G. Microbial terroir and food innovation: The case of yeast biodiversity in wine. *Microbiol. Res.* **2015**, *181*, 75–83. [CrossRef] [PubMed]

110. Thongekkaew, J.; Kongsanthia, J. Screening and identification of cellulase producing yeast from Rongkho Forest, Ubon Ratchathani University. *Bioeng. Biosci.* **2016**, *4*, 29–33. [CrossRef]

111. Williams, P.J.; Cynkar, W.; Francis, I.L.; Gray, J.D.; Iland, P.G.; Coombe, B.G. Quantification of glycosides in grapes, juices, and wines through a determination of glycosyl glucose. *J. Agric. Food Chem.* **1995**, *43*, 121–128. [CrossRef]

112. Winterhalter, P.; Skouroumounis, G.K. Glycoconjugated aroma compounds: Occurrence, role and biotechnological transformation. *Adv. Biochem. Eng. Biotechnol.* **1997**, *55*, 73–105. [PubMed]

113. Rosi, I.; Vinella, M.; Domizio, P. Characterization of β-glucosidase activity in yeasts of oenological origin. *J. Appl. Bacteriol.* **1994**, *77*, 519–527. [CrossRef] [PubMed]

114. Cordero Otero, R.R.; Ubeda Iranzo, J.F.; Briones-Perez, A.I.; Potgieter, N.; Villena, M.A.; Pretorius, I.S.; van Rensburg, P. Characterization of the β-glucosidase activity produced by enological strains of non-*Saccharomyces* yeasts. *Food Microbiol. Saf.* **2003**, *68*, 2564–2569. [CrossRef]

115. Hock, R.; Benda, I.; Schreier, P. Formation of terpenes by yeasts during alcoholic fermentation. *Zeitschrift für Leb. Und-forsch.* **1984**, *179*, 450–452. [CrossRef]

116. Jolly, N.P.; Augustyn, O.H.P.; Pretorius, I.S. The effect of non-*Saccharomyces* yeasts on fermentation and wine quality. *S. Afr. J. Enol. Vitic.* **2003**, *24*, 55–62. [CrossRef]

117. Acosta, N.; Beldarraín, A.; Rodríguez, L.; Alonso, Y. Characterization of recombinant invertase expressed in methylotrophic yeasts. *Biotechnol. Appl. Biochem.* **2000**, *32*, 179–187. [CrossRef] [PubMed]

118. Lee, W.; Huang, C. Modeling of ethanol fermentation using *Zymomonas mobilis* ATCC 10988 grown on the media containing glucose and fructose. *Biochem. Eng. J.* **2000**, *4*, 217–227. [CrossRef]

119. Uma, C.; Gomathi, D.; Ravikumar, G.; Kalaiselvi, M.; Palaniswamy, M. Production and properties of invertase from a *Cladosporium cladosporioides* in SmF using pomegranate peel waste as substrate. *Asian Pac. J. Trop. Biomed.* **2012**, S605–S611. [CrossRef]

120. Pataro, C.; Guerra, J.B.; Gomes, F.C.O.; Neves, M.J.; Pimentel, P.F.; Rosa, C.A. Trehalose accumulation, invertase activity and physiological characteristics of yeasts isolates from 24 h fermentative cycles during the production of artisanal Brazilian cachaça. *Braz. J. Microbiol.* **2002**, *33*, 202–208. [CrossRef]

121. Gargel, C.A.; Baffi, M.A.; Gomes, E.; Da-Silva, R. Invertase from a *Candida stellata* strain isolated from grape: Production and physico-chemical characterization. *J. Microbiol. Biotechnol. Food Sci.* **2014**, *4*, 24–28. [CrossRef]

122. Van Bogaert, I.N.A.; Saerens, K.; De Muynck, C.; Develter, D.; Soetaert, W.; Vandamme, E.J. Microbial production and application of sophorolipids. *Appl. Microbiol. Biotechnol.* **2007**, *76*, 23–34. [CrossRef] [PubMed]

123. Kurtzman, C.P.; Price, N.P.J.; Ray, K.J.; Kuo, T.M. Production of sophorolipid biosurfactants by multiple species of the *Starmerella (Candida) bombicola* yeast clade. *FEMS Microbiol. Lett.* **2010**, *311*, 140–146. [CrossRef] [PubMed]

124. Parekh, V.J.; Pandit, A.B. Optimization of fermentative production of sophorolipid biosurfactant by *Starmerella bombicola* NRRL Y-17069 using response surface methodology. *Int. J. Pharm. Biol. Sci.* **2011**, *1*, 103–116.

125. Pekin, G.; Vardar-Sukan, F.; Kosaric, N. Production of sophorolipids from *Candida bombicola* ATCC 22214 using Turkish corn oil and honey. *Eng. Life Sci.* **2005**, *5*, 357–362. [CrossRef]

126. Cavalero, D.A.; Cooper, D.G. The effect of medium composition on the structure and physical state of sophorolipids produced by *Candida bombicola* ATCC 22214. *J. Biotechnol.* **2003**, *103*, 31–41. [CrossRef]

127. De Waele, S.; Vandenberghe, I.; Laukens, B.; Planckaert, S.; Verweire, S.; Van Bogaert, I.N.A.; Soetaert, W.; Devreese, B.; Ciesielska, K. Optimized expression of the *Starmerella bombicola* lactone esterase in Pichia pastoris through temperature adaptation, codon-optimization and co-expression with HAC1. *Protein Expr. Purif.* **2018**, *143*, 62–70. [CrossRef] [PubMed]

128. Andorrà, I.; Berradre, M.; Mas, A.; Esteve-Zarzoso, B.; Guillamón, J.M. Effect of mixed culture fermentations on yeast populations and aroma profile. *LWT-Food Sci. Technol.* **2012**, *49*, 8–13. [CrossRef]

129. Sadoudi, M.; Tourdot-Maréchal, R.; Rousseaux, S.; Steyer, D.; Gallardo-Chacón, J.J.; Ballester, J.; Vichi, S.; Guérin-Schneider, R.; Caixach, J.; Alexandre, H. Yeast-yeast interactions revealed by aromatic profile analysis of Sauvignon Blanc wine fermented by single or co-culture of non-*Saccharomyces* and *Saccharomyces* yeasts. *Food Microbiol.* **2012**, *32*, 243–253. [CrossRef] [PubMed]

130. Aguilar-Uscanga, B.; François, J.M. A study of the yeast cell wall composition and structure in response to growth conditions and mode of cultivation. *Lett. Appl. Microbiol.* **2003**, *37*, 268–274. [CrossRef] [PubMed]

131. Doco, T.; Quellec, N.; Moutounet, M.; Pellerin, P. Polysaccharide patterns during the ageing of Carignan noir red wines. *Am. J. Enol. Vitic.* **1999**, *50*, 25–32.

132. Dufrechou, M.; Doco, T.; Poncet-Legrand, C.; Sauvage, F.X.; Vernhet, A. Protein/Polysaccharide interactions and their impact on haze formation in white wines. *J. Agric. Food Chem.* **2015**, *63*, 10042–10053. [CrossRef] [PubMed]

133. Marchal, R.; Jeandet, P. Use of enological additives for colloid and tartrate salt stabilization in white wines and for improvement of sparkling wine foaming properties. In *Wine Chemistry and Biochemistry*; Moreno-Arribas, M.V., Polo, M.C., Eds.; Springer: New York, NY, USA, 2009; pp. 127–158. ISBN 9780387741161.

134. Chalier, P.; Angot, B.; Delteil, D.; Doco, T.; Gunata, Z. Interactions between aroma compounds and whole mannoprotein isolated from *Saccharomyces cerevisiae* strains. *Food Chem.* **2007**, *100*, 22–30. [CrossRef]

135. Pérez-Magariño, S.; Martínez-Lapuente, L.; Bueno-Herrera, M.; Ortega-Heras, M.; Guadalupe, Z.; Ayestarán, B. Use of commercial dry yeast products rich in mannoproteins for white and rosé sparkling wine elaboration. *J. Agric. Food Chem.* **2015**, *63*, 5670–5681. [CrossRef] [PubMed]

136. Vidal, S.; Francis, L.; Williams, P.; Kwiatkowski, M.; Gawel, R.; Cheynier, V.; Waters, E. The mouth-feel properties of polysaccharides and anthocyanins in a wine like medium. *Food Chem.* **2004**, *85*, 519–525. [CrossRef]

137. Guilloux-Benatier, M.; Guerreau, J.; Feuillat, M. Influence of initial colloid content on yeast macromolecule production and on the metabolism of wine microorganisms. *Am. J. Enol. Vitic.* **1995**, *46*, 486–492.

138. Domizio, P.; Romani, C.; Comitini, F.; Gobbi, M.; Lencioni, L.; Mannazu, I.; Ciani, M. Potential spoilage non-*Saccharomyces* yeasts in mixed cultures with *Saccharomyces* cerevisiae. *Ann. Microbiol.* **2011**, *61*, 137–144. [CrossRef]

139. Giovani, G.; Rosi, I.; Bertuccioli, M. Quantification and characterization of cell wall polysaccharides released by non-*Saccharomyces* yeast strains during alcoholic fermentation. *Int. J. Food Microbiol.* **2012**, *160*, 113–118. [CrossRef] [PubMed]

140. González-Royo, E.; Pascual, O.; Kontoudakis, N.; Esteruelas, M.; Esteve-Zarzoso, B.; Mas, A.; Canals, J.M.; Zamora, F. Oenological consequences of sequential inoculation with non-*Saccharomyces* yeasts (*Torulaspora delbrueckii* or *Metschnikowia pulcherrima*) and *Saccharomyces cerevisiae* in base wine for sparkling wine production. *Eur. Food Res. Technol.* **2015**, *240*, 999–1012. [CrossRef]

141. Suzuki, M.; Suh, S.O.; Sugita, T.; Nakase, T.A. phylogenetic study on galactose-containing *Candida* species based on 18S ribosomal DNA sequences. *J. Gen. Appl. Microbiol.* **1999**, *45*, 229–238. [CrossRef] [PubMed]

142. Cominiti, F.; Gobbi, M.; Domizio, P.; Romani, C.; Lencioni, L.; Mannazzu, I.; Ciani, M. Selected non-*Saccharomyces* wine yeasts in controlled multistarter fermentations with *Saccharomyces cerevisiae*. *Food Microbiol.* **2011**, *28*, 873–882. [CrossRef] [PubMed]

143. Domizio, P.; Liu, Y.; Bisson, L.F.; Barile, D. Use of non-*Saccharomyces* wine yeasts as novel sources of mannoproteins in wine. *Food Microbiol.* **2014**, *43*, 5–15. [CrossRef] [PubMed]

144. Ballou, C. Structure and biosynthesis of the mannan component of the yeast cell envelope. *Adv. Microb. Physiol.* **1976**, *14*, 93–158. [CrossRef] [PubMed]

fermentation

MDPI

Review

The Multiple and Versatile Roles of *Aureobasidium pullulans* in the Vitivinicultural Sector

Despina Bozoudi and Dimitrios Tsaltas *

Cyprus University of Technology, 3036 Lemesos, Cyprus; despoina.bozoudi@cut.ac.cy
* Correspondence: dimitris.tsaltas@cut.ac.cy; Tel.: +357-25-00-2545

Received: 27 August 2018; Accepted: 4 October 2018; Published: 9 October 2018

Abstract: The saprophytic yeast-like fungus *Aureobasidium pullulans* has been well documented for over 60 years in the microbiological literature. It is ubiquitous in distribution, being found in a variety of environments (plant surfaces, soil, water, rock surfaces and manmade surfaces), and with a worldwide distribution from cold to warm climates and wet/humid regions to arid ones. Isolates and strains of *A. pullulans* produce a wide range of natural products well documented in the international literature and which have been regarded as safe for biotechnological and environmental applications. Showing antagonistic activity against plant pathogens (especially post-harvest pathogens) is one of the major applications currently in agriculture of the fungus, with nutrient and space competition, production of volatile organic compounds, and production of hydrolytic enzymes and antimicrobial compounds (antibacterial and antifungal). The fungus also shows a positive role on mycotoxin biocontrol through various modes, with the most striking being that of binding and/or absorption. *A. pullulans* strains have been reported to produce very useful industrial enzymes, such as β-glucosidase, amylases, cellulases, lipases, proteases, xylanases and mannanases. Pullulan (poly-α-1,6-maltotriose biopolymer) is an *A. pullulans* trademark product with significant properties and biotechnological applications in the food, cosmetic and pharmaceutical industries. Poly (β-L-malic acid), or PMA, which is a natural biopolyester, and liamocins, a group of produced heavy oils and siderophores, are among other valuable compounds detected that are of possible biotechnological use. The fungus also shows a potential single-cell protein source capacity with high levels of nucleic acid components and essential amino acids, but this remains to be further explored. Last but not least, the fungus has shown very good biocontrol against aerial plant pathogens. All these properties are of major interest in the vitivinicultural sector and are thoroughly reviewed under this prism, concluding on the importance that *A. pullulans* may have if used at both vineyard and winery levels. This extensive array of properties provides excellent tools for the viticulturist/farmer as well as for the oenologist to combat problems in the field and create a high-quality wine.

Keywords: *Aureobasidium pullulans*; biotechnological applications; viticulture; enzymes; non-Saccharomyces yeasts

1. Introduction

The genus *Aureobasidium* includes members of a ubiquitous nature that are able to survive in a diverse range of habitats. *Aureobasidium pullulans* is one of the common organisms readily found in most phyllospheric habitats including grapevines, with high morphological and genetic diversity [1]. *A. pullulans* is a yeast-like fungus (Figure 1) frequently isolated from the phyllosphere and carposphere of fruits and vegetables crops [2], and is associated with the endophyte population of many plant species possessing high antagonistic activity [3]. *A. pullulans* is one of the predominant yeast species isolated from grape berries at all stages of maturity [4] and other vine tissues from both diseased and healthy vines [2]. This observed abundance led many scientists to explore its biocontrol potential for important grape diseases such as Botrytis grey mould [5], and for bunch rot caused by species of *Aspergillus* [6]. Interestingly, Dimakopoulou and coworkers [7] found that isolate of *A. pullulans* was as effective as commercial fungicides for bunch rots. *A. pullulans* may also degrade and detoxify ochratoxin A, preventing wine contamination [8]. Nowadays, *A. pullulans'* diverse habitats, environmental conditions with a repertoire of biochemical characteristics, make it a first -lass source for biotechnological uses even across boundaries. The biosafety of *A. pullulans* has been explored as well, although most studies are related to immunocompromised individuals undergoing surgical treatments, for severe injuries with open wounds or those suffering serious diseases (AIDS, pulmonary infections and chronic diseases). Reports describe the infections as serious due to their severity and difficulties in treatment, although the isolates were not exhibiting resistance. In addition, the fungus shows strong affinity to synthetic materials and surgically implanted Silastic devices [9]. Although outside the scope of this review, it is worth mentioning the potential role of pullulan in biomedical applications reviewed by Singh and coworkers [10].

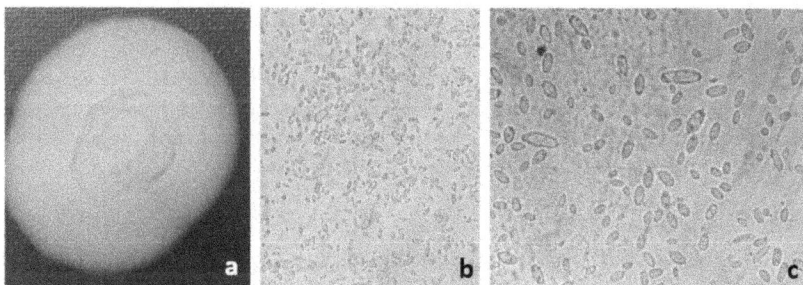

Figure 1. *Aureobasidium pullulans* (**a**) colony on Sabouraud Dextrose Agar (**b**) microscopic view of yeast-like cells with characteristic pseudomycelium, (**c**) characteristic microscopic view of yeast-like cells of various shapes and sizes.

2. Distribution and Diversity

As already mentioned, *A. pullulans* is characterized by its vast habitat presence. In the following, we will only refer to the presence of the fungus on the vine and must, although it is well practiced today that biotechnological applications could be across very isolated boundaries in order to make use of unique useful traits.

It has been recorded that soil, grape variety and grape growing practices influence the microbial ecosystem [11–13]. Microbial species present on the surface of grape berries at harvest play an important role in winemaking, and thus, counting and identifying them is of great importance. Studying several regions in the Bordeaux area, Renouf et al. [13] found that *A. pullulans*, the most widespread yeast species at the berry set, was never detected at harvest. Its number fell significantly at veraison, as it was superseded by fermentative yeasts, and was finally undetectable at harvest.

On the contrary, *A. pullulans* was found at significant high levels at studies conducted in Italy [14], Spain [15], Canada [16], Australia [4] and South Africa [17], while it often was isolated from Brazil, France, New Zealand, Greece and Slovenia, [18–22] as reviewed by Bozoudi and Tsaltas [12].

A. pullulans was also isolated from the grapes of the indigenous Cypriot varieties Xinisteri and Maratheftiko at low rates (6.29%; Bozoudi et al., unpublished data). Work from Zalar et al. [23], on *A. pullulans* diversity, describes that the fungus occurs particularly in the phyllosphere. Although *A. pullulans* is one of the most abundant microorganisms on grape berries and other vine tissues, the diversity of *Aureobasidium* spp. on vine tissues has not been explored.

Rathnayake et al. [24] reported that the diversity of *Aureobasidium* isolates from different tissue types was greater than on a regional scale. The authors reported that the vineyards treated with no fungicides were having differences in colonization, having higher genetic variation in the *Aureobasidium* isolates observed. Additionally, they show that the *Aureobasidium* population may have been established for a long period of time, and being adapted to the climatic conditions. Alternatively, the introduction of rootstocks could have co-introduced new isolates. Therefore, it is possible that the genetic variation expressed by the *Aureobasidium* isolates from different vineyards in close proximity may be the result of evolution of these isolates over time in order to cope with different environmental selection pressures [24].

A. *pullulans* is characterized by high genetic variability [25]. Morphological and cultural characteristics alone are not sufficient to assess interspecific variability and to differentiate closely related strains. Thus, RAPD–PCR and other PCR techniques were used to successfully differentiate A. *pullulans* populations and to obtain information about the genetic complexity of this microorganism [26]. Small groups of strains of A. *pullulans* were described as varieties in the literature [23]. In 2014, [1] with the publication of four species' genome sequences, we cleared up significantly the knowledge of the genus and the discrimination of the species A. *pullulans*, A. *melanogenum*, A. *subglaciale* and A. *namibiae*. By comparison of their genomic data, Gostincar and coworkers showed that the differences between these "varieties" were large enough to justify their redefinition as four separate *Aureobasidium* species. These new data help address and explain the differences between strains and "varieties", which are of course attributed to different genetic material, coding for potentially different traits from the proteins that they encode. This work redefined clearly that the opportunistic human pathogens belong only to A. *melanogenum*, and we can now have a more clear understanding of the molecular background of *Aureobasidium* spp.

3. Products

A. *pullulans* has been known since 1891, as reported in the work of Cooke [27]. This allowed the scientific community to have gathered a substantial amount of information on the lifestyle and physiology of this fungus. A wide array of products have been isolated, characterized and tested for various biological and nonbiological functions. Antimicrobials, enzymes, polysaccharides, siderophores, polyesters and heavy oils are among the most prominent and will be reported analytically below.

3.1. Antimicrobials

Bacteria and yeasts are most likely to show a mutualistic behavior to each other in order to efficiently colonize the berries' surfaces. During these interactions on the berry surfaces, they may have increased nutrient-capture capabilities, resisting environmental stresses and interacting also with other categories of microorganisms such as moulds as well as viruses. Among yeast and bacteria, some species are known to have an antagonistic effect on mould development. A. *pullulans* is known to possess antagonistic properties towards other yeasts and fungi, and it can be speculated that it may influence the overall grape ecology [4].

It has been reported that A. *pullulans* exhibits reduction of *Botrytis cinerea* growth on the surface of table grape berries [13]. The proposed mechanisms explaining this antifungal activity were exclusion by bacteria and yeasts of fungal adhesion sites [28], competition for nutrients [29] and production of antagonistic metabolites or lytic enzymes. It is described that A. *pullulans* secretes chitinase and glucanase enzymes able to hydrolyse moulds [30]. In addition, an A. *pullulans* strain was found to produce antimicrobial compounds that were inhibitory towards the Gram-negative *Pseudomonas fluorescens* and Gram-positive *Staphylococcus aureus* bacteria [31]. The antibacterial activity of A. *pullulans* strains was attributed to 2-propylacrylic acid, 8,9-dihydroxy-2-methyl-4H,5H-pyrano [3,2-c]-chromon-4-one, 2-methylenesuccinic acid and hexane-1,2,3,5,6-hexol [32]. More work on antifungal properties was performed and a group of antifungals was named as aureobasidins. Aureobasidins are derivatives of cyclic deosipeptides (molar mass ranging 1070–1148 Da). Depending on their structures, aureobasidins are designated with the letters A to R [33]. Aureobasidin A seems to be reported in most cases, while work on factors affecting their production and activity has shown that glucose increases antifungal activity, and that the culture medium's amino acid composition has a variable role in some cases [34].

3.2. Enzymes

As reported above, enzymes may play a role as antimicrobials, but enzyme production and enzymatic activity have important roles in various biotechnological applications. *A. pullulans* is reported to produce amylases [35], cellulases [36], lipases [37], xylanases [38,39], proteases [40–42], laccase [43] and mannanases [44]. Currently in the wine industry, pectinases, glucanases, xylanases and proteases are used to improve the clarification and processing of wine. In addition, glycosidase is used for the release of varietal aromas from precursor compounds, urease for the reduction of ethyl carbamate formation, and glucose oxidase for the reduction of alcohol levels [45].

β-glucosidase has been also detected in *A. pullulans* [36,46,47] as well as glucose oxidase [48]. Urease activity has not been reported to the best of our knowledge. The work of Baffi et al. [49] is characteristic of the potential of non-Saccharomyces yeasts and their role in wine aroma, since they observed a notably increased amount of monoterpenes. Secreting cold-active pectinolytic activity has been also documented [50,51] and has good potential in winemaking as well.

Lastly, of interest is the indirect role of enzymes in the possible microbial relations on the grape berries, because intact grape berry surfaces are likely to be poor in carbon. *A. pullulans* may well be a slowly rotting machine orchestrating the degradation of epidermal cells via pectolytic or cellulolytic activities, necessary to degrade pectin and cellulose, the most important plant cell constituents. *A. pullulans* produces extracellular pectolytic enzymes while growing on medium containing pectin as sole carbon source [52]. Also, pectinases are inducible in carbon starvation conditions according to Biely et al., [53] and pectinolytic activity of *A. pullulans* is maximum when pectin is the sole carbon source.

3.3. Pullulan and Other Polysaccharides

A. pullulans produces an extracellular and unbranched homopolysaccharide: the pullulan, which consists of α-(1→6) linkages of α-(1→4)-linked maltotriose units [54]. This flexible and sticky polymer can form an oxygen-impermeable film, a property which is especially interesting to the understanding of the presence of several anaerobic bacteria on the berry. Moreover, the pullulan envelope may facilitate the adhesion of the bacterial cells to the berry surface. In order to preserve cell populations, the microorganisms need nutritive sources. The biofilm may act as a nutrient trap [55].

Pullulan production during initial fermentation stages by the fungus may help with must stabilization and improving mouth feel of wine due to the molecule's rheological properties in both aqueous and/or ethanolic media. Polysaccharides may also improve aroma and flavor delivery and perception also due to their physicochemical reactions with the aromatic compounds. In addition, polysaccharides (and pullulan) can retain better the colour and the antioxidant capacity of red wine.

Recently, other properties of pullulan have been explored, such as applications in medical sciences, particularly drug delivery, as well as the interaction of the molecule with various types of cells (liver, cancer cells) [56,57]. Such properties could be very interesting in the enhancement of the antioxidant role of red wine, as well as in the investigation of the beneficial role of wine in human health in general.

Other interesting extracellularly produced polysaccharides by *A. pullulans* include soluble β-glucan, consisting of a β-(1,3)-linked glucose main chain, and β-(1,6)-linked glucose branches. β-glucan exhibits immune stimulatory activity, and is consumed as a supplement in many countries. Also, *A. pullulans* culture supernatant is believed to exhibit beneficial effects in delaying the onset of a number of diseases, and has been reported to exhibit antitumor, antiallergy and anti-infectious disease activities in mouse models [58–68]. An interesting review by Li et al. [69] on *Aureobasidium* spp. and biosynthesis and regulation of their extracellular polymers should be read by anyone interested in the field.

Lastly, there is interest in wine waste as a substrate for *A. pullulans* growth and pullulan production [70–72]. Grape skin pulp is considered as one of the best substrates for pullulan production, especially hot water extracts of the pulp. The product is of higher molecular weight and rather pure.

3.4. PMA

Poly (β-L-malic acid), or PMA, is a natural biopolyester produced by many microorganisms including *A. pullulans*. The interest in this molecule derives from its properties being biodegradable, water soluble and biocompatible, and its uses in the pharmaceutical industry [73–76]. No applications in the wine industry have been reported, but possible relationships can be explored via wine waste as substrate for PMA production and PMA as a coating for grape postharvest protection.

3.5. Liamocins

Back in 1994, Kurosawa et al. [77] discovered the production of heavy oils in the culture medium of *Aureobasidium* sp. In 2013, Price et al. [78] named them liamocins, and further clarified their molecular structure consisting of a single mannitol headgroup that is partially O-acetylated with 3,5-dihydroxy-decanoic ester groups. Liamocins showed immediately their interesting biological activities as antimicrobials [79,80] and anticancer agents [81,82].

Liamocins' role in plant disease control (grapes) and their role in wine in a technological aspect, as well as towards human health, remain to be explored.

3.6. Siderophores

Siderophores are low-molecular-weight high-affinity iron-chelating molecules that are produced by many microbes (fungi and bacteria) living under iron-depleted environments. The molecules help to sequester and solubilize the iron (Fe^{3+} and Fe^{2+} ions). Siderophores are very interesting molecules for medical, agricultural and environmental applications, and have been of interest in biotechnology. Although only one strain of *A. pullulans* (HN6.2) has been reported in the literature to produce siderophores, studies of *A. pullulans* isolates report the effect and role of siderophores in biocontrol processes of plant and human pathogens [1,44,65,83–88].

4. Single-Cell Protein

With the world population reaching 9 billion by 2050, there is strong evidence that agriculture will not be able to meet the demand for food, and particularly protein, and as a result, food security is under serious threat. Common agriculture has serious drawbacks such as high water footprint, high land use, biodiversity loss, soil erosion, and contribution to climate change of a third of all greenhouse gases. For these reasons, food out of microbes is considered a sustainable way to proceed.

The biomass and protein extracted from cultures of fungi, bacteria and algae may be used as an ingredient or a substitute for protein-rich foods. Single-cell proteins (SCPs) refer to edible unicellular or multicellular microorganisms. The products are of high value, suitable for animal and human consumption, and efforts to grow SCP on agricultural and food waste as well as autotrophically are quite successful.

A. pullulans has not been explored for its use as a source of SCP, but fungi proteins have more advantages than those obtained from bacteria and algae [89]. Work from Chi et al. [90] showed that *A. pullulans* isolates had high levels of nucleic acid components and essential amino acids. Such properties could be helpful in wine fermentation if these components can be used for feeding the alcohol- and aroma-producing yeasts. Alternatively, dried, lysed *A. pullulans* cells can be a source of feed for wine fermenting yeasts [91,92].

5. Biocontrol Agent

A. pollulans' biocontrol capabilities have been explored for many years for diseases in both the field and post-harvest. In addition, diseases of the phylloplane and the carposphere, as well as diseases of the internal tissues, have been combated less or more successfully. The antagonistic feature of fungi may be attributed to competition for nutrients and space, parasitism on the fungal pathogens, secretion of antifungal compounds, attachment and biofilm formation, production of volatile organic compounds, as well as the induction of host plant resistance [93].

As a fast-growing yeast-like fungus, *A. pullulans* competes for nutrients as well as space. Extracellular polysaccharides, enzymes as well as other secreted molecules (liamocins, aureobasidins etc.) require significant amounts of carbon and nitrogen sources, as well as other micronutrients that are soon depleted from the environment and their competitors. In addition, pullulan and/or other high-MW molecules take space while at the same time creating a less favorable or even hostile environment for plant pathogens.

5.1. Competition for Nutrients

In 2006, Bencheqroum et al. [94] presented their first data that application of high amounts of exogenous amino acids, vitamins or sugars on apple wounds significantly reduced the protective level of *A. pullulans*, and in 2007, [95] the authors confirmed with in-vitro and in-situ evidence that competition for apple nutrients, most particularly amino acids, may be a main mechanism of the biocontrol activity of *A. pullulans*.

5.2. Competition for Space

Competition for space is amongst the most common but efficient ways in which biocontrol agents operate. Speedy growth helps a microorganism to dominate the space over slow growers. In addition, certain microbes occupy extra space with copious amounts of secreted polysaccharides that have both direct (occupying space) and indirect (attachment inhibitors, growth inhibitors etc.) roles on the growth of competitors. Schena and collaborators [26,96] have tested various isolates from different sources (epiphytic and endophytic) of *A. pullulans*, and showed good results in biocontrol of various postharvest diseases of fruits and vegetables.

5.3. Production of Volatile Organic Compounds

Volatile organic compounds (VOCs) could play an essential role in the antagonistic activity of *A. pullulans* against postharvest pathogens. Mari et al. [97] suggested first that *A. pullulans* L1 and L8 strains could be considered as good candidates for the development of biofungicides. Compounds such as 2-phenyl, 1-butanol-3-methyl, 1-butanol-2-methyl and 1-propanol-2-methyl belonging to the group of alcohols are mainly produced from *A. pullulans* within 3–4 days of growth. 2-Phenethyl alcohol was determined as the most active, with EC_{50} values lower than 0.8 µL ml^{-1}, responsible for reduction of vegetative growth and sporulation, and also reducing ochratoxin A (OTA) production and OTAbiosynthetic gene expression [98]. Similar results have been recently confirmed for other yeasts as well, [99] so it is worthwhile to revisit the *A. pullulan* isolates' VOCs capacity.

5.4. Production of Hydrolytic Enzymes

Hydrolytic enzymes were always considered first in biological control modes of action against pathogens. Chitinase and glucanase are amongst the most prominent enzymes having a role in biological agents' biocontrol activity. In addition, killer toxins have been attributed to have a role in fungal–fungal interactions [100,101].

5.5. OTA Biodegradation, Detoxification and Absorption

Mycotoxins could be decomposed, transformed or absorbed by microorganisms [102]. Their microbial degradation or transformation with specific attention to the actual detoxification is an important feature of various microorganisms [103]. De Felice and coworkers [104] showed that *A. pullulans* can transform OTA to OTAα on berries.

Yarrowia lipolytica Y-2 has the capacity to biodegrade OTA to OTAα through the hydrolytic activity of carboxypeptidases [105]. The same authors also support that, in addition, many proteins of *Y. lipolytica* Y-2 involved in stress response and reactive O_2 species elimination also play a role in OTA degradation. In the case of *A. pullulans*, carboxylpeptidases should be specifically explored for a similar role, though there are some toxicity issues regarding the use of enzymes to degrade OTA in wine because of their undesirable effects on must fermenting microbes [106]. More information can be found in an excellent review by Zang et al. [107]

6. Aromatic Properties

Microorganisms of enological interest have been grouped into three main classes: (a) easily controllable species without the ability to spoil wine when good manufacturing practices are applied, (b) fermenting species responsible for sugar and malic acid conversion, and (c) spoilage species [2]. As previously mentioned, *A. pullulans* holds a dominant position in most grapevine terroirs studied, and is classified in the first group. In grape microbiome reported work, *A. pullulans* emits typical, well-known flavour components of red wine (i.e., 2-methylbutanoic acid, 3-methyl-1-butanol and ethyl octanoate) [108]. It is not yet reported whether endophytic microorganisms have a role on grape aromatic compounds, but grapevine endophyte studies have progressed, and are very likely to identify such interplay in the near future [109,110]. Also, as referred to earlier (Section 3.2), β-glucosidase and pectinases have been involved in aroma production [49–51,111,112].

7. Conclusions and Future Perspectives

Aureobasidium pullulans' cosmopolitan presence has been well documented in the past 100 years. The list of properties (Table 1) of this yeast-like fungus is still growing. Here, we have presented what has been documented in relation to the microorganism and fruits, other microorganisms, as well as wine fermentation. *A. pullulans* has a vast potential in biotechnological uses, and in particular, in the vitivinicultural sector. In addition, new exotic isolates from extreme environments are likely to enhance significantly the repertoire of properties. Enzymes and metabolites of these isolates are very likely to help us resolve many technological problems requiring extreme solutions. In our conclusion, two major research directions are currently suggested, and these are: (1) reexamining all isolates in laboratory collections with the current knowledge of properties and molecular analysis tools (DNA/RNA level and proteins, including phylogenetics), and (2) exploring all known products of *A. pullulans* for novel uses and functions in the vitivinicultural sector, as described earlier.

Table 1. *Aureobasidium pullulans* main properties.

Main Property	Specific Property	Strain #	Reference
Antimicrobials		FRR4800, WH9	[28–30]
	Antibacterial (Aureobasidins)	NRRL 58561, NRRL 58562, NRRL 58563, NRRL 58514, NRRL 58536, NRRL 58516, NRRL 58517, NRRL 58520	[31–34]
Enzymes	Lytic enzymes		
	Amylases	Cau19	[113]
	Cellulases	ER-16	[36,114]
	Lipases	HN2-3	[37,115]
	Xylanases	ATCC20524	[38,39]
	Proteases	HN2-3, 10, PLS	[40–42]
	Laccase	NRRL50381	[43,116]
	Mannanase		[44,117,118]
	β-glucosidase	NRRL Y-12974, Ap-beta-gl	[36,46,49]
	Pectinolytic	GM-R-22, LV-10	[50–53,111]
Pullulan		CGMCC1234, P56, CH1, ATCC 201253, HP2001	[54,56,119–122]
β-glucan		SM2001	[58,59,61,63,64,66–68,123–125]
PMA		CCTCCM2012223	[73–76]
Liamocins		NRRL 50380	[78,126]
Siderophores		HN6.2, Y-1	[83–87]
Single-Cell Protein		G7b, 4#2	[90]
Biocontrol		SL250, SL236, L47, Ach1-1, 533, 547, L1, L8, ACBL77, LS30, AU34-2	[3,26,94–97,100,101,104,127,128]
Aromatic Compounds		T4B1c.17-P	[108]

Author Contributions: Conceptualization, D.T.; Writing-Original Draft Preparation, D.B.; Writing-Review & Editing, D.T.; Funding Acquisition, D.T.

Funding: This work received no external funding.

References

1. Gostincar, C.; Ohm, R.A.; Kogej, T.; Sonjak, S.; Turk, M.; Zajc, J.; Zalar, P.; Grube, M.; Sun, H.; Han, J.; et al. Genome Sequencing of Four *Aureobasidium pullulans* Varieties: Biotechnological Potential, Stress Tolerance, and Description of New Species. *BMC Genom.* **2014**, *15*, 549. [CrossRef] [PubMed]
2. Barata, A.; Malfeito-Ferreira, M.; Loureiro, V. The Microbial Ecology of Wine Grape Berries. *Int. J. Food Microbiol.* **2012**, *153*, 243–259. [CrossRef] [PubMed]
3. De Curtis, F.; De Felice, D.V.; Ianiri, G.; De Cicco, V.; Castoria, R. Environmental Factors Affect the Activity of Biocontrol Agents against Ochratoxigenic Aspergillus Carbonarius on Wine Grape. *Int. J. Food Microbiol.* **2012**, *159*, 17–24. [CrossRef] [PubMed]
4. Prakitchaiwattana, C.J.; Fleet, G.H.; Heard, G.M. Application and Evaluation of Denaturing Gradient Gel Electrophoresis to Analyse the Yeast Ecology of Wine Grapes. *FEMS Yeast Res.* **2004**, *4*, 865–877. [CrossRef] [PubMed]
5. Parafati, L.; Vitale, A.; Restuccia, C.; Cirvilleri, G. Biocontrol Ability and Action Mechanism of Food-Isolated Yeast Strains against Botrytis Cinerea Causing Post-Harvest Bunch Rot of Table Grape. *Food Microbiol.* **2015**, *47*, 85–92. [CrossRef] [PubMed]
6. Pantelides, I.S.; Christou, O.; Tsolakidou, M.-D.; Tsaltas, D.; Ioannou, N. Isolation, Identification and In Vitro Screening of Grapevine Yeasts for the Control of Black Aspergilli on Grapes. *Biol. Control* **2015**, *88*, 46–53. [CrossRef]
7. Dimakopoulou, M.; Tjamos, S.E.; Antoniou, P.P.; Pietri, A.; Battilani, P.; Avramidis, N.; Markakis, E.A.; Tjamos, E.C. Phyllosphere Grapevine Yeast *Aureobasidium pullulans* Reduces Aspergillus Carbonarius (Sour Rot) Incidence in Wine-Producing Vineyards in Greece. *Biol. Control* **2008**, *46*, 158–165. [CrossRef]

8. De Curtis, F.; De Cicco, V.; Lima, G. Efficacy of Biocontrol Yeasts Combined with Calcium Silicate or Sulphur for Controlling Durum Wheat Powdery Mildew and Increasing Grain Yield Components. *Field Crops Res.* **2012**, *134*, 36–46. [CrossRef]

9. Hawkes, M.; Rennie, R.; Sand, C.; Vaudry, W. *Aureobasidium pullulans* Infection: Fungemia in an Infant and a Review of Human Cases. *Diagn. Microbiol. Infect. Dis.* **2005**, *51*, 209–213. [CrossRef] [PubMed]

10. Singh, R.S.; Kaur, N.; Rana, V.; Kennedy, J.F. Pullulan: A Novel Molecule for Biomedical Applications. *Carbohydr. Polym.* **2017**, *171*, 102–121. [CrossRef] [PubMed]

11. Bokulich, N.A.; Thorngate, J.H.; Richardson, P.M.; Mills, D.A. Microbial Biogeography of Wine Grapes Is Conditioned by Cultivar, Vintage, and Climate. *Proc. Natl. Acad. Sci. USA* **2014**, *111*, E139–E148. [CrossRef] [PubMed]

12. Bozoudi, D.; Tsaltas, D. Grape Microbiome: Potential and Opportunities as a Source of Starter Cultures. In *Grape and Wine Biotechnology*; InTech: Rijeka, Croatia, 2016.

13. Renouf, V.; Claisse, O.; Lonvaud-Funel, A. Understanding the Microbial Ecosystem on the Grape Berry Surface through Numeration and Identification of Yeast and Bacteria. *Aust. J. Grape Wine Res.* **2005**, *11*, 316–327. [CrossRef]

14. Francesca, N.; Chiurazzi, M.; Romano, R.; Aponte, M.; Settanni, L.; Moschetti, G. Indigenous Yeast Communities in the Environment of "Rovello Bianco" Grape Variety and Their Use in Commercial White Wine Fermentation. *World J. Microbiol. Biotechnol.* **2010**, *26*, 337–351. [CrossRef]

15. Clavijo, A.; Calderón, I.L.; Paneque, P. Diversity of Saccharomyces and Non-Saccharomyces Yeasts in Three Red Grape Varieties Cultured in the Serranía de Ronda (Spain) Vine-Growing Region. *Int. J. Food Microbiol.* **2010**, *143*, 241–245. [CrossRef] [PubMed]

16. Subden, R.E.; Husnik, J.I.; Van Twest, R.; Van Der Merwe, G.; Van Vuuren, H.J.J. Autochthonous Microbial Population in a Niagara Peninsula Icewine Must. *Food Res. Int.* **2003**, *36*, 747–751. [CrossRef]

17. Setati, M.E.; Jacobson, D.; Bauer, F.F. Sequence-Based Analysis of the *Vitis vinifera* L. Cv Cabernet Sauvignon Grape Must Mycobiome in Three South African Vineyards Employing Distinct Agronomic Systems. *Front. Microbiol.* **2015**, *6*, 1358. [CrossRef] [PubMed]

18. Baffi, M.A.; dos Santos Bezerra, C.; Arévalo-Villena, M.; Briones-Pérez, A.I.; Gomes, E.; Da Silva, R. Isolation and Molecular Identification of Wine Yeasts from a Brazilian Vineyard. *Ann. Microbiol.* **2011**, *61*, 75–78. [CrossRef]

19. David, V.; Terrat, S.; Herzine, K.; Claisse, O.; Rousseaux, S.; Tourdot-Maréchal, R.; Masneuf-Pomarede, I.; Ranjard, L.; Alexandre, H. High-Throughput Sequencing of Amplicons for Monitoring Yeast Biodiversity in Must and during Alcoholic Fermentation. *J. Ind. Microbiol. Biotechnol.* **2014**, *41*, 811–821. [CrossRef] [PubMed]

20. Gayevskiy, V.; Goddard, M.R. Geographic Delineations of Yeast Communities and Populations Associated with Vines and Wines in New Zealand. *ISME J.* **2012**, *6*, 1281. [CrossRef] [PubMed]

21. Nisiotou, A.A.; Nychas, G.-J.E. Yeast Populations Residing on Healthy or Botrytis-Infected Grapes from a Vineyard in Attica, Greece. *Appl. Environ. Microbiol.* **2007**, *73*, 2765–2768. [CrossRef] [PubMed]

22. Raspor, P.; Milek, D.M.; Polanc, J.; Možina, S.S.; Čadež, N. Yeasts Isolated from Three Varieties of Grapes Cultivated in Different Locations of the Dolenjska Vine-Growing Region, Slovenia. *Int. J. Food Microbiol.* **2006**, *109*, 97–102. [CrossRef] [PubMed]

23. Zalar, P.; Gostincar, C.; de Hoog, G.S.; Ursic, V.; Sudhadham, M.; Gunde-Cimerman, N. Redefinition of *Aureobasidium pullulans* and Its Varieties. *Stud. Mycol.* **2008**, *61*, 21–38. [CrossRef] [PubMed]

24. Rathnayake, R.; Savocchia, S.; Schmidtke, L.M.; Steel, C.C. Characterisation of *Aureobasidium pullulans* Isolates from *Vitis vinifera* and Potential Biocontrol Activity for the Management of Bitter Rot of Grapes. *Eur. J. Plant Pathol.* **2018**, *151*, 593–611. [CrossRef]

25. Yurlova, N.A.; Mokrousov, I.V.; de Hoog, G.S. Intraspecific Variability and Exopolysaccharide Production in *Aureobasidium pullulans*. *Anton. Leeuwenhoek* **1995**, *68*, 57–63. [CrossRef]

26. Schena, L.; Ippolito, A.; Zahavi, T.; Cohen, L.; Nigro, F.; Droby, S. Genetic Diversity and Biocontrol Activity of *Aureobasidium pullulans* Isolates against Postharvest Rots. *Postharvest Biol. Technol.* **1999**, *17*, 189–199. [CrossRef]

27. Cooke, W.B. An Ecological Life History of *Aureobasidium pullulans* (De Bary) Arnaud. *Mycopathol. Mycol. Appl.* **1959**, *12*, 1–45. [CrossRef] [PubMed]

28. Benbow, J.M.; Sugar, D. Fruit Surface Colonization and Biological Control of Postharvest Diseases of Pear by Preharvest Yeast Applications. *Plant Dis.* **1999**, *83*, 839–844. [CrossRef]

29. Vero, S.; Mondino, P.; Burgueno, J.; Soubes, M.; Wisniewski, M. Characterization of Biocontrol Activity of Two Yeast Strains from Uruguay against Blue Mold of Apple. *Postharvest Biol. Technol.* **2002**, *26*, 91–98. [CrossRef]

30. Ippolito, A.; El Ghaouth, A.; Wilson, C.L.; Wisniewski, M. Control of Postharvest Decay of Apple Fruit by *Aureobasidium pullulans* and Induction of Defense Responses. *Postharvest Biol. Technol.* **2000**, *19*, 265–272. [CrossRef]

31. McCormack, P.J.; Wildman, H.G.; Jeffries, P. Production of Antibacterial Compounds by Phylloplane-Inhabiting Yeasts and Yeastlike Fungi. *Appl. Environ. Microbiol.* **1994**, *60*, 927–931. [PubMed]

32. Zain, M.E.; Awaad, A.S.; Razak, A.A.; Maitland, D.J.; Khamis, N.E.; Sakhawy, M.A. Secondary Metabolites of *Aureobasidium pullulans* Isolated from Egyptian Soil and Their Biological Activity. *J. Appl. Sci. Res.* **2009**, *5*, 1582–1591.

33. Takesako, K.; Ikai, K.; Haruna, F.; Endo, M.; Shimanaka, K.; Sono, E.; Nakamura, T.; Kato, I.; Yamaguchi, H. Aureobasidins, New Antifungal Antibiotics. Taxonomy, Fermentation, Isolation, and Properties. *J. Antibiot.* **1991**, *44*, 919–924. [CrossRef] [PubMed]

34. Prasongsuk, S.; Ployngam, S.; Wacharasindhu, S.; Lotrakul, P.; Punnapayak, H. Effects of Sugar and Amino Acid Supplementation on *Aureobasidium pullulans* NRRL 58536 Antifungal Activity against Four Aspergillus Species. *Appl. Microbiol. Biotechnol.* **2013**, *97*, 7821–7830. [CrossRef] [PubMed]

35. Manitchotpisit, P.; Skory, C.D.; Leathers, T.D.; Lotrakul, P.; Eveleigh, D.E.; Prasongsuk, S.; Punnapayak, H. Alpha-Amylase Activity during Pullulan Production and Alpha-Amylase Gene Analyses of *Aureobasidium pullulans*. *J. Ind. Microbiol. Biotechnol.* **2011**, *38*, 1211–1218. [CrossRef] [PubMed]

36. Leite, R.S.; Bocchini, D.A.; Martins Eda, S.; Silva, D.; Gomes, E.; Da Silva, R. Production of Cellulolytic and Hemicellulolytic Enzymes from Aureobasidium Pulluans on Solid State Fermentation. *Appl. Biochem. Biotechnol.* **2007**, *137–140*, 281–288.

37. Leathers, T.D.; Rich, J.O.; Anderson, A.M.; Manitchotpisit, P. Lipase Production by Diverse Phylogenetic Clades of *Aureobasidium pullulans*. *Biotechnol. Lett.* **2013**, *35*, 1701–1706. [CrossRef] [PubMed]

38. Manitchotpisit, P.; Leathers, T.D.; Peterson, S.W.; Kurtzman, C.P.; Li, X.L.; Eveleigh, D.E.; Lotrakul, P.; Prasongsuk, S.; Dunlap, C.A.; Vermillion, K.E.; et al. Multilocus Phylogenetic Analyses, Pullulan Production and Xylanase Activity of Tropical Isolates of *Aureobasidium pullulans*. *Mycol. Res.* **2009**, *113 Pt 10*, 1107–1120. [CrossRef] [PubMed]

39. Ohta, K.; Fujimoto, H.; Fujii, S.; Wakiyama, M. Cell-Associated Beta-Xylosidase from *Aureobasidium pullulans* ATCC 20524: Purification, Properties, and Characterization of the Encoding Gene. *J. Biosci. Bioeng.* **2010**, *110*, 152–157. [CrossRef] [PubMed]

40. Chi, Z.; Ma, C.; Wang, P.; Li, H.F. Optimization of Medium and Cultivation Conditions for Alkaline Protease Production by the Marine Yeast *Aureobasidium pullulans*. *Bioresour. Technol.* **2007**, *98*, 534–538. [CrossRef] [PubMed]

41. Ma, C.; Ni, X.; Chi, Z.; Ma, L.; Gao, L. Purification and Characterization of an Alkaline Protease from the Marine Yeast *Aureobasidium pullulans* for Bioactive Peptide Production from Different Sources. *Mar. Biotechnol.* **2007**, *9*, 343–351. [CrossRef] [PubMed]

42. Ni, X.; Chi, Z.; Ma, C.; Madzak, C. Cloning, Characterization, and Expression of the Gene Encoding Alkaline Protease in the Marine Yeast *Aureobasidium pullulans* 10. *Mar. Biotechnol.* **2008**, *10*, 319–327. [CrossRef] [PubMed]

43. Rich, J.O.; Leathers, T.D.; Anderson, A.M.; Bischoff, K.M.; Manitchotpisit, P. Laccases from *Aureobasidium pullulans*. *Enzym. Microb. Technol.* **2013**, *53*, 33–37. [CrossRef] [PubMed]

44. Chi, Z.; Wang, F.; Chi, Z.; Yue, L.; Liu, G.; Zhang, T. Bioproducts from *Aureobasidium pullulans*, a Biotechnologically Important Yeast. *Appl. Microbiol. Biotechnol.* **2009**, *82*, 793–804. [CrossRef] [PubMed]

45. Van Oort, M.; Canal-Llaubères, R.-M.; Law, B.A. Enzymes in Wine Production. In *Enzymes in Food Technology*; Sheffield Academic Press: Sheffield, UK, 2002; pp. 76–90.

46. Iembo, T.; Da-Silva, R.; Pagnocca, F.C.; Gomes, E. Production, Characterization and Properties of Beta-Glucosidase and Beta-Xylosidase from a Strain of *Aureobasidium* Sp. *Prikl. Biokhim. Mikrobiol.* **2002**, *38*, 639–643. [PubMed]

47. Leite, R.S.R.; Alves-Prado, H.F.; Cabral, H.; Pagnocca, F.C.; Gomes, E.; Da-Silva, R. Production and Characteristics Comparison of Crude β-Glucosidases Produced by Microorganisms Thermoascus Aurantiacus e *Aureobasidium pullulans* in Agricultural Wastes. *Enzym. Microb. Technol.* **2008**, *43*, 391–395. [CrossRef]
48. Smotrova, N.G.; Kremenchutskii, G.N. Isolation from the Environment of Strains of Microorganisms with Glucose Oxidase Activity. *Mikrobiol. Zh.* **2002**, *64*, 28–34.
49. Baffi, M.A.; Tobal, T.; Lago, J.H.; Boscolo, M.; Gomes, E.; Da-Silva, R. Wine Aroma Improvement Using a Beta-Glucosidase Preparation from *Aureobasidium pullulans*. *Appl. Biochem. Biotechnol.* **2013**, *169*, 493–501. [CrossRef] [PubMed]
50. Merin, M.G.; Mendoza, L.M.; Farias, M.E.; Morata de Ambrosini, V.I. Isolation and Selection of Yeasts from Wine Grape Ecosystem Secreting Cold-Active Pectinolytic Activity. *Int. J. Food Microbiol.* **2011**, *147*, 144–148. [CrossRef] [PubMed]
51. Merin, M.G.; Morata de Ambrosini, V.I. Kinetic and Metabolic Behaviour of the Pectinolytic Strain *Aureobasidium pullulans* GM-R-22 during Pre-Fermentative Cold Maceration and Its Effect on Red Wine Quality. *Int. J. Food Microbiol.* **2018**, *285*, 18–26. [CrossRef] [PubMed]
52. Manachini, P.L.; Parini, C.; Fortina, M.G. Pectic Enzymes from *Aureobasidium pullulans* LV 10. *Enzym. Microb. Technol.* **1988**, *10*, 682–685. [CrossRef]
53. Biely, P.; Heinrichová, K.; Kru, M. Induction and Inducers of the Pectolytic System in *Aureobasidium pullulans*. *Curr. Microbiol.* **1996**, *33*, 6–10. [CrossRef]
54. Seo, H.P.; Son, C.W.; Chung, C.H.; Jung, D.I.; Kim, S.K.; Gross, R.A.; Kaplan, D.L.; Lee, J.W. Production of High Molecular Weight Pullulan by *Aureobasidium pullulans* HP-2001 with Soybean Pomace as a Nitrogen Source. *Bioresour. Technol.* **2004**, *95*, 293–299. [CrossRef] [PubMed]
55. Filonow, A.B.; Vishniac, H.S.; Anderson, J.A.; Janisiewicz, W.J. Biological Control of Botrytis Cinerea in Apple by Yeasts from Various Habitats and Their Putative Mechanisms of Antagonism. *Biol. Control* **1996**, *7*, 212–220. [CrossRef]
56. Fundueanu, G.; Constantin, M.; Ascenzi, P. Preparation and Characterization of PH-and Temperature-Sensitive Pullulan Microspheres for Controlled Release of Drugs. *Biomaterials* **2008**, *29*, 2767–2775. [CrossRef] [PubMed]
57. Prajapati, V.D.; Jani, G.K.; Khanda, S.M. Pullulan: An Exopolysaccharide and Its Various Applications. *Carbohydr. Polym.* **2013**, *95*, 540–549. [CrossRef] [PubMed]
58. Ikewaki, N.; Fujii, N.; Onaka, T.; Ikewaki, S.; Inoko, H. Immunological Actions of Sophy Beta-Glucan (Beta-1,3-1,6 Glucan), Currently Available Commercially as a Health Food Supplement. *Microbiol. Immunol.* **2007**, *51*, 861–873. [CrossRef] [PubMed]
59. Jung, M.Y.; Kim, J.W.; Kim, K.Y.; Choi, S.H.; Ku, S.K. Polycan, a Beta-Glucan from *Aureobasidium pullulans* SM-2001, Mitigates Ovariectomy-Induced Osteoporosis in Rats. *Exp. Ther. Med.* **2016**, *12*, 1251–1262. [CrossRef] [PubMed]
60. Aoki, S.; Iwai, A.; Kawata, K.; Muramatsu, D.; Uchiyama, H.; Okabe, M.; Ikesue, M.; Maeda, N.; Uede, T. Oral Administration of the *Aureobasidium pullulans*-Derived Beta-Glucan Effectively Prevents the Development of High Fat Diet-Induced Fatty Liver in Mice. *Sci. Rep.* **2015**, *5*, 10457. [CrossRef] [PubMed]
61. Kataoka-Shirasugi, N.; Ikuta, J.; Kuroshima, A.; Misaki, A. Antitumor Activities and Immunochemical Properties of the Cell-Wall Polysaccharides from *Aureobasidium pullulans*. *Biosci. Biotechnol. Biochem.* **1994**, *58*, 2145–2151. [CrossRef] [PubMed]
62. Kawata, K.; Iwai, A.; Muramatsu, D.; Aoki, S.; Uchiyama, H.; Okabe, M.; Hayakawa, S.; Takaoka, A.; Miyazaki, T. Stimulation of Macrophages with the Beta-Glucan Produced by *Aureobasidium pullulans* Promotes the Secretion of Tumor Necrosis Factor-Related Apoptosis Inducing Ligand (TRAIL). *PLoS One* **2015**, *10*, e0124809. [CrossRef] [PubMed]
63. Ku, S.K.; Kim, J.W.; Cho, H.R.; Kim, K.Y.; Min, Y.H.; Park, J.H.; Kim, J.S.; Park, J.H.; Seo, B.I.; Roh, S.S. Effect of Beta-Glucan Originated from *Aureobasidium pullulans* on Asthma Induced by Ovalbumin in Mouse. *Arch. Pharm. Res.* **2012**, *35*, 1073–1081. [CrossRef] [PubMed]
64. Muramatsu, D.; Iwai, A.; Aoki, S.; Uchiyama, H.; Kawata, K.; Nakayama, Y.; Nikawa, Y.; Kusano, K.; Okabe, M.; Miyazaki, T. Beta-Glucan Derived from *Aureobasidium pullulans* Is Effective for the Prevention of Influenza in Mice. *PLoS ONE* **2012**, *7*, e41399. [CrossRef] [PubMed]

65. Prasongsuk, S.; Lotrakul, P.; Ali, I.; Bankeeree, W.; Punnapayak, H. The Current Status of *Aureobasidium pullulans* in Biotechnology. *Folia Microbiol.* **2018**, *63*, 129–140. [CrossRef] [PubMed]

66. Tanaka, K.; Tanaka, Y.; Suzuki, T.; Mizushima, T. Protective Effect of Beta-(1,3 → 1,6)-D-Glucan against Irritant-Induced Gastric Lesions. *Br. J. Nutr.* **2011**, *106*, 475–485. [CrossRef] [PubMed]

67. Zhang, W.; Yu, X.; Kwak, M.; Xu, L.; Zhang, L.; Yu, Q.; Jin, J.O. Maturation of Dendritic Cells by Pullulan Promotes Anti-Cancer Effect. *Oncotarget* **2016**, *7*, 44644–44659. [CrossRef] [PubMed]

68. Lim, J.-M.; Do, E.; Park, D.-C.; Jung, G.-W.; Cho, H.-R.; Lee, S.-Y.; Shin, J.W.; Baek, K.M.; Choi, J.-S. Ingestion of Exopolymers from *Aureobasidium pullulans* Reduces the Duration of Cold and Flu Symptoms: A Randomized, Placebo-Controlled Intervention Study. *Evid.-Based Complement. Altern. Med.* **2018**, *2018*. [CrossRef] [PubMed]

69. Li, Y.; Chi, Z.; Wang, G.Y.; Wang, Z.P.; Liu, G.L.; Lee, C.F.; Ma, Z.C.; Chi, Z.M. Taxonomy of *Aureobasidium* Spp. and Biosynthesis and Regulation of Their Extracellular Polymers. *Crit. Rev. Microbiol.* **2015**, *41*, 228–237. [CrossRef] [PubMed]

70. Arapoglou, D.; Israilides, C.J.; Bocari, M.; Scanlon, B.; Smith, A.; Thessaly, U.; Thessaloniki, A.U. A Novel Approach to Grape Waste Treatment. In *Protection and Restoration of the Environment VI: Proceedings of the International Conference*; Center for Environmental Systems at Stevens Institute of Technology: Skiathos, Greece, 2002; p. 469.

71. Arvanitoyannis, I.S.; Ladas, D.; Mavromatis, A. Potential Uses and Applications of Treated Wine Waste: A Review. *Int. J. Food Sci. Technol.* **2006**, *41*, 475–487. [CrossRef]

72. Israilides, C.; Scanlon, B.; Smith, A.; Harding, S.E.; Jumel, K. Characterization of Pullulans Produced from Agro-Industrial Wastes. *Carbohydr. Polym.* **1994**, *25*, 203–209. [CrossRef]

73. Feng, J.; Yang, J.; Li, X.; Guo, M.; Wang, B.; Yang, S.T.; Zou, X. Reconstruction of a Genome-Scale Metabolic Model and in Silico Analysis of the Polymalic Acid Producer *Aureobasidium pullulans* CCTCC M2012223. *Gene* **2017**, *607*, 1–8. [CrossRef] [PubMed]

74. Feng, J.; Yang, J.; Yang, W.; Chen, J.; Jiang, M.; Zou, X. Metabolome- and Genome-Scale Model Analyses for Engineering of *Aureobasidium pullulans* to Enhance Polymalic Acid and Malic Acid Production from Sugarcane Molasses. *Biotechnol. Biofuels* **2018**, *11*, 94. [CrossRef] [PubMed]

75. Leathers, T.D.; Manitchotpisit, P. Production of Poly(Beta-L-Malic Acid) (PMA) from Agricultural Biomass Substrates by *Aureobasidium pullulans*. *Biotechnol. Lett.* **2013**, *35*, 83–89. [CrossRef] [PubMed]

76. Manitchotpisit, P.; Skory, C.D.; Peterson, S.W.; Price, N.P.; Vermillion, K.E.; Leathers, T.D. Poly(Beta-L-Malic Acid) Production by Diverse Phylogenetic Clades of *Aureobasidium pullulans*. *J. Ind. Microbiol. Biotechnol.* **2012**, *39*, 125–132. [CrossRef] [PubMed]

77. Kurosawa, T.; Sakai, K.; Nakahara, T.; Oshima, Y.; Tabuch, T. Extracellular Accumulation of the Polyol Lipids, 3,5-Dihydroxydecanoyl and 5-Hydroxy-2-Decenoyl Esters of Arabitol and Mannitol, by *Aureobasidium* Sp. *Biosci. Biotechnol. Biochem.* **1994**, *58*, 2057–2060. [CrossRef]

78. Price, N.P.; Manitchotpisit, P.; Vermillion, K.E.; Bowman, M.J.; Leathers, T.D. Structural Characterization of Novel Extracellular Liamocins (Mannitol Oils) Produced by *Aureobasidium pullulans* Strain NRRL 50380. *Carbohydr. Res.* **2013**, *370*, 24–32. [CrossRef] [PubMed]

79. Bischoff, K.M.; Leathers, T.D.; Price, N.P.J.; Manitchotpisit, P. Liamocin Oil from *Aureobasidium pullulans* Has Antibacterial Activity with Specificity for Species of Streptococcus. *J. Antibiot.* **2015**, *68*, 642. [CrossRef] [PubMed]

80. Kim, J.S.; Lee, I.K.; Yun, B.S. A Novel Biosurfactant Produced by *Aureobasidium pullulans* L3-GPY from a Tiger Lily Wild Flower, Lilium Lancifolium Thunb. *PLoS ONE* **2015**, *10*, e0122917. [CrossRef] [PubMed]

81. Manitchotpisit, P.; Price, N.P.J.; Leathers, T.D.; Punnapayak, H. Heavy Oils Produced by *Aureobasidium pullulans*. *Biotechnol. Lett.* **2011**, *33*, 1151–1157. [CrossRef] [PubMed]

82. Manitchotpisit, P.; Watanapokasin, R.; Price, N.P.; Bischoff, K.M.; Tayeh, M.; Teeraworawit, S.; Kriwong, S.; Leathers, T.D. *Aureobasidium pullulans* as a Source of Liamocins (Heavy Oils) with Anticancer Activity. *World J. Microbiol. Biotechnol.* **2014**, *30*, 2199–2204. [CrossRef] [PubMed]

83. Chi, Z.; Wang, X.X.; Ma, Z.C.; Buzdar, M.A.; Chi, Z.M. The Unique Role of Siderophore in Marine-Derived *Aureobasidium pullulans* HN6.2. *Biometals* **2012**, *25*, 219–230. [CrossRef] [PubMed]

84. Murugappan, R.; Karthikeyan, M.; Aravinth, A.; Alamelu, M. Siderophore-Mediated Iron Uptake Promotes Yeast-Bacterial Symbiosis. *Appl. Biochem. Biotechnol.* **2012**, *168*, 2170–2183. [CrossRef] [PubMed]

85. Wang, W.; Chi, Z.; Liu, G.; Buzdar, M.A.; Chi, Z.; Gu, Q. Chemical and Biological Characterization of Siderophore Produced by the Marine-Derived *Aureobasidium pullulans* HN6.2 and Its Antibacterial Activity. *Biometals* **2009**, *22*, 965–972. [CrossRef] [PubMed]

86. Wang, W.L.; Chi, Z.M.; Chi, Z.; Li, J.; Wang, X.H. Siderophore Production by the Marine-Derived *Aureobasidium pullulans* and Its Antimicrobial Activity. *Bioresour. Technol.* **2009**, *100*, 2639–2641. [CrossRef] [PubMed]

87. Tjamos, E.C.; Tjamos, S.E.; Antoniou, P.P. Biological Management of Plant Diseases: Highlights on Research and Application. *J. Plant Pathol.* **2010**, S17–S21.

88. Blakeman, J.P.; Fokkema, N.J. Potential for Biological Control of Plant Diseases on the Phylloplane. *Annu. Rev. Phytopathol.* **1982**, *20*, 167–190. [CrossRef]

89. Money, N.P. Fungi and Biotechnology. In *The Fungi*, 3rd ed.; Elsevier: Amsterdam, The Netherlands, 2015; pp. 401–424.

90. Chi, Z.; Yan, K.; Gao, L.; Li, J.; Wang, X.; Wang, L. Diversity of Marine Yeasts with High Protein Content and Evaluation of Their Nutritive Compositions. *J. Mar. Biol. Assoc. UK* **2008**, *88*, 1347–1352. [CrossRef]

91. Mina, M.; Tsaltas, D. Contribution of Yeast in Wine Aroma and Flavour. In *Yeast-Industrial Applications*; InTech: Rijeka, Croatia, 2017.

92. Fugelsang, K.C.; Edwards, C.G. *Wine Microbiology: Practical Applications and Procedures*; Springer Science & Business Media: Berlin, Germany, 2007.

93. Liu, J.; Sui, Y.; Wisniewski, M.; Droby, S.; Liu, Y. Utilization of Antagonistic Yeasts to Manage Postharvest Fungal Diseases of Fruit. *Int. J. Food Microbiol.* **2013**, *167*, 153–160. [CrossRef] [PubMed]

94. Bencheqroun, S.K.; Bajji, M.; Massart, S.; Bentata, F.; Labhilili, M.; Achbani, H.; El Jaafari, S.; Jijakli, M.H. Biocontrol of Blue Mold on Apple Fruits by *Aureobasidium pullulans* (Strain Ach 1-1): In Vitro and in Situ Evidence for the Possible Involvement of Competition for Nutrients. *Commun. Agric. Appl. Biol. Sci.* **2006**, *71* Pt 3B, 1151–1157.

95. Bencheqroun, S.K.; Bajji, M.; Massart, S.; Labhilili, M.; El Jaafari, S.; Jijakli, M.H. In Vitro and in Situ Study of Postharvest Apple Blue Mold Biocontrol by *Aureobasidium pullulans*: Evidence for the Involvement of Competition for Nutrients. *Postharvest Biol. Technol.* **2007**, *46*, 128–135. [CrossRef]

96. Schena, L.; Nigro, F.; Pentimone, I.; Ligorio, A.; Ippolito, A. Control of Postharvest Rots of Sweet Cherries and Table Grapes with Endophytic Isolates of *Aureobasidium pullulans*. *Postharvest Biol. Technol.* **2003**, *30*, 209–220. [CrossRef]

97. Mari, M.; Martini, C.; Spadoni, A.; Rouissi, W.; Bertolini, P. Biocontrol of Apple Postharvest Decay by *Aureobasidium pullulans*. *Postharvest Biol. Technol.* **2012**, *73*, 56–62. [CrossRef]

98. Di Francesco, A.; Ugolini, L.; Lazzeri, L.; Mari, M. Production of Volatile Organic Compounds by *Aureobasidium pullulans* as a Potential Mechanism of Action against Postharvest Fruit Pathogens. *Biol. Control* **2015**, *81*, 8–14. [CrossRef]

99. Farbo, M.G.; Urgeghe, P.P.; Fiori, S.; Marcello, A.; Oggiano, S.; Balmas, V.; Hassan, Z.U.; Jaoua, S.; Migheli, Q. Effect of Yeast Volatile Organic Compounds on Ochratoxin A-Producing Aspergillus Carbonarius and A. Ochraceus. *Int. J. Food Microbiol.* **2018**, *284*, 1–10. [CrossRef] [PubMed]

100. Di Francesco, A.; Roberti, R.; Martini, C.; Baraldi, E.; Mari, M. Activities of *Aureobasidium pullulans* Cell Filtrates against Monilinia Laxa of Peaches. *Microbiol. Res.* **2015**, *181*, 61–67. [CrossRef] [PubMed]

101. Ferraz, L.P.; Cunha, T.D.; da Silva, A.C.; Kupper, K.C. Biocontrol Ability and Putative Mode of Action of Yeasts against Geotrichum Citri-Aurantii in Citrus Fruit. *Microbiol. Res.* **2016**, *188–189*, 72–79. [CrossRef] [PubMed]

102. Pfliegler, W.P.; Pusztahelyi, T.; Pócsi, I. Mycotoxins–prevention and Decontamination by Yeasts. *J. Basic Microbiol.* **2015**, *55*, 805–818. [CrossRef] [PubMed]

103. Vanhoutte, I.; Audenaert, K.; De Gelder, L. Biodegradation of Mycotoxins: Tales from Known and Unexplored Worlds. *Front. Microbiol.* **2016**, *7*, 561. [CrossRef] [PubMed]

104. De Felice, D.V.; Solfrizzo, M.; De Curtis, F.; Lima, G.; Visconti, A.; Castoria, R. Strains of *Aureobasidium pullulans* Can Lower Ochratoxin A Contamination in Wine Grapes. *Phytopathology* **2008**, *98*, 1261–1270. [CrossRef] [PubMed]

105. Zhang, X.; Yang, H.; Apaliya, M.T.; Zhao, L.; Gu, X.; Zheng, X.; Hu, W.; Zhang, H. The Mechanisms Involved in Ochratoxin A Elimination by Yarrowia Lipolytica Y-2. *Ann. Appl. Biol.* **2018**, *163*, 164–174. [CrossRef]

106. Amézqueta, S.; González-Peñas, E.; Murillo-Arbizu, M.; de Cerain, A.L. Ochratoxin A Decontamination: A Review. *Food Control* **2009**, *20*, 326–333. [CrossRef]
107. Zhang, H.; Apaliya, M.T.; Mahunu, G.K.; Chen, L.; Li, W. Control of Ochratoxin A-Producing Fungi in Grape Berry by Microbial Antagonists: A Review. *Trends Food Sci. Technol.* **2016**, *51*, 88–97. [CrossRef]
108. Verginer, M.; Leitner, E.; Berg, G. Production of Volatile Metabolites by Grape-Associated Microorganisms. *J. Agric. Food Chem.* **2010**, *58*, 8344–8350. [CrossRef] [PubMed]
109. Campisano, A.; Pancher, M.; Puopolo, G.; Puddu, A.; Lòpez-Fernàndez, S.; Biagini, B.; Yousaf, S.; Pertot, I. Diversity in Endophytic Populations Reveals Functional and Taxonomic Diversity between Wild and Domesticated Grapevines. *Am. J. Enol. Vitic.* **2014**, *66*, 12–21. [CrossRef]
110. Campisano, A.; Antonielli, L.; Pancher, M.; Yousaf, S.; Pindo, M.; Pertot, I. Bacterial Endophytic Communities in the Grapevine Depend on Pest Management. *PLoS ONE* **2014**, *9*, e112763. [CrossRef] [PubMed]
111. Merin, M.G.; Martin, M.C.; Rantsiou, K.; Cocolin, L.; de Ambrosini, V.I. Characterization of Pectinase Activity for Enology from Yeasts Occurring in Argentine Bonarda Grape. *Braz. J. Microbiol.* **2015**, *46*, 815–823. [CrossRef] [PubMed]
112. Merin, M.G.; Mendoza, L.M.; Morata de Ambrosini, V.I. Pectinolytic Yeasts from Viticultural and Enological Environments: Novel Finding of Filobasidium Capsuligenum Producing Pectinases. *J. Basic Microbiol.* **2014**, *54*, 835–842. [CrossRef] [PubMed]
113. Mulay, Y.R.; Deopurkar, R.L. Purification, Characterization of Amylase from Indigenously Isolated *Aureobasidium pullulans* Cau 19 and Its Bioconjugates with Gold Nanoparticles. *Appl. Biochem. Biotechnol.* **2018**, *184*, 644–658. [CrossRef] [PubMed]
114. Kudanga, T.; Mwenje, E. Extracellular Cellulase Production by Tropical Isolates of *Aureobasidium pullulans*. *Can. J. Microbiol.* **2005**, *51*, 773–776. [CrossRef] [PubMed]
115. Kudanga, T.; Mwenje, E.; Mandivenga, F.; Read, J.S. Esterases and Putative Lipases from Tropical Isolates of *Aureobasidium pullulans*. *J. Basic Microbiol.* **2007**, *47*, 138–147. [CrossRef] [PubMed]
116. Martins, M.A.; Lima, N.; Silvestre, A.J.; Queiroz, M.J. Comparative Studies of Fungal Degradation of Single or Mixed Bioaccessible Reactive Azo Dyes. *Chemosphere* **2003**, *52*, 967–973. [CrossRef]
117. Kremnicky, L.; Slavikova, E.; Mislovicova, D.; Biely, P. Production of Extracellular Beta-Mannanases by Yeasts and Yeast-like Microorganisms. *Folia Microbiol.* **1996**, *41*, 43–47. [CrossRef]
118. Augustin, J. Polysaccharide Hydrolases of *Aureobasidium pullulans*. *Folia Microbiol.* **2000**, *45*, 143–146. [CrossRef]
119. Sheng, L.; Tang, G.; Su, P.; Zhang, J.; Xiao, Q.; Tong, Q.; Ma, M. Understanding the Influence of Tween 80 on Pullulan Fermentation by *Aureobasidium pullulans* CGMCC1234. *Carbohydr. Polym.* **2016**, *136*, 1332–1337. [CrossRef] [PubMed]
120. Roukas, T.; Serris, G. Effect of the Shear Rate on Pullulan Production from Beet Molasses by *Aureobasidium pullulans* in an Airlift Reactor. *Appl. Biochem. Biotechnol.* **1999**, *80*, 77–89. [CrossRef]
121. Lazaridou, A.; Biliaderis, C.G.; Roukas, T.; Izydorczyk, M. Production and Characterization of Pullulan from Beet Molasses Using a Nonpigmented Strain of *Aureobasidium pullulans* in Batch Culture. *Appl. Biochem. Biotechnol.* **2002**, *97*, 1–22. [CrossRef]
122. Roukas, T. Pullulan Production from Deproteinized Whey by *Aureobasidium pullulans*. *J. Ind. Microbiol. Biotechnol.* **1999**, *22*, 617–621. [CrossRef] [PubMed]
123. Muramatsu, D.; Okabe, M.; Takaoka, A.; Kida, H.; Iwai, A. *Aureobasidium pullulans* Produced Beta-Glucan Is Effective to Enhance Kurosengoku Soybean Extract Induced Thrombospondin-1 Expression. *Sci. Rep.* **2017**, *7*, 2831. [CrossRef] [PubMed]
124. Aoki, S.; Iwai, A.; Kawata, K.; Muramatsu, D.; Uchiyama, H.; Okabe, M.; Ikesue, M.; Maeda, N.; Uede, T. Oral Administration of the β-Glucan Produced by *Aureobasidium pullulans* Ameliorates Development of Atherosclerosis in Apolipoprotein E Deficient Mice. *J. Funct. Foods* **2015**, *18*, 22–27. [CrossRef]
125. Muramatsu, D.; Kawata, K.; Aoki, S.; Uchiyama, H.; Okabe, M.; Miyazaki, T.; Kida, H.; Iwai, A. Stimulation with the *Aureobasidium pullulans*-Produced Beta-Glucan Effectively Induces Interferon Stimulated Genes in Macrophage-like Cell Lines. *Sci. Rep.* **2014**, *4*, 4777. [CrossRef] [PubMed]
126. Leathers, T.D.; Price, N.P.; Bischoff, K.M.; Manitchotpisit, P.; Skory, C.D. Production of Novel Types of Antibacterial Liamocins by Diverse Strains of *Aureobasidium pullulans* Grown on Different Culture Media. *Biotechnol. Lett.* **2015**, *37*, 2075–2081. [CrossRef] [PubMed]

127. Di Francesco, A.; Ugolini, L.; D'Aquino, S.; Pagnotta, E.; Mari, M. Biocontrol of Monilinia Laxa by *Aureobasidium pullulans* Strains: Insights on Competition for Nutrients and Space. *Int. J. Food Microbiol.* **2017**, *248*, 32–38. [CrossRef] [PubMed]

128. Di Francesco, A.; Mari, M.; Ugolini, L.; Baraldi, E. Effect of *Aureobasidium pullulans* Strains against Botrytis Cinerea on Kiwifruit during Storage and on Fruit Nutritional Composition. *Food Microbiol.* **2018**, *72*, 67–72. [CrossRef] [PubMed]

fermentation

MDPI

Review

Zygosaccharomyces rouxii: Control Strategies and Applications in Food and Winemaking

Carlos Escott *, Juan Manuel del Fresno *, Iris Loira, Antonio Morata and José Antonio Suárez-Lepe

Chemistry and Food Technology Department, Polytechnic University of Madrid, Avenida Complutense S/N, 28040 Madrid, Spain; iris.loira@upm.es (I.L.); antonio.morata@upm.es (A.M.); joseantonio.suarez.lepe@upm.es (J.A.S.-L.)

* Correspondence: c.escott@alumnos.upm.es (C.E.); juan.fresno.florez@alumnos.upm.es (J.M.d.F.)

Received: 30 July 2018; Accepted: 17 August 2018; Published: 22 August 2018

Abstract: The genus *Zygosaccharomyces* is generally associated to wine spoilage in the winemaking industry, since a contamination with strains of this species may produce re-fermentation and CO_2 production in sweet wines. At the same time, this capacity might be useful for sparkling wines production, since this species may grow under restrictive conditions, such as high ethanol, low oxygen, and harsh osmotic conditions. The spoilage activity of this genus is also found in fruit juices, soft drinks, salad dressings, and other food products, producing besides package expansion due to gas production, non-desired compounds such as ethanol and esters. Despite these drawbacks, *Zygosaccharomyces* spp. produces high ethanol and acetoin content in wines and may play an important role as non-*Saccharomyces* yeasts in differentiated wine products. Control strategies, such as the use of antimicrobial peptides like Lactoferricin B (Lfcin B), the use of dimethyl dicarbonate (DMDC) or non-thermal sterilization techniques may control this spoilage genus in the food industry.

Keywords: wine; *Zygosaccharomyces rouxii*; re-fermentation; spoilage-control; non-*Saccharomyces*; high-ethanol

1. The Genus *Zygosaccharomyces*

There were nine species accepted in the genus *Zygosaccharomyces* at the beginning of the century [1], and by then, the genus also included *Zygosaccharomyces rouxii* (*Z. rouxii*). Nonetheless, the number of species of the genus *Zygosaccharomyces* has increased rapidly over the past years and the classification of the genus by 2014 included the following species: *Z. bailii*, *Z. bisporus*, *Z. gambellarensis*, *Z. kombuchaensis*, *Z. lentus*, *Z. machadoi*, *Z. mellis*, *Z. parabaillii*, *Z. pseudobailii*, *Z. pseudorouxii*, *Z. rouxii*, *Z. sapae*, and *Z. siamensi* [2]. Out of the recently isolated species, *Z. pseudorouxii* is closely related to the species *Z. sapae*. Some of these species affect the food and beverage industries as spoiling microorganisms, and others are associated with fermentations and sweet foodstuff like honey. From the aforementioned species, only *Z. rouxii* and *Z. bailii* have their genome sequenced [2]. In this way, from the osmophilic yeasts, being the first cause of fruit juice spoilage, the genus *Zygosaccharomyces* is the most frequently described [3].

The genus *Zygosaccharomyces* is related to an important genus in winemaking, *Saccharomyces*, and at the same time, is a genus involved in food and beverage spoilage [4]. This genus is considered a spoiling microorganism since it shows high tolerance to osmotic stress, and therefore, the *Zygosaccharomyces* species can grow in harsh environments with high sugars concentration. In this regard, contamination by these microorganisms is often seen in fruit juices, sauces, carbonated soft drinks, salad dressings, ketchup [1], sugar syrups, candied fruit, jams and preserves, tomato sauce, and wines [5]. Wine is subjected to spoilage by this genus, since it is also capable of growing at very low pH values.

This genus can also resist extreme conditions in the presence of organic acids, low oxygen levels, and high concentration of permitted preservatives [6]. These preservatives, commonly used in the food production industry, comprise the use of sorbic acid, benzoic acid, acetic acid, and ethanol [5] and up to 200 mg/L SO_2 in winemaking.

The physiology, the metabolism, and the spoilage/industrial activity of the species *Z. rouxii*, where the morphology is shown in Figure 1, will be described in the following sections.

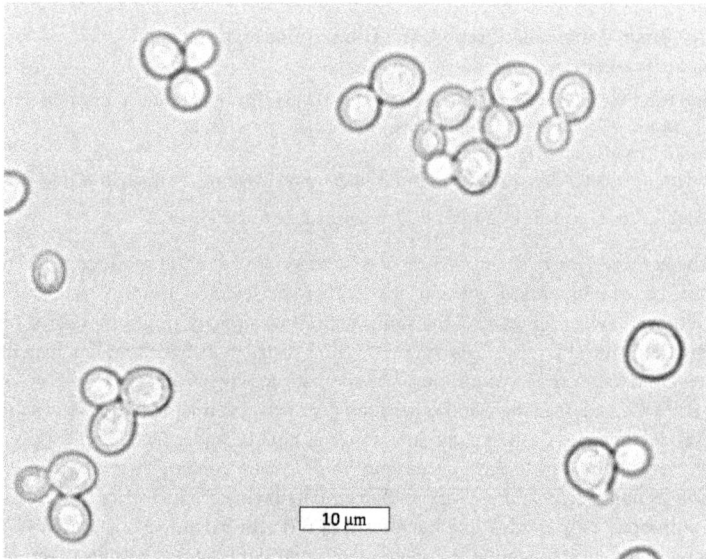

Figure 1. Optical microscopic picture of the species *Z. rouxii* (100× magnification).

2. Physiology and Metabolism of *Z. rouxii*

The species *Z. rouxii* and *Z. bailii* are associated with food spoilage, especially affecting those products with a high concentration of sugar and/or salt, low pH, or week-organic acids content.

From these species, *Z. rouxii* is able to endure very low water activity (a_w) environments, and due to this, it is one of the most xerophilic organisms known. It can grow in food with up to 70% glucose in their composition, and some strains survive to even higher concentrations, 5 M (>90% w/v) and 5.5 M glucose (saturated glucose solution) [7,8]. Moreover, it opts to consume fructose over glucose, being the reason why it is considered a fructophilic yeast. As a fructophilic yeast, *Z. rouxii* possesses genes FFZ that encode specific fructose facilitators and proteins, which have been characterized for both species, *Z. rouxii* [9] and *Z. bailii* [10]. The fructose transporter systems mediate the uptake of hexoses via a facilitated diffusion mechanism. Figure 2 is a schematic representation of such transporters during ethanol fermentation. ZrFfz1 and ZrFsy1 are fructose transporters, Hxt is a hexose transporter, and ZrFfz2 is a fructose/glucose transporter. The intermediate reactions in the production of fructose 1,6-bisphosphate from glucose and fructose, double phosphorylation in glucose, and phosphorylation in fructose, are not shown.

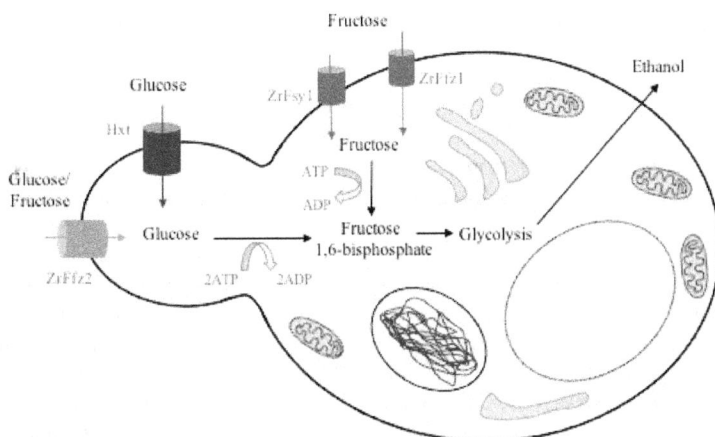

Figure 2. Fructose and glucose transporters during *Z. rouxii* ethanol fermentation.

The yeast species *Zygosaccharomyces rouxii* is usually haploid and heterothallic [11]. Besides being osmophilic and xerophilic, it is also considered acid-tolerant. The species can adapt and grow in acidic media at pH values of 2.2 or even as low as 1.8 [7,11], being the reason why this species could either spoil products, such as grape juice concentrates [12] or be used industrially in the production of soy sauce and miso paste [11]. Total inhibition is achieved at pH values below 1.7, although this might be difficult to achieve at an industrial scale [13]. Assuring pH 2.2, food products such as concentrated grape juice, could extend their shelf-life significantly for storage or shipping overseas.

In terms of inhibition temperature, 47 °C is needed to reduce log2 CFU/mL in inoculums of spoilage yeast species *Z. rouxii*, *P. guillermondii*, and *Z. lentus* [7]; nonetheless, some strains require higher temperatures, between 55 °C and 60 °C, to inhibit their growth.

Water activity tolerance is one physiological difference between the species *Zygosaccharomyces rouxii* and *Zygosaccharomyces bailii*. The species *Z. rouxii* can tolerate low water activity (a_w) environments, whilst *Z. bailii* requires environments with (a_w) of at least 0.85 [7]. This characteristic makes it difficult for *Z. bailii* to survive in high sugar foods, such as syrups and candied fruits.

The yeast species *Zygosaccharomyces rouxii*, as well as other osmotolerant microorganisms, adjusts its internal osmotic pressure to tolerate high concentrations of salt (NaCl) of about 3–4 M (ca. 20% w/v) [14]. The mechanism is the efflux of sodium cations (Na^+) from cells under high concentration of salt [11]. A change in the fluidity of the lipidic cell membranes has been observed when exposing yeast cells to 15% NaCl [14]. The lipid composition of the cell membrane and the plasma membranes changed by means of a decrease in the degree of saturation, an increase in ergosterol concentration, and a decrease in the phospholipid to protein ratio. The accumulation of glycerol as a compatible solute is a mechanism that *Z. rouxii* follows to survive to high osmolarity. This protects the cell against lysis [11].

Another metabolic feature of this yeast is the production of volatiles in high-sugar food matrices at an early stage; this would indicate the presence of spoilage yeast *Z. rouxii*. Ethanol, acetone, ethyl acetate, acetaldehyde, or 3-methyl-1-butanol could be detected by analytical techniques and even before the human nose is able to [15].

3. Food Spoilage Activity

As already described, the physiologies of *Zygosaccharomyces rouxii*, comprising the osmotolerance, the xerophilic ability, the fructophilic capacity, and the weak-acid tolerance, are responsible for causing food spoilage. The food products prone to growing spoilage microorganisms include juice concentrates, sugar syrups, honey, jams, confectionary products, and dried fruits [16]. Among several fruit concentrate juices, those with a higher incidence of *Z. rouxii*, even after 16 months stored at −18 °C, are cherry and orange juices; grape concentrated juice grew populations of *C. stellate, K, thermotolerans, P. anomala, S. cerevisiae, and Z. rouxii* [17]. Despite the presence of other yeast species in non-spoiled juices, 100% of the yeasts isolated in spoiled concentrated grape juice, belonged to the species *Zygosaccharomyces rouxii* [18].

One of the most obvious effects observed in food products after the contamination of spoilage yeasts, is the production of excess gas. This gas compromises the integrity of the food package as it can swell containers, and it could also be responsible for "blown" cans or exploding glass bottles. This excess gas is the result of the fermentation of sugars by yeasts, during the product's shelf-life. The volume of gas produced is variable and so is the pressure inside the food package; this effect depends on the fermentative yeast species and their fermentation power. In this matter, three *Zygosaccharomyces* species (*Z. lentus, Z. bailii,* and *Z. rouxii*) produced larger amounts of gas in comparison with other spoilage yeast genera (*P. guillermondii, C. halophila* and *C. magnolia*), as evidence of the high fermenting capacity of *Zygosaccharomyces* genus in food with high sugar content [7].

Taking all this into consideration, the efforts in controlling the spoilage yeasts should not just focus on the conservation alternatives using preservatives, but also in the production facilities in terms of hygienic practices to avoid the contamination of piping, containers, and any other equipment and machinery. The quick detection of spoilage microorganisms, would solve contamination in the initial stage and save related costs.

4. Detection

Food spoilage caused by *Zygosaccharomyces rouxii* strains, is perceived by consumers due to the formation of non-desired odors affecting the products. In products where the aroma profile is synonymous of quality, like apple juice, the presence of such aromas may contribute to increased product waste. Electronic noses (e-nose) may detect the contamination by *Zygosaccharomyces rouxii* strains, even at populations as low as log2 CFU/mL, during the production stage or in the product's shelf life [19]. The electronic noses work with gas sensory array technology and are able to detect changes in the volatile pattern associated with microorganism spoilage [20]. Organic acids and esters, are compounds detectable by e-noses which indicate microorganism activity. The production of such volatiles is then related to a certain microorganism; this characteristic makes this technique even able to distinguish contamination coming from different species [21]. No sample preparation is needed when using the e-nose technique, but a series of sensors able to distinguish among aromatic compounds, nitrogen oxide, ammonia, alkanes, methane, Sulphur compounds, alcohol, etc. are required [15]. These electronic noses could be coupled with chemometric analysis, for the early diagnosis of the contamination by *Z. rouxii* strains in apple juice [19]. Other analytical techniques, more expensive than e-noses, but quite extended on the identification and quantification of spoilage yeast metabolites among other compounds, are gas chromatography and liquid chromatography. Gas chromatography coupled with mass-spectrometry, detects concentrations of guaiacol, 2,6-dibromophenol, and 2,6-dichlorophenol in kiwi juices [22]; ultra-high-performance liquid chromatography coupled with tandem mass spectrometry (UHPLC-MS/MS), is suitable to identify molecules from different mycotoxins present in cereal syrups [23]. The selection of the appropriate analytical technique is also related to the stability and solubility of the target molecules, and no less important is the sample preparation, which may include solid phase extraction, partitioning via salting-out interfaces between aqueous and organic solvent layers, etc.

Although several analytical techniques might help in identifying the presence of *Zygosaccharomyces* genus in food products and in winemaking by the detection of molecules associated to its metabolic activity, the use of primers with plasmid DNA in multiplex polymerase chain reaction (PCR) would also allow the identification between *Z. rouxii* and *Z. bailii*, as well as to differentiate this genus from the species *S. cerevisiae* [24]. This method was used in the late twentieth century and it later allowed the identification of the genes TPS1 and TPS2, encoding trehalose-6-phosphate synthase and trehalose-6-phosphate phosphatase, respectively [25], for the synthesis of trehalose; or the identification of the nucleotide sequence of the genes Nq$^+$/H$^+$-antiporter (ZSOD2 and ZSOD22) related to the salt tolerance of this yeast species [26]. Lately, the use of pre-treatment methodologies in PCR analysis, has allowed the identification of the food-spoilage yeast *Z. rouxii* in real apple juice samples. Double washing dielectrophoretic (DEP) manipulation of yeast cells, is one of such pre-treatments [27]. The DEP device washes out PCR inhibitors and improves the analysis.

The use of selective high-sugar medium, such as PYGF broths with 300 g/L glucose and 300 g/L fructose, may also help in isolating osmotolerant and fructophilic yeast strains present in food matrices, for further identification during microbiological controls and detection [15].

5. Control Strategies

Zygosaccharomyces bailii species has shown high resistance to food preservatives, such as sorbic acid, benzoic acid, acetic acid, cinnamic acid, ethanol, and to heat. On the other hand, they lacked resistance to peracetic acid or hypochlorite, suggesting the possibility of using biocidal cleaning agents to control their population in production facilities [7]. Regarding sanitization practices, it is interesting to pay attention to the materials used in the food industry. Frisón, Chiericatti, Aríngoli, Basílico, and Basílico [28] have compared the effect of different sanitizing materials, such as peracetic acid, monochloramine, iodophor, and quaternary ammonium compounds, in a variety of surfaces, including wood, glass, PVC plastic, and stainless steel, against the yeast species *Zygosaccharomyces rouxii*. The results obtained revealed that peracetic acid was effective to avoid contamination by *Z. rouxii*, and it was preferred over the rest of the products tested due to its higher safety. Stainless steel was completely sanitized with all the compounds tested, being the reason why it shall be more appropriate to use this material in the diverse food industry sectors. Regarding limiting growing conditions for *Z. rouxii*, these have been determined as glucose concentration above 5M, temperature above 46.5 °C, and pH lower than 2.2 [7]. This yeast could also produce 31.5 mL of gas in substrates with 2% glucose, and up to 102 mL of gas in substrates with 18% glucose. Table 1 summarizes the comparison of different preservatives and disinfectants, and the minimum concentration for their inhibitory effect.

Zygosaccharomyces rouxii, similar to *Z. bailii*, has a high resistance to different chemical compounds used as food preservatives. Hydroxycinnamic acids, such as caffeic acid and *p*-coumaric acid, have a rather low inhibitory effect of around 15%, whilst preservatives like potassium sorbate, sodium benzoate, dimethyldicarbonate, and vanillin can inhibit the growth of this yeast species up to 40% [29]. The acetic acid has an impact in the respiratory activity of the halo-tolerant yeast *Zygosaccharomyces rouxii* R-1, and it also inhibits the formation of cytochromes. *Z. rouxii* was significantly inhibited, and its growth was considerably reduced, in the presence of 0.5% acetic acid and also in media containing NaCl above 18% [30].

The use of thermo-sonication, ultrasounds in combination with heat, to inactivate *Zygosaccharomyces rouxii* strains at different pH and water activity conditions [31], results in a reduction of >log5 CFU/mL of yeast population. The higher the temperature of sonication, the greater the effect of the temperature in the inactivation. The use of ultrasounds under these conditions produces irreversible cell damage contributing to yeast inactivation; the synergetic contribution of lower water activity (a_w) and pH decreases with lower temperatures of sonication. Therefore, this approach might be useful as an alternative to traditional pasteurization of fruit juices.

Table 1. Minimum inhibitory concentration of preservatives and disinfectants against *Zygosaccharomyces rouxii* strains.

Yeast Strain	Ferulic Acid (mg/L)	p-Coumaric Acid (mg/L)	Potassium Sorbate (mg/L)	Sodium Benzoate (mg/L)	Sorbic Acid (mg/L)	Benzoic Acid (mg/L)	Acetic Acid (g/L)	Cinnamic Acid (mg/L)	Ethanol (mg/L)	SO₂ (mg/L)	DMDC (mg/L)	Peracetic Acid (mg/L)	Sodium Hypochlorite (mg/L)	H₂O₂ (mg/L)
CECT 12003 [1]	-	-	-	-	320	> 439.6	6.6	> 446.7	73.7	217	227.9	247.2	204.7	459
CECT 12004 [1]	-	-	-	-	314	372	6.4	369.6	78.3	262	227.9	247.2	189.8	544
MC8, MC9, MC10 [2]	229	707	28.5	33.8	-	-	-	-	-	-	6.03	-	-	-
R1 [3]	-	-	-	-	-	-	26	-	-	-	-	-	-	-
Z. rouxii [4]	-	-	-	-	-	-	-	-	-	-	-	169.6	55.1	-

[1] [7]; [2] [29]; [3] [30]; [4] Isolated *Z. rouxii* [28].

Dielectric barrier discharge (DBD) plasma at 90W for 140 s, have shown to reduce log5 viable *Z. rouxii* cells in apple juice [6]. The DBD plasma have produced alterations in the permeability of the *Z. rouxii* cell membranes, and as a result, the release of intracellular macromolecules, such as nucleic acids and proteins. The disruptions caused in the cell membrane are observable with scanning electron microscopy (SEM) imaging. During DBD plasma processing, reactive species like H_2O_2 and NO_2 were produced, and these reactive compounds would contribute to the inactivation of *Z. rouxii* together with the alterations produced in the membrane permeability. A drawback on the use of DBD plasma, is the negative effect on color parameters of apple juices treated. The juices reported higher acidic values, whilst on the other hand, the content of reducing sugars, total soluble solids, and total phenolics remained practically without change. DBD plasma might then be used as effective control technique to inactivate *Z. rouxii* in apple juice.

Other control strategies may involve the use of antimicrobial peptides such as Lactoferricin B (Lfcin B) and the use of killer toxins.

Lfcin B is a peptide produced after the gastric digestion of protein lactoferrin from bovine origin, and according to Escott, Loira, Morata, Bañuelos, and Suárez-Lepe [32], it has antibacterial and antifungal properties, besides being considered a peptide with antiviral, antitumor, anti-inflammatory, and immunoregulatory properties. These properties have shown to have an effective antimicrobial effect against spoilage yeasts, such as *Dekkera bruxellensis* and *Zygosaccharomyces* spp. in wine production.

Killer toxins, such as *Pichia membranifaciens* killer toxins (PMKT), may interact with other antimicrobial agents like metabisulphite, to avoid the spoilage by *Zygosaccharomyces* spp. in beverages with a high sugar concentration [33]. The interaction effect would reduce the amount of metabisulphite needed as antimicrobial, and therefore, reducing the potential negative effect on health and to the environment. Both yeasts, *Saccharomyces* spp. and non-*Saccharomyces* spp., have shown to possess the ability to produce killer toxins during spontaneous wine fermentation, as it is, in the case of the toxins K1, K2, K28, and Klus from *S. cerevisiae*. In some cases, the activity observed in killer toxins is due to β-glucanases. These β-glucanases are used to produced synthetic preparations as antimicrobial agents against spoilage yeasts *Dekkera bruxellensis* and *Zygosaccharomyces bailii* [34].

To assess the effect of β-glucanases on the inhibition of different spoilage yeasts species, including the species *Zygosaccharomyces rouxii*, an experiment was carried out by Escott et al. [32], and the results are shown in Figure 3. It could be observed that all control growing media have fermented as there was CO_2 production in the tubes. After using β-glucanase 1, the optical density (OD) did not change for *S. ludwigii*, suggesting a complete inhibition effect, while *D. bruxellensis* and *W. anomalus* had slightly increased OD suggesting a high inhibition effect. Finally, lesser effects were observed in *Z. rouxii* as there was CO_2 production and higher OD, suggesting partial inhibition effect. On the other hand, the use of β-glucanase 2 has successfully inhibited the growth of all spoilage yeast strains evaluated.

Sulphur dioxide (SO_2) has two purposes in the winemaking industry, to avoid microbiological spoilage of musts and wines, and to act as an antioxidant of wines, especially to avoid browning of white wines [35]. The role of SO_2 in red wines could be detrimental towards the anthocyanin content. Oenologists prefer the use of SO_2 to preserve wines, rather than using it during winemaking. According to research, the minimal concentration of free SO_2 needed to inhibit the growth of *Z. rouxii* populations is strain related and goes from 160–185 mg/L [36] to 217–262 mg/L (Table 1).

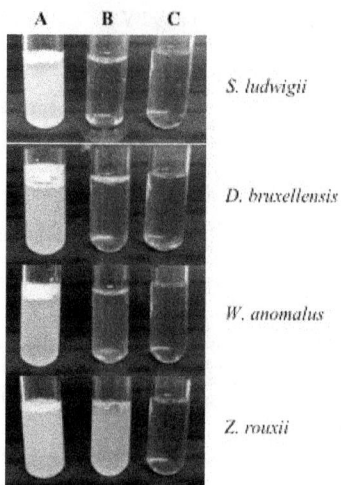

Figure 3. Use of β-glucanases as yeasts inhibitors. The tubes contain the yeasts species *S. ludwigii*, *D. bruxellensis*, *W. anomalus* and *Z. rouxii* in yeast extract peptone dextrose (YEPD) liquid growing media with same optical density. (**A**) control, (**B**) β-glucanase 1 and (**C**) β-glucanase 2.

6. Food Applications

Soy sauce is probably the main product elaborated with the use of *Zygosaccharomyces rouxii*. The species *Z. rouxii*, as osmotolerant yeast, makes feasible the production of soy sauce and miso paste industrially [11]. This species contributes to enhancing the flavor of soy sauce during its production, since this yeast is able to increase the concentration of certain aromatic volatile compounds. These compounds comprise the formation of larger amounts of 3-methyl-1-butanol (isoamyl alcohol), 2-methil-1-butanol (amyl alcohol), and 2-methyl-1-propanol (isobutyl alcohol) [37]. Some strains are even prone to forming larger amounts of acetoin than others [4], contributing to the overall flavor formation.

Parallel to the production of soy sauce and miso paste, *Z. rouxii* is used industrially in the production of other salted condiments, such as balsamic vinegar [38,39].

In addition, *Z. rouxii* has also been used for the production of certain compounds of interest. Hecquet, Sancelme, Bolte and Demuynck [40] showed that *Z. rouxii* produced 4-hydroxy-2,5-dimethyl-3(2H)-furanone when this yeast grows aerobically with D-fructose 1,6-bisphosphate (10%) as precursor. 4-hydroxy-2,5-dimethyl-3(2H)-furanone is used in the food industry as an additive; it exhibits caramel-like odors and has a relatively low perception threshold. Similarly, Saha, Sakakibara and Cotta [41] isolated a strain of *Z. rouxii*, that produced d-arabitol as the main metabolic product from glucose. In this way, this yeast shows potential to be used for production of xylitol from glucose via the d-arabitol route. Xylitol is a five-carbon sugar alcohol, used as a natural food sweetener. In addition, *Z. rouxii* has been used in solid-state fermentation, to produce extra-cellular L-glutaminase [42].

Other uses of *Zygosaccharomyces rouxii* strains are given in a patent application in the United States of America, that proposes the commercial utilization of a novel yeast strain of the species *Zygosaccharomyces rouxii* and its fermented metabolites as probiotics, as well as antioxidant and antimicrobial agents in foods and cosmetics [43].

Finally, the use of genome shuffling technique was successfully used to improve the flavor formation. This improvement impacted the formation of flavor components and amino acid nitrogen, with the result of enhancing the quality of soy sauce [44].

7. Alcohol-Fermentation Applications

Latest research, has shown the potential of using strains from the species *Z. rouxii* in the production of low-alcohol beer [45]. The use of these nonconventional yeast strains, as well as yeasts from the species *S. ludwigii*, could become an alternative to current practices in the production of beer with ethanol content between 0.5 and 1.2% *v/v*. The use of *Z. rouxii* strains, compared to *S. ludwigii* strains, has produced ethanol and diacetyl in larger amounts, above the taste threshold in beers.

Regarding wine production, there is scarce research on the potential use of *Zygosaccharomyces rouxii* strains in winemaking. Although some metabolic features of strains from the genus *Zygosaccharomyces*, may be interesting in the production of wines.

Species from the yeast *Zygosaccharomyces* genus, found in grapes and musts are able to increase the production of higher alcohols, at the same time that acetoin is reduced [4]. Although the contribution of acetoin to wine aroma profile is difficult to assess and its threshold is rather high (150 mg/L) [46], concentration above 300 mg/L is expected to produce butter aromas not pleasant in wine [47]. *Zygosaccharomyces* spp. generally produce a lower amount of acetoin in comparison with high-acetoin producer yeasts, such as the genera *Kloeckera* and *Hanseniaspora*. Higher alcohols, or fusel alcohols, like isoamyl alcohol, amyl alcohol, and isobutyl alcohol, are also found during the production of soy sauce by *Z. rouxii* [37]. These volatile compounds may contribute to the aroma profile of wines.

Wines trials with lower ethanol concentration were produced at laboratory scale, with yeast strains of the species *Torulaspora delbrueckii* and *Zygosaccharomyces bailii*; and *S. cerevisiae* was used to ensure completion of fermentation after 50% of sugars were consumed by non-*Saccharomyces* yeasts. These strains allowed the production of wines with less ethanol concentration, under limited aerobic conditions (5 to 10 mL/min). The ethanol reduction depended on the aeration regime. Both strains, *T. delbrueckii* and *Z. bailii*, were able to reduce ethanol in 1.5% (*v/v*) and 2.0% (*v/v*), respectively, in comparison to anaerobic fermentation carried out with *S. cerevisiae* as control [48]. The media used in this analysis comprised the preparation of chemically defined grape juice having 100 g glucose, 100 g fructose, 0.2 g citric acid, 3 g malic acid, inorganic salts and nitrogen sources.

One drawback of having uncontrolled *Zygosaccharomyces* spp. yeasts in sweet wine production is the potential re-fermentation, producing turbidity and CO_2 production [33], although this effect does not produce off-characters to wine. This, and the high production of acetic acid, has limited the possibilities of using this yeast species in winemaking production in the past years [49].

On the other hand, other *Zygosaccharomyces* species studied like *Z. fermentati* and *Z. bailii*, produce lower levels of H_2S and malic acid degradation, respectively [50]. These characteristics might be useful for wine production, where mixed fermentations with these and other spoilage non-*Saccharomyces* yeasts, and *S. cerevisiae* could be performed. Laboratory testing has shown that most detrimental metabolites produced by spoilage yeasts in pure culture or spoiled juices, are reduced in mixed fermentations; the production of polysaccharides increased improving body of wines and this also had a positive effect on aroma and protein stability [51].

The main concern in the winemaking industry, comes from the fact that these strains have high stress tolerance and may produce off-metabolites. Nonetheless, there are commercial *Zygosaccharomyces* yeast products prepared for stuck fermentations and potentially suitable for musts from riper grapes where the concentration of sugars, particularly fructose, is higher, and this could limit the implantation of other fermentative strains, such as *S. cerevisiae* [52].

8. Conclusions

Although the presence of yeast strains of the genus *Zygosaccharomyces* in many food products may represent a quality control danger and negative economic impact, the controlled used of some strains may positively contribute to enhancing organoleptic parameters of a particular range of products in the food industry. The potential use of these strains in winemaking is still controversial for their high spoilage activity, but it might also be an alternative to current technologically challenging conditions, such as stuck fermentations or the use of high fructose riper grape musts. Further studies on the impact of using the species *Zygosaccharomyces rouxii*, a yeast species which might not always endanger wine production, at all levels in the winemaking industry are promising.

Author Contributions: C.E. and J.M.d.F. performed the bibliographic revision and drafted the manuscript, I.L. and A.M. coordinated the bibliographic work and revised the manuscript and J.A.S.-L. revised and corrected the manuscript.

Conflicts of Interest: The authors declare no conflict of interest.

References

1. Esteve-Zarzoso, B.; Zorman, T.; Belloch, C.; Quero, A. Molecular Characterisation of the Species of the Genus *Zyosaccharomyces*. *Syst. Appl. Microbiol.* **2003**, *26*, 404–411. [CrossRef] [PubMed]
2. Hulin, M.; Wheals, A. Rapid Identification of *Zygosaccharomyces* with Genus-Specific Primers. *Int. J. Food Microbiol.* **2014**, *173*, 9–13. [CrossRef] [PubMed]
3. Wang, H.; Hu, Z.; Long, F.; Guo, C.; Niu, C.; Yuan, Y.; Yue, T. Combined Effect of Sugar Content and PH on the Growth of a Wild Strain of *Zygosaccharomyces Rouxii* and Time for Spoilage in Concentrated Apple Juice. *Food Control.* **2016**, *59*, 298–305. [CrossRef]
4. Romano, P.; Suzzi, G. Higher Alcohol and Acetoin Production by *Zygosaccharomyces* Wine Yeasts. *J. Appl. Bacteriol.* **1993**, *75*, 541–545. [CrossRef]
5. Steels, H.; James, S.A.; Roberts, I.N.; Stratford, M. *Zygosaccharomyces Lentus*: A Significant New Osmophilic, Preservative-Resistant Spoilage Yeast, Capable of Growth at Low Temperature. *J. Appl. Microbiol.* **1999**, *87*, 520–527. [CrossRef] [PubMed]
6. Xiang, Q.; Liu, X.; Li, J.; Liu, S.; Zhang, H.; Bai, Y. Effects of Dielectric Barrier Discharge Plasma on the Inactivation of *Zygosaccharomyces Rouxii* and Quality of Apple Juice. *Food Chem.* **2018**, *254*, 201–207. [CrossRef] [PubMed]
7. Martorell, P.; Stratford, M.; Steels, H.; Fernández-Espinar, M.; Querol, A. Physiological Characterization of Spoilage Strains of *Zygosaccharomyces Bailii* and *Zygosaccharomyces Rouxii* Isolated from High Sugar Environments. *Int. J. Food Microbiol.* **2007**, *114*, 234–242. [CrossRef] [PubMed]
8. Dakal, T.; Solieri, L.; Giudici, P. Adaptive Response and Tolerance to Sugar and Salt Stress in the Food Yeast *Zygosaccharomyces Rouxii*. *Int. J. Food Microbiol.* **2014**, *185*, 140–157. [CrossRef] [PubMed]
9. Leandro, M.J.; Sychrová, H.; Prista, C.; Loureiro-Dias, M.C. The Osmotolerant Fructophilic Yeast *Zygosaccharomyces Rouxii* Employs Two Plasma-Membrane Fructose Uptake Systems Belonging to a New Family of Yeast Sugar Transporters. *Microbiology* **2011**, *157*, 601–608. [CrossRef] [PubMed]
10. Pina, C.; Gonçalves, P.; Prista, C.; Loureiro-Dias, M.C. Ffz1, a New Transporter Specific for Fructose from *Zygosaccharomyces Bailii*. *Microbiology* **2004**, *150*, 2429–2433. [CrossRef] [PubMed]
11. Gordon, J.L.; Wolfe, K.H. Recent Allopolyploid Origin of *Zygosaccharomyces Rouxii* Strain ATCC 42981. *Yeast* **2008**, *25*, 449–456. [CrossRef] [PubMed]
12. Rojo, M.C.; Torres Palazzolo, C.; Cuello, R.; González, M.; Guevara, F.; Ponsone, M.L.; Mercado, L.A.; Martínez, C.; Combina, M. Incidence of Osmophilic Yeasts and *Zygosaccharomyces Rouxii* during the Production of Concentrate Grape Juices. *Food Microbiol.* **2017**, *64*, 7–14. [CrossRef] [PubMed]
13. Rojo, M.C.; Arroyo López, F.N.; Lerena, M.C.; Mercado, L.; Torres, A.; Combina, M. Effects of pH and Sugar Concentration in *Zygosaccharomyces Rouxii* Growth and Time for Spoilage in Concentrated Grape Juice at Isothermal and Non-Isothermal Conditions. *Food Microbiol.* **2014**, *38*, 143–150. [CrossRef] [PubMed]
14. Hosono, K. Effect of Salt Stress on Lipid Composition and Membrane Fluidity of the Salttolerant Yeast *Zygosaccharomyces Rouxii*. *J. Gen. Microbiol.* **1992**, *138*, 91–96. [CrossRef]
15. Wang, H.; Hu, Z.; Long, F.; Guo, C.; Yuan, Y.; Yue, T. Detection of *Zygosaccharomyces rouxii* and *Candida tropicals* in a High Sugar Medium by a Metal Oxide Sensor-Based Electronic Nose and Comparison with Test Panel Evaluation. *J. Food Prot.* **2015**, *78*, 2052–2063. [CrossRef] [PubMed]
16. Fleet, G. Yeast Spoilage of Foods and Beverages. In *The Yeasts*; Elsevier: Amsterdam, The Netherlands, 2010; pp. 53–63.
17. Deak, T.; Beuchat, L.R. Yeasts Associated with Fruit Juice Concentrates. *J. Food Prot.* **1993**, *56*, 777–782. [CrossRef]

18. Combina, M.; Daguerre, C.; Massera, A.; Mercado, L.; Sturm, M.E.; Ganga, A.; Martinez, C. Yeast Identification in Grape Juice Concentrates from Argentina. *Lett. Appl. Microbiol.* **2007**, *46*, 192–197. [CrossRef] [PubMed]

19. Wang, H.; Hu, Z.; Long, F.; Guo, C.; Yuan, Y.; Yue, T. Early Detection of *Zygosaccharomyces Rouxii*—Spawned Spoilage in Apple Juice by Electronic Nose Combined with Chemometrics. *Int. J. Food Microbiol.* **2016**, *217*, 68–78. [CrossRef] [PubMed]

20. Fujioka, K.; Arakawa, E.; Kita, J.I.; Aoyama, Y.; Manome, Y.; Ikeda, K.; Yamamoto, K. Detection of Aeromonas Hydrophila in Liquid Media by Volatile Production Similarity Patterns, Using a FF-2A Electronic Nose. *Sensors* **2013**, *13*, 736–745. [CrossRef] [PubMed]

21. Gobbi, E.; Falasconi, M.; Concina, I.; Mantero, G.; Bianchi, F.; Mattarozzi, M.; Musci, M.; Sberveglieri, G. Electronic Nose and *Alicyclobacillus* spp. Spoilage of Fruit Juices: An Emerging Diagnostic Tool. *Food Control* **2010**, *21*, 1374–1382. [CrossRef]

22. Zhang, J.; Yue, T.; Yuan, Y. Alicyclobacillus Contamination in the Production Line of Kiwi Products in China. *PLoS ONE* **2013**, *8*, e67704. [CrossRef] [PubMed]

23. Arroyo-Manzanares, N.; Huertas-Pérez, J.F.; Gámiz-Gracia, L.; García-Campaña, A.M. Simple and Efficient Methodology to Determine Mycotoxins in Cereal Syrups. *Food Chem.* **2015**, *177*, 274–279. [CrossRef] [PubMed]

24. Pearson, B.M.; McKee, R.A. Rapid Identification of *Saccharomyces cerevisiae*, *Zygosaccharomyces bailii* and *Zygosaccharomyces rouxii*. *Int. J. Food Microbiol.* **1992**, *16*, 63–67. [CrossRef]

25. Kwon, H.; Yeo, E.; Hahn, S.; Bae, S.; Kim, D.; Byun, M. Cloning and Characterization of Genes Encoding Trehalose-6-Phosphate Synthase (TPS1) and Trehalose-6-Phosphate Phosphatase (TPS2) from *Zygosaccharomyces rouxii*. *FEMS Yeast Res.* **2003**, *3*, 433–440. [CrossRef]

26. Iwaki, T.; Higashida, Y.; Tsuji, H.; Tamai, Y.; Watanabe, Y. Characterization of a Second Gene (*ZSOD22*) of Na$^+$/H$^+$ Antiporter from Salt-Tolerant Yeast *Zygosaccharomyces rouxii* and Functional Expression of ZSOD2 and ZSOD22 in *Saccharomyces cerevisiae*. *Yeast* **1998**, *14*, 1167–1174. [CrossRef]

27. Jaramillo, M.C.; Huttener, M.; Alvarez, J.M.; Homs-Corbera, A.; Samitier, J.; Torrens, E.; Juárez, A. Dielectrophoresis Chips Improve PCR Detection of the Food-Spoiling Yeast *Zygosaccharomyces rouxii* in Apple Juice. *Electrophoresis* **2015**, *36*, 1471–1478. [CrossRef] [PubMed]

28. Frisón, L.N.; Chiericatti, C.A.; Aríngoli, E.E.; Basílico, J.C.; Basílico, M.Z. Effect of Different Sanitizers against *Zygosaccharomyces Rouxii*. *J. Food Sci. Technol.* **2014**, *52*, 4619–4624. [CrossRef] [PubMed]

29. Rojo, M.C.; Arroyo López, F.N.; Lerena, M.C.; Mercado, L.; Torres, A.; Combina, M. Evaluation of Different Chemical Preservatives to Control *Zygosaccharomyces Rouxii* Growth in High Sugar Culture Media. *Food Control* **2014**, *50*, 349–355. [CrossRef]

30. Kusumegi, K.; Yoshida, H.; Tomiyama, S. Inhibitory Effects of Acetic Acid on Respiration and Growth of *Zygosaccharomyces Rouxii*. *J. Ferment. Bioeng.* **1998**, *85*, 213–217. [CrossRef]

31. Kirimli, S.; Kunduhoglu, B. Inactivation of Zygosaccharomyces Rouxii Using Ultrasound at Different Temperatures, PH and Water Activity Conditions. *Ital. J. Food Sci.* **2016**, *28*, 64–72.

32. Escott, C.; Loira, I.; Morata, A.; Bañuelos, M.; Suárez-Lepe, J. Wine Spoilage Yeasts: Control Strategy. In *Yeast-Industrial Applications*; Morata, A., Loira, I., Eds.; InTech: Rijeka, Croatia, 2017; pp. 89–116.

33. Alonso, A.; Belda, I.; Santos, A.; Navascués, E.; Marquina, D. Advances in the Control of the Spoilage Caused by *Zygosaccharomyces* Species on Sweet Wines and Concentrated Grape Musts. *Food Control* **2015**, *51*, 129–134. [CrossRef]

34. Enrique, M.; Ibáñez, A.; Marcos, J.F.; Yuste, M.; Martínez, M.; Vallés, S.; Manzanares, P. β-Glucanases as a Tool for the Control of Wine Spoilage Yeasts. *J. Food Sci.* **2010**, *75*, M41–M45. [CrossRef] [PubMed]

35. Usseglio-Tomasset, L. Properties and Use of Sulphur Dioxide. *Food Addit. Contam.* **1992**, *9*, 399–404. [CrossRef] [PubMed]

36. Warth, A.D. Resistance of Yeast Species to Benzoic and Sorbic Acids and to Sulfur Dioxide. *J. Food Prot.* **1985**, *48*, 564–569. [CrossRef]

37. Jansen, M.; Veurink, J.; Euverink, G. Growth of the Salt-Tolerant Yeast *Zygosaccharomyces Rouxii* in Microtiter Plates: Effects of NaCl, PH and Temperature on Growth and Fusel Alcohol Production From. *FEMS Yeast Res.* **2003**, *3*, 313–318. [PubMed]

38. Solieri, L.; Cassanelli, S.; Giudici, P. A New Putative *Zygosaccharomyces* Yeast Species Isolated from Traditional Balsamic Vinegar. *Yeast* **2007**, *24*, 403–417. [CrossRef] [PubMed]

39. Solieri, L.; Giudici, P. Yeasts Associated to Traditional Balsamic Vinegar: Ecological and Technological Features. *Int. J. Food Microbiol.* **2008**, *125*, 36–45. [CrossRef] [PubMed]

40. Hecquet, L.; Sancelme, M.; Bolte, J.; Demuynck, C. Biosynthesis of 4-Hydroxy-2,5-Dimethyl-3(2*H*)-Furanone by *Zygosaccharomyces Rouxii*. *J. Agric. Food Chem.* **1996**, *44*, 1357–1360. [CrossRef]

41. Saha, B.C.; Sakakibara, Y.; Cotta, M.A. Production of D-Arabitol by a Newly Isolated Zygosaccharomyces Rouxii. *J. Ind. Microbiol. Biotechnol.* **2007**, *34*, 519–523. [CrossRef] [PubMed]

42. Kashyap, P.; Sabu, A.; Pandey, A.; Szakacs, G.; Soccol, C.R. Extra-Cellular l-Glutaminase Production by *Zygosaccharomyces Rouxii* under Solid-State Fermentation. *Process Biochem.* **2002**, *38*, 307–312. [CrossRef]

43. Ok, T. Method of Utilization of *Zygosaccharomyces Rouxii*. U.S. Patent US20030219456A1, 21 May 2002.

44. Cao, X.; Hou, L.; Lu, M.; Wang, C.; Zeng, B. Genome Shuffling of *Zygosaccharomyces Rouxii* to Accelerate and Enhance the Flavour Formation of Soy Sauce. *J. Sci. Food Agric.* **2010**, *90*, 281–285. [CrossRef] [PubMed]

45. De Francesco, G.; Turchetti, B.; Sileoni, V.; Marconi, O.; Perretti, G. Screening of New Strains of *Saccharomycodes ludwigii* and *Zygosaccharomyces rouxii* to Produce Low-Alcohol Beer. *J. Inst. Brew.* **2015**, *121*, 113–121. [CrossRef]

46. Romano, P.; Suzzi, G. MINIREVIEW Origin and Production of Acetoin during Wine Yeast Fermentation. *Appl. Environ. Microbiol.* **1996**, *62*, 309–315. [PubMed]

47. Romano, P.; Suzzi, G.; Zironi, R.; Comi, G. Biometric Study of Acetoin Production in *Hanseniaspora guilliermondii* and *Kloeckera apiculata*. *Appl. Environ. Microbiol.* **1993**, *59*, 1838–1841. [PubMed]

48. Contreras, A.; Hidalgo, C.; Schmidt, S.; Henschke, P.A.; Curtin, C.; Varela, C. The Application of Non-*Saccharomyces* Yeast in Fermentations with Limited Aeration as a Strategy for the Production of Wine with Reduced Alcohol Content. *Int. J. Food Microbiol.* **2015**, *205*, 7–15. [CrossRef] [PubMed]

49. Loureiro, V.; Malfeito-Ferreira, M. Spoilage Yeasts in the Wine Industry. *Int. J. Food Microbiol.* **2003**, *86*, 23–50. [CrossRef]

50. Romano, P.; Suzzi, G. Potential Use for *Zygosaccharomyces* Species in Winemaking. *J. Wine Res.* **1993**, *4*, 87–94. [CrossRef]

51. Domizio, P.; Romani, C.; Lencioni, L.; Comitini, F.; Gobbi, M.; Mannazzu, I.; Ciani, M. Outlining a Future for Non-*Saccharomyces* Yeasts: Selection of Putative Spoilage Wine Strains to Be Used in Association with *Saccharomyces cerevisiae* for Grape Juice Fermentation. *Int. J. Food Microbiol.* **2011**, *147*, 170–180. [CrossRef] [PubMed]

52. Jolly, N.P.; Varela, C.; Pretorius, I.S. Nor Your Ordinary Yeast: Non-*Saccharomyces* Yeasts in Wine Production Uncovered. *FEMS Yeast Res.* **2014**, *14*, 215–237. [CrossRef] [PubMed]

fermentation

MDPI

Review

Saccharomycodes ludwigii, Control and Potential Uses in Winemaking Processes

Ricardo Vejarano

Faculty of Engineering, Universidad Privada del Norte (UPN), Av. Del Ejército 920, 13001 Trujillo, Peru; ricardo.vejarano@upn.edu.pe; Tel.: +51-(44)60-6200-4298

Received: 25 July 2018; Accepted: 25 August 2018; Published: 27 August 2018

Abstract: Non-*Saccharomyces* yeasts are becoming important because most of them are considered as spoilage species in winemaking processes, among them the species *Saccharomycodes ludwigii*. This species is frequently isolated at the end of the fermentation process and/or during storage of the wine, i.e., it can to grow in the presence of high levels of ethanol. Besides, this species is adaptable to unfavorable conditions such as high concentrations of SO_2 and is characterized by its capacity to produce high amounts of undesirable metabolites as acetoin, ethyl acetate or acetic acid. To the present, physical (gamma irradiation and continuous pulsed electric fields), chemical (inhibitory compounds such as chitosan and dimethyl dicarbonate) and biological (antagonistic biocontrol by killer yeasts) treatments have been developed in order to control the growth of this spoilage yeast in wines and other fruit derivatives. Therefore, this review is focused on the most relevant studies conducted to control contamination by *S. ludwigii*. Moreover, potential applications of *S. ludwigii* in alternative winemaking techniques, for example for ageing-on-lees and stabilization of red wines, and improvement of aromatic profile are also examined.

Keywords: non-*Saccharomyces* yeast; *Saccharomycodes ludwigii*; *S. ludwigii*; spoilage yeasts' control; ageing-on-lees

1. Introduction

In addition to spoilage bacteria that cause problems in the wine industry, detection and control of spoilage yeasts are vital [1], especially those capable of growing under conditions of low water activity (A_w) and high ethanol content and acidity as well as in the presence of chemical preservatives [2–4], conditions in which other microorganisms are not completely viable.

Grapes and the presence of vectors (insects) that transport microorganisms to the interior of wineries are considered to be the main sources of contamination [5]. Yeasts such as *Dekkera/Brettanomyces* spp., *Zygosaccharomyces bailii* and *Saccharomycodes ludwigii* are considered detrimental to the winemaking process [4]. The presence of these yeasts is indicated by the appearance of superficial films and the production of gases in stored wine, turbidity, sediments, as well as undesirable odors and flavors [6,7].

2. *Saccharomycodes ludwigii*

S'codes ludwigii is known for its ability to contaminate fruit juices and fermented beverages such as wines and cider. Morphologically, it appears as elongated cells with bipolar apiculation (budding yeast) and swelling in the middle (Figure 1), and it presents asexual reproduction by bipolar budding [8,9].

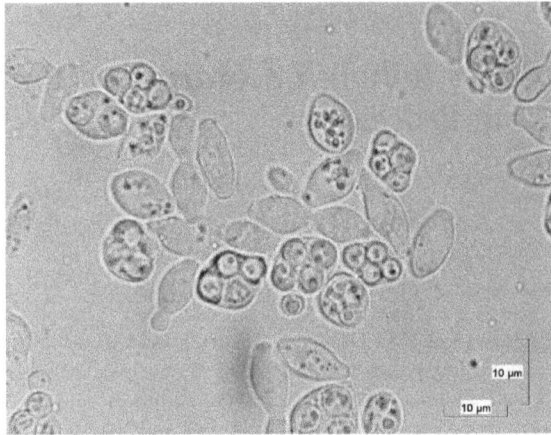

Figure 1. Apiculated cells of *Saccharomycodes ludwigii* at a magnification of 600×.

S'codes ludwigii causes serious problems in the industry due to its high tolerance to sulfur dioxide (SO_2) [10] and is commonly referred to as the "winemaker's nightmare" due to the difficulty in eradicating it from contaminated environments [11–13]. It has also shown resistance to pressurized carbon dioxide (CO_2) with the ability to deteriorate carbonated beverages [6]. It has also been isolated from sweet wines, thus demonstrating its tolerance to high sugar levels [3,4]. Together with *Z. bailii* and some *Saccharomyces cerevisiae* strains, *S'codes ludwigii* is among the spoilage species in bottling lines, especially wines, the process of which uses additives such as SO_2 or sorbic acid [14].

Regarding its fermentative capacity, *S'codes ludwigii* can produce up to 12% *v/v* of ethanol [2,10,15] and acetic acid, in most cases at concentrations <1.0 g/L [2,3]. Some strains have shown acetic acid yields of 0.3–0.5 g/L, similar to some selected strains of *S. cerevisiae* [10]. In addition, this yeast is characterized by its high production of secondary metabolites, such as isobutanol (20.0–200 mg/L), amyl alcohol (32.0–58.0 mg/L), isoamyl alcohol (75.0–190 mg/L), acetaldehyde (46.7–124 mg/L), acetoin (104–478 mg/L) and ethyl acetate (141–580 mg/L) [2,3,10,11,16], which can confer negative undertones to the wine upon exceeding their respective thresholds of perception. However, some strains have shown high production of metabolites such as succinic acid (up to 1.4 g/L) and glycerol (up to 11.7 g/L) [2,3].

The high yield of isobutanol [11] and acetaldehyde [10] can be considered discriminant characteristics of *S'codes ludwigii*; however, Romano et al. [11] obtained low yield of acetaldehyde with several strains, in contrast to the high production of this metabolite by *S'codes ludwigii* as reported in the literature. Other differential characteristics of this yeast are its great capacity to release polysaccharides and its high production of ethyl acetate [10].

Regarding its sugar consumption, *S'codes ludwigii* can ferment glucose, sucrose and raffinose, although it cannot ferment maltose, galactose and lactose [6,17]. It is also capable of assimilating glycerin, cadaverine and ethylamine, although it does not assimilate nitrates [17].

S'codes ludwigii has been reported to increase its production of glycerin, acetic acid and ethyl acetate in media with high concentrations of sugar [3], a phenomenon that can be related to the mechanism of adaptation to osmotic stress to prevent dehydration. Glycerin synthesis involves the oxidation of NADH to NAD^+, and acetic acid synthesis allows NADH to be regenerated [18]. A similar response has been observed in yeasts against toxins, such as the *Pichia membranifaciens* killer toxin (PMKT) [19], so that the osmophilic media protect the yeasts from the action of the PMKT toxin.

2.1. Sources of Contamination by Saccharomycodes ludwigii

Most spoilage yeasts come directly from the surface of grapes (Figure 2) and of equipment and cellar installations [20]. Commonly, *S'codes ludwigii* has been isolated in cases of stuck or sluggish fermentations or during storage of wines [3,7,10,21]. It has also been detected in fruit juices and their fermented derivatives [9,22], tequila and mezcal [23], in soil samples [24], insects [5] and tree secretions [8,25,26].

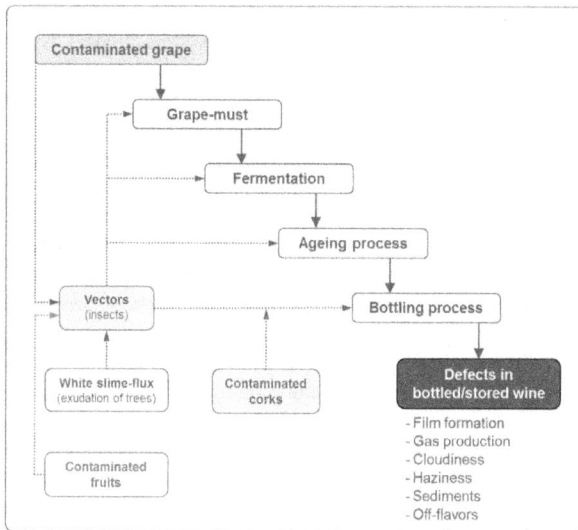

Figure 2. Main sources of *Saccharomycodes ludwigii*, as a spoilage yeast in wines.

In the case of tree secretions, *S'codes ludwigii*, together with other microorganisms, would be transported from "sick trees" to the wineries by insects [5,6]. Sick specimens of trees such as oak, birch, poplar, beech, willow, maple and ash can produce the so-called "alcoholic flux" or "white slime flux", which is characterized by its high content of microorganisms and its smell of beer, malic ester and vinegar [25]. Cases of contamination by *S'codes ludwigii* have also been reported in corks that were inadequately treated with SO_2 before packing [4].

Another source of *S'codes ludwigii*, as well as species such as *S. cerevisiae* and *Z. bailii*, is palm sap, from which a fermented drink known as "palm wine" is obtained in Cameroon [26]. *S'codes ludwigii* is the dominant species at the beginning of the fermentation process, and as the fermentation proceeds, its population decreases in favor of *S. cerevisiae*.

2.2. Detection of Saccharomycodes ludwigii

S'codes ludwigii has been proven to have a high polluting capacity, starting from only one or two cells per liter [27]. The limitations in its proper detection and control are the same as those in the case of other spoilage yeasts, such as the short incubation periods of traditional methods and the use of media for counting "total molds and yeasts" [4].

Among other alternatives, detection based on biomarkers, such as the low content of long chain fatty acids (C18:2 and C18:3) characteristic of *S'codes ludwigii* [7], can be applied; however, its application at the industrial level requires access to databases that allow the interpretation of these molecular profiles in real time to take immediate corrective actions. Another alternative is the use of chemical and organoleptic indicators, similar to 4-ethylphenol produced by *Dekkera/Brettanomyces* spp. [28]. Isobutanol, acetaldehyde, ethyl acetate and acetoin can be used as aromatic indicators of *S'codes ludwigii* [2,3,10,11,16].

2.3. Disadvantages of "Sulfiting" and Resistance of Saccharomycodes ludwigii

SO_2 is generally recognized as a safe additive and is used as an antioxidant and preservative in the control of spoilage bacteria, molds and yeasts. According to Stratford et al. [12], the three forms of SO_2 in solution as a function of pH are called "sulfites"; the molecular form (SO_2) predominates at pH values of <1.80, the HSO_3^- form at pH values of 1.80–7.20 and the SO_3^{2-} form at pH values of >7.20. Of these, SO_2 has the greatest antimicrobial effect [29]. The International Organization of Vine and Wine [30] establishes maximum levels of sulfites according to the type of wine (red, white or rosé), with a higher dose of SO_2 at higher levels of reducing sugars. In Europe, the presence of sulfites must be stated on the bottle when they exceed 10 mg/L (European Union Regulation No. 1991/2004).

Some of the disadvantages of using sulfites are the resistance of *S'codes ludwigii* [12], as well as the dependence of the effect of SO_2 on pH, the generation of undesirable odors and flavors and binding of >50% of the added dose to certain grape-must/wine molecules [12,31], thus losing its antiseptic and antioxidant activity [1]; therefore, its use in the established doses does not always ensure total protection. In addition, sulfites can generate health problems in consumers, such as headaches, allergic reactions and respiratory difficulties in asthmatic individuals, both in its free and linked form [32,33].

Therefore, there is a growing interest in the search for alternative treatments to SO_2, in line with consumers' growing preference for products free of chemical additives [34,35].

Regarding the resistance to SO_2, Stratford et al. [12] required doses of up to 7.8 mM of free sulfites to inactivate *S'codes ludwigii*, which is considerably higher than that required to inactivate *S. cerevisiae* (1.56 mM free sulfites). They also obtained a high yield of acetaldehyde with *S'codes ludwigii* in the presence of SO_2. This response would be a defense mechanism, as is the case with other yeasts, through which SO_2 joins acetaldehyde and other molecules such as pyruvate and 2-oxoglutarate ("sulfite-binding compounds").

The resistance of glyceraldehyde-3-phosphate dehydrogenase (GPDH), especially that of *S'codes ludwigii*, to sulfites has also been proposed as a defense mechanism, without affecting the production of cellular ATP [12,36]. Likewise, only the SO_2 form crosses the cell membrane [37]. *S'codes ludwigii* has a higher C18:1 fatty acid content in its cell membrane [7], which may give greater fluidity to the diffusion of SO_2 toward the outside [38] and may palliate its toxic effect. This would add an additional mechanism related to low intracellular pH in *S'codes ludwigii*, which would favor SO_2 remaining as such, allowing it to flow to the outside without accumulating in the cytoplasm [37].

In Table 1, a summary of studies related to the control of *S'codes ludwigii* in grape-must, wine and fruit juices is presented, which will be described in more detail in the subsequent sections.

Table 1. Applied treatments for the control of *Saccharomycodes ludwigii*.

Sample	Applied Treatment	Reference
Grape must	DMDC	[31,39]
	DMDC + SO$_2$	[39,40]
	DMDC + sorbic acid	[39]
	Toxin of *Pichia anomala* WC65	[41]
	Toxin KpKt	[42]
	Biological control: *Metschnikowia pulcherrima*	[43]
Apple juice	Chitosan	[22]
Mango pulp	Gamma radiation	[44]
	Gamma radiation + steaming	
Wine	DMDC in red wine	[21]
	DMDC in semi-sweet wine	[40]
	PEF	[45]

DMDC: dimethyl dicarbonate. KpKt: *Kluyveromyces phaffii* killer toxin. PEF: pulsed electric fields.

3. Control by Chemical Treatments

3.1. Dimethyl Dicarbonate

Dimethyl dicarbonate (DMDC), also known as dimethyl pyrocarbonate, can be used to partially replace and help reduce SO$_2$ doses [31], with the advantage that it does not generate odors or unpleasant flavors in wine [40], even at the maximum dose of 200 mg/L, authorized by the International Organization of Vine and Wine (OIV) [46].

In aqueous solutions, DMDC is rapidly hydrolyzed mainly to CO$_2$ and methanol at concentrations considered to be safe [47], and its hydrolysis rate increases with temperature; for example, at 10, 20 and 30 °C, it is hydrolyzed in 4, 2 and 1 h, respectively [40]. Its rapid hydrolysis gives it effectiveness as an oenological additive, capable of disinfecting grape-must/wine without leaving toxic residues, unlike its "cousin" diethyl dicarbonate (DEDC), which generates ethyl carbamate, having a carcinogenic potential [48].

The antimicrobial activity of DMDC is favored by low microbial population, low pH values, high ethanol and SO$_2$ contents and temperatures of 20–30 °C. Higher doses are needed to sterilize grape-musts and dealcoholized wines than wines [21,39,40]; DMDC has a greater effect on yeasts than on bacteria [21,31] possibly due to the denaturation of the enzymes GPDH and alcohol dehydrogenase [49].

DMDC acts quickly after dosing, although its period of action is short, unlike SO$_2$, which acts progressively and action is durable; thus, the effectiveness of both preservatives lies in their simultaneous use [50]. Low pH values, which would allow for a high molecular SO$_2$ concentration, are desirable [29,50].

Terrell et al. [39] evaluated the antimicrobial capacity of 0.2, 0.4 and 0.8 mM of potassium metabisulfite (as SO$_2$), potassium sorbate and DMDC in fermentation with *S. cerevisiae* at different levels of inoculum (2, 200 and 20,000 CFU/mL) and temperatures of 21 and 31 °C. The pure DMDC showed an inhibitory effect, and its combinations showed an inhibitory effect to a lesser extent. At 31 °C, the effectiveness of DMDC and its combinations at a dose of 0.8 mM at all inoculum levels increased. No significant differences were observed between SO$_2$, sorbate and SO$_2$ + sorbate at different temperatures and at different preservative concentrations, at all inoculum levels.

Threlfall and Morris [40] evaluated the growth and fermentation capacity of *Saccharomyces bayanus* in grape-must and semi-sweet wine at 20 °C and at different pH values. Certain combinations of SO$_2$ and DMDC were only effective at pH values of 3.0 and 3.2. The minimum doses to completely inhibit microbial growth and fermentation were 200 mg/L of DMDC in grape-must and 50 mg/L of SO$_2$ or 100 mg/L of DMDC in semi-sweet wine, whereas the most effective minimum combinations were

50 mg/L of SO$_2$ + 100 mg/L of DMDC in grape-must and 10 mg/L of SO$_2$ + 50 mg/L of DMDC in semi-sweet wine at any pH value.

On the other hand, Delfini et al. [31] evaluated the inhibitory effect of DMDC (50–10,000 mg/L) in fermentation with grape-must. The dose of 400 mg/L was sufficient to inhibit *S'codes ludwigii* and other species, such as *Hanseniaspora osmophila*, *S. pombe* and *Z. bailii*. Higher doses were required to inhibit bacteria such as *Acetobacter aceti* and *Lactobacillus* sp. (1000 and 500 mg/L, respectively). The authors also concluded that in grape-must treated with 200 mg/L of DMDC (maximum authorized dose), it is recommended to inoculate with *S. cerevisiae* for at least 12 h after dosing to ensure complete hydrolysis and antimicrobial action.

A dose of 200 mg/L would be recommended to confer prolonged stability [21,50]. However, DMDC cannot be used to replace SO$_2$, so its use can only help minimize the doses of the latter [40]. Therefore, during barrel aging, the addition combined with SO$_2$ would be the best alternative, considering that SO$_2$, in addition to microbicide acts as an antioxidant, and thus, the loss of color in red wine in the presence of pure DMDC due to oxidation is avoided [50].

Regarding the rapid hydrolysis of DMDC during barrel aging, periodic dosages of low concentrations (25 mg/L) can help maintain the microbiological quality of wine and lower the doses of SO$_2$. However, it should be emphasized that the maximum dose allowed by the OIV (200 mg/L) is more effective against yeast than against bacteria, especially those producing lactic acid and acetic acid, requiring >500 mg/L doses of DMDC [21,31].

3.2. Chitosan

Chitosan is a deacetylated derivative of chitin, which is a part of the structure of many organisms. Chitosan is considered to act as a chelator of minerals such as Ca and Fe from the fermentation medium, affecting their availability for microbial growth [51,52]. The loss of protein compounds and UV radiation-absorbing material from the cell membrane has also been proposed [53,54].

Due to its polycationic nature (high presence of NH$_2$$^+$ groups), it can interact with negatively-charged groups present in cell surface molecules, such as proteins, anionic polysaccharides, fatty acids and phospholipids, among others, affecting the cell functions and transport of essential nutrients to the inside of the yeast [55,56]. The most commonly-used chemical forms include chitosan glutamate and chitosan lactate [22], the latter having an effect on *S. cerevisiae* at a concentration of 1.0 g/L [57].

Roller and Covill [22] evaluated the effectiveness of different doses of chitosan in apple juice sterilized by ultra-high temperature (UHT) and without additives, against a strain of *S'codes ludwigii* (isolated from contaminated cider). Total inhibition was achieved at a dose of 5.0 g/L of chitosan, whereas a dose of 1.0 g/L only induced a delay in the start of fermentation, without affecting the end of the fermentation process.

Evidently, there is little literature regarding the application of chitosan in the control of *S'codes ludwigii*, which leaves open the possibility of future research to better understand the potential of this polymer in controlling this yeast.

4. Biological Controllers

Another alternative is biological controllers, specifically the so-called killer yeasts, which have an antimicrobial effect on *S'codes ludwigii*, for example some species of *Pichia*, *Kluyveromyces* and *Metschnikowia*.

S'codes ludwigii has shown sensitivity to the microbial toxin produced by *Pichia anomala* WC65. Sawant et al. [41] observed good stability of this toxin at pH values of 2.0–5.0, the usual range in wines. At high concentrations, however, the toxin showed a tendency to aggregate, with loss of activity against *S'codes ludwigii* and other yeasts and blocking of recognition sites being the possible cause for this loss of activity [58]. Therefore, low concentrations could be useful for treatment, although no more studies have been reported.

Another genus, the toxin of which has antimicrobial activity, is *Kluyveromyces*. Palpacelli et al. [42] evaluated the killer activity of the species *Kluyveromyces phaffii*, *Kluyveromyces lactis* and *Kluyveromyces vanudenii*. All showed antimicrobial activity against *S'codes ludwigii*. *K. phaffii* also showed activity against *Kloeckera apiculata* and *Zygosaccharomyces rouxii*. According to the authors, the toxin involved is *Kluyveromyces phaffii* killer toxin (KpKt). Nevertheless, the authors pointed out the need to apply the procedure at the fermentation level, considering that the study was conducted at the laboratory level (plate cultures).

The yeast *Metschnikowia pulcherrima* also has antimicrobial activity and has been used as a biological controller of fungi that cause diseases in fruits [59]. This yeast can grow in mature and overgrown grapes, botrytized grapes and grapes used to make so-called ice wines [60]. The activity of *M. pulcherrima*, in addition to the killer phenomenon [61], would be mainly related to the production of the pulcherrimin pigment by the chelation of Fe in the medium [62], thus decreasing the availability of this mineral for the development of other microorganisms.

Oro et al. [43] evaluated the antimicrobial activity of *M. pulcherrima* against different yeasts and did not observe any effect on *S. cerevisiae*, but did see an effect on *Pichia*, *Brettanomyces/Dekkera*, *Hanseniaspora* and especially on *S'codes ludwigii*.

An interesting alternative for controlling spoilage yeasts during the fermentation process could be the use of mixed inocula with *S. cerevisiae*, which is not affected by *M. pulcherrima* by regulating the absence of Fe in the fermentative medium [63], in addition to taking advantage of other benefits of *M. pulcherrima*, such as its ability to produce aromatic compounds [64].

Although the literature does not report any cases, the potential killer of other yeasts could also be studied for the control of *S'codes ludwigii*; for example, *Candida pyralidae*, producer of the *C. pyralidae* killer toxin (CpKT), with activity against *Brettanomyces bruxellensis*. This toxin has shown stability at pH values of 3.5–4.5 and at temperatures of 15–25 °C, i.e., it is compatible with the winemaking conditions and is not affected by the sugar and ethanol levels present in grape-must/wine. In addition, it has not shown effects on *S. cerevisiae* or on lactic acid bacteria, which would not affect the normal red winemaking process [65]. In the same way, in winemaking conditions (pH values of 3.0–4.5 and at temperatures of 15–25 °C), *Ustilago maydis* fungus has shown killer activity against *B. bruxellensis* [66].

Another potential killer yeast against *S'codes ludwigii* is *P. membranifaciens*, the PMKT toxin of which has shown antifungal activity, with mechanisms that include the alteration of plasma membrane permeability, alteration of cell cycle and induction of cellular apoptosis [19,67]. *S. cerevisiae* has not shown sensitivity to PMKT, but *Z. rouxii* has shown high sensitivity, as well as *Z. bailii*, to a lesser degree [68].

In addition, PMKT can synergistically increase the effect of SO_2. In this regard, Alonso et al. [68] evaluated combinations of PMKT and SO_2 in a medium with high glucose content (60% w/v), showing an inhibitory effect against *Z. rouxii*, although the mechanism of synergistic action PMKT-SO_2 is not fully understood. Pure SO_2 showed no inhibitory effect. Therefore, PMKT could also be used in the control of *S'codes ludwigii*, considering that it shares similar characteristics with *Zygosaccharomyces*, such as the capacity to grow in media with high acidity and low A_W and resistance to osmotic stress and to SO_2, in addition to contaminating concentrated grape-musts, sweet wines and other wines with high residual sugar content [3,4].

Despite the previously-described studies, several killer toxins from *Saccharomyces* and non-*Saccharomyces* yeasts have not yet been characterized. Therefore, further studies are needed in order to identify their genetic origin, mode of action and how to employ them at the industrial level in the control of spoilage yeast, especially *S'codes ludwigii*.

Finally, another interesting strategy to reduce or prevent both the growth of *S'codes ludwigii* and its production of undesirable metabolites in the wine could be the use of starter cultures of yeasts and lactic acid bacteria, as biocontrol agents during alcoholic and malolactic fermentations, similarly to biocontrol processes tested in *B. bruxellensis* [69]. This strategy could ensure a fast and complete fermentation, limiting the available nutrients for growing of spoilage yeasts.

5. Control by Physical Treatments

Out of the various available options, only pulsed electric fields (PEFs) and gamma radiation (γ) have been studied in specific cases with *S'codes ludwigii*. Other technologies, such as high hydrostatic pressure (HHP), ultrasound (US), pulsed light and e-beam radiation, have been applied to inactivate and reduce total populations of yeasts and bacteria in grape-musts and wines [70].

5.1. Pulsed Electric Fields

PEFs cause cell damage through a mechanism related to electroporation or electrical disruption of the membrane, altering the permeability [71]; thus, yeasts become more sensitive (larger size and oval shape) than bacteria [45,72,73] and without the disadvantage of modifying the physicochemical properties and sensorial attributes of grape-must/wine [72,74].

The effectiveness of PEFs varies depending on ethanol content, acidity and temperature, among other factors. *Z. bailii* has shown greater sensitivity in the presence of ethanol [75]. *S. cerevisiae* in a treatment at 45 kV/cm, 46.3 pulses and 70 μs in beer with alcoholic degrees of 0%, 5.2% and 7.0% showed logarithmic reductions of 0.2, 0.7 and 2.2, respectively [76]. Even in the last study, a greater effect was observed at 40–50 °C during treatments in the order of micro- to milli-seconds. Similar results were obtained by Timmermans et al. [73]. However, the optimization of the applicable dose and temperature is required for the purposes of seeking the industrial applicability of PEF as some constituents of grape-must/wine are thermosensitive.

On the other hand, Puértolas et al. [72] managed to reduce the contaminating flora by 99% in grape-must and wine at 186 kJ/kg and 29.0 kV/cm, with greater effectiveness on yeast. There were no significant changes in the color and odor of must and wine treated, even at high doses of PEF. Likewise, all microorganisms were more sensitive in wine than in must, an effect that was attributed to the ethanol content of wine, in accordance with previous results [75,76].

Of the limited experience with the specific application of PEF with *S'codes ludwigii*, only González-Arenzana et al. [45] evaluated a semi-industrial continuous flow system (13.75 L/h) for the control of artificially-contaminated wine, *S'codes ludwigii* being the microorganism that showed greater sensitivity to a specific energy of 60 kJ/kg (103 μs).

In this sense, PEF could allow for a significant reduction in the doses of SO_2 through combined treatment, or in the best of cases, to dispense with its use. In spite of the scarce background, PEF would be an interesting alternative for the control of *S'codes ludwigii* considering its elongated cell morphology [72]. However, one aspect to be taken into account is the tolerance to ethanol shown by *S'codes ludwigii*; therefore, the study of the combined effect of different treatments could better elucidate control pathways during the fermentation process.

5.2. Gamma Radiation

Ionizing radiation, or in combination with conventional chemical and thermal treatments, has been proposed as a replacement alternative. Youssef et al. [44] studied the effect of a combined treatment with steam and γ radiation on the microbiological quality in mango pulp, obtaining a considerable increase in the shelf-life of the product (270 days) compared with irradiated samples without pretreatment with steam (90 days) and with controls without any treatment (15 days). No defects of a chemical, rheological or sensory nature were found.

In addition, six strains of *S'codes ludwigii* were isolated from the untreated pulp, which were inhibited in a medium based on mango pulp at a D_{10}-dose of 2.23 kGy of γ radiation (D_{10}: dose necessary to inactivate 90% of the microbial population), whereas a greater effect was observed in saline solution ($D_{10} = 1.75$ kGy). This indicates that the effect of the γ radiation is influenced by interactions with solids in the medium, which make higher doses of radiation necessary. Therefore, more studies will contribute to improving its application, without producing chemical and sensory changes.

6. Other Applications of *Saccharomycodes ludwigii*

Traditionally, *S'codes ludwigii* and other non-*Saccharomyces* have been considered spoilage yeasts, which is a concept that has changed in recent decades thanks to several works that demonstrate their advantages in the production of wine and other beverages.

6.1. Aromatic Profile Improvement in Wines

It has been noted that most of the secondary metabolites produced by pure cultures of non-*Saccharomyces* do not reach the thresholds of perception when they are made in mixed fermentation with *S. cerevisiae*, since the latter can modulate the metabolism of the former [10,77]. Although *S'codes ludwigii* produces high levels of ethyl acetate and acetic acid, it is possible to modulate this production in mixed cultures, in addition to improving the yield of esters with a positive impact on the wine's aromatic profile.

One case is the Sd64 strain studied by Domizio et al. [10], which in mixed cultures with *S. cerevisiae* (ratio 10^3:10^7 cells/mL), increased the production of glycerin (up to 21.8%), isoamyl acetate (up to 20.8%) and 2-phenylethanol (>200%) compared with a pure culture of *S. cerevisiae*, besides producing low volatile acidity (0.32 g/L), lower than other non-*Saccharomyces* and the pure culture of *S. cerevisiae*.

However, in mixed culture, high levels of acetaldehyde (up to 33% higher) and ethyl acetate (up to 10-times higher) were also obtained compared with the pure culture of *S. cerevisiae*, which could be improved with the selection of strains with low production of these metabolites and with the optimization of fermentative parameters that regulate their production. Granchi et al. [3] obtained a lower yield of acetaldehyde, acetoin and ethyl acetate at 25 °C, when compared with that obtained during fermentation at 15 °C. Conditions also compatible with the β-glucosidase activity of some strains of *S'codes ludwigii*, 46% higher than *S. cerevisiae* at 30 °C [78], favor the release of aromatic compounds from non-aromatic precursors of grapes [79,80], thus improving the wine's aroma. Of course, this improvement would be advisable only in white wines, since β-glucosidase or anthocyanase generates the hydrolysis of anthocyanins [81], so that its applicability would not be viable in red wines due to the loss of color.

6.2. Reduction of Alcohol Content in Wine

High temperatures in vineyards induce changes in the chemical composition of grapes, mainly an increase in sugar and decrease in acids and anthocyanins, which results in wines with a higher concentration of ethanol and alteration in the mouthfeel, flavor and aroma, or even an increase in the sensation of astringency, bitterness and roughness [82], to which we must add the consequences of high doses of ethanol on the consumer's health.

S'codes ludwigii can lower the production of ethanol in mixed cultures with *S. cerevisiae*, as obtained by Domizio et al. [10] with the Sd64 strain (previously mentioned), with which they achieved a reduction in alcoholic degree of up to 1.74% *v/v* in mixed culture compared with the pure culture of *S. cerevisiae*. Therefore, *S'codes ludwigii* can also be considered as a potential yeast to lower the alcoholic degree in mixed fermentations, a field not studied so far.

6.3. Release of Polysaccharides in Red Wines

Several studies have demonstrated the feasibility of using *S'codes ludwigii* for the release of polysaccharides in wine not only in the traditional aging-on-lees (AOL), as a result of cellular autolysis [83,84], but also during growth and alcoholic fermentation [10,15,85] due to the controlled hydrolysis of cell walls (β-glucanase activity) to allow cell budding [86].

The yeast *S'codes ludwigii* has shown a high capacity to release polysaccharides during the fermentation process—up to 300% more than *S. cerevisiae* [10,15]—while rates of release during AOL are >200% compared with *S. cerevisiae* [84,85].

The most abundant polysaccharides are the mannoproteins (Table 2), located in the outer layer of the cell wall, linked by β-1,6 glucan, β-1,3 glucan and chitin chains [84,87]. Generally speaking, they contain 85%–90% of carbohydrates, mainly mannose, and 10%–15% of proteins [84,85,87].

Table 2. Composition of cell walls of *Saccharomycodes ludwigii*, *Schizosaccharomyces pombe* and *Saccharomyces cerevisiae*.

Component (%)	*S'codes ludwigii*	*S. pombe*	*S. cerevisiae*	Reference
Proteins	12	11	24 *	
Mannose	93	55	88 *	[85]
Glucose	7	22	12 *	
Galactose	-	23	-	
α (1-3) glucan		Yes	No	
β (1-3) glucan		Yes	Yes	[84] **
β (1-6) glucan		Yes	Yes	
Chitin (% of dry weight)		0.5	0.1	

(*) Average for three strains. (**) Not reported for *S'codes ludwigii*.

Polysaccharides, especially in red wines, can improve the mouthfullness and body [88], sweetness and roundness [89], aromatic persistence [90], protein and tartaric stability [91,92], interaction with tannins and reduction of astringency [93] and protection of phenolic compounds against oxidation, making it possible to maintain antioxidant and anti-inflammatory capacity [94].

Polysaccharides also interact with tertiary aromatic compounds [95], which may confer a lower perception of wood aromas in long-aged wines, in addition to stimulating the malolactic fermentation [96], as well as improving the quality of foam in sparkling wines [97] and adsorbing undesirable compounds such as ochratoxin A [98], the presence of which in wine leads to risks to the consumer's health [99].

Palomero et al. [84] obtained a high release of polysaccharides by *S'codes ludwigii* (110.51 mg/L) and by *S. pombe* (103.61 mg/L), with respect to *S. cerevisiae* (36.65 mg/L), in a hydro-alcoholic medium. In the case of *Saccharomycodes* and *Schizosaccharomyces*, polysaccharides were of a larger molecular size, with a potential positive impact on the wine's palatability. These yeasts' high capacity for releasing polysaccharides is related to the chemical composition and structure of their cell walls (Table 2), mainly glucans and mannoproteins [85].

Likewise, Palomero et al. [84] evaluated the effect of lees in red wine (Garnacha), observing a loss of color due to the weak and reversible interaction between monomer anthocyanins and polysaccharides [100]. Lower loss was observed in pyranoanthocyanins due to the presence of the fourth heteroaromatic ring in its structure [101]. However, the loss of color was lower with *S'codes ludwigii* and *S. pombe* than with *S. cerevisiae*.

No significant effect was observed on the volatile fraction, whereas the sensory analysis in the wine treated with lees from *S'codes ludwigii* showed low astringency and bitterness and greater body. However, with this yeast, the perception of the reduction aroma was high, which indicates the need for more work in the selection of strains that confer this characteristic to the treated wine to a lesser extent.

6.4. Combined Treatments: Aging-on-Lees with Ultrasound

The coupling of AOL with US is possible because of the cavitation generated in the cell wall by the creation of localized areas with high temperature (up to 5000 °C) and high pressure (up to 50,000 kPa) [102], in addition to the formation of hydroxyl radicals (OH) that act on the cell wall altered by US waves [103], thus improving the release of polysaccharides.

A research work was conducted on this topic by Kulkarni et al. [83] with *S'codes ludwigii*, *S. pombe*, *M. pulcherrima*, *S. cerevisiae* and other yeasts in a hydro-alcoholic medium (seven weeks of AOL at 23 °C, applying US at a dose of 50 kHz for 10 min a day). *S'codes ludwigii* showed a high rate of release of polysaccharides from the third week around 460 mg/L.

The authors also applied AOL in red wine, observing a decrease in the anthocyanin content, without affecting the stability of pyranoanthocyanins (vitisins and vinylphenols), in accordance with Morata et al. [101] and Palomero et al. [84], especially with the lees of *S'codes ludwigii*. A decrease in the content of proanthocyanidins was also observed, particularly with the lees of *S'codes ludwigii*, contributing to a decrease in the astringency and bitterness of wine (sensory analysis). Regarding aroma, esters were the main group released, especially ethyl lactate, which could be related to the esterase activity during autolysis.

Finally, AOL implies economic impacts due to the investment necessary to store wines in wine cellars, as well as the potential risk of organoleptic and microbiological alterations in these wines. It is thus necessary to optimize the time and conditions under which AOL is carried out in addition to optimizing the time and intensity of US doses, which in addition to accelerating the process, minimizes the degradation of polysaccharides by the action of US waves [104].

6.5. Non-Wine Fermentations

Another interesting alternative for the use of *S'codes ludwigii* is the elaboration of "fruit wines", in which the high production of aromatic compounds and organic acids can be exploited.

This type of drink is traditionally made with a poor aromatic profile in different parts of the world, mainly because *S. cerevisiae* is used [105,106]. Mixed or sequential fermentations could contribute to improving the sensory profile of these beverages, which constitutes an opportunity for the use of *S'codes ludwigii* due to its high production of ethyl acetate, isoamyl acetate and amyl, isoamyl or 2-phenylethyl alcohols [10,78].

7. Future Perspectives

7.1. Adaptation to Harsh Conditions

It is known that the most studied spoilage yeast is *B. bruxellensis*, which can be used as a reference to know how much progress has been made and what is possible to improve, allowing the design of effective strategies for spoilage yeast control in wines. Like *S'codes ludwigii*, *B. bruxellensis* is capable of surviving and proliferating after alcoholic fermentation is completed [107], even in the presence of SO_2 [108].

According to Smith and Divol [109], the factors that allow these spoilage yeasts to be better adapted to unfavorable environments could either be internal (genotypic) or external (nutritional, phenotypic) in nature or both. In this regard, many studies have been performed in order to investigate genetic bases that allow these yeasts to adapt to unfavorable conditions in which other microorganisms are not completely viable, for example high ethanol levels. *B. bruxellensis* is well adapted to these conditions including its ability to utilize ethanol as a carbon source [110].

On the other hand, to control the proliferation of most of the spoilage yeasts, SO_2 is commonly employed, and many studies have been carried out, especially with *B. bruxellensis*, in order to explore the relationship between SO_2 tolerance and genotype. The identification of susceptible or resistant strains to sulfite could help to develop appropriate antimicrobial techniques and efficient spoilage prevention [111]. Capozzi et al. [112] observed the expression of genes involved in carbohydrate metabolism and encoding heat shock proteins, as well as enriched categories including amino acid transport and transporter activity in the presence of SO_2. Moreover, geographical origin has shown a significant influence on the biodiversity of spoilage yeast such as *B. bruxellensis*, displaying variation in tolerance to SO_2 [113].

Evidently, there is little literature regarding the genotypic and phenotypic characterization of *S'codes ludwigii*, which would lead to better understanding of its mechanisms of adaptation to unfavorable conditions. This aspect leaves open the possibility of future research to better design strategies for effective control of this yeast in winemaking processes.

7.2. Emerging Technologies for Controlling S'codes ludwigii

Emerging technologies, such as HHP and PEF, are interesting alternatives to reduce the doses of antimicrobial agents and antioxidants such as SO_2 [114], especially in red wines, which are less susceptible to oxidation than white wines. It is also possible to produce SO_2-free red wines by applying UV or e-beam irradiation if hygienic conditions during the process are adequate [70], allowing, among other advantages, for the proper implantation of starter cultures during fermentation, apart from contributing to improving the extraction of phenolic and aromatic compounds.

However, the scaling up of technologies such as PEF at the industrial level is still a pending issue, since most studies have been carried out with small sample volumes and in static systems [115] and occasionally in continuous flow systems at the laboratory level [116]; therefore, it is necessary to conduct more studies that allow for its application in large volumes and in continuous flow systems to implement this technology in the winery, such as the one developed by González-Arenzana et al. [45].

On the other hand, the antimicrobial effect of radiation can be altered due to its interactions with the components of food samples, as observed by Youssef et al. [44], requiring a greater dose of gamma radiation to reduce (by 90%) the population of *S'codes ludwigii* in mango pulp (2.23 kGy) compared to a saline solution (1.75 kGy). No studies have been reported (review in ScienceDirect) on the application of this radiation in grape-must.

7.3. Considerations about Chemical Preservatives

An important point made by Roller and Covill [22] is the need to evaluate the effect of parameters such as pH, temperature, yeast strains, presence of other preservatives and food composition on the microbicidal capacity of potential preservatives such as chitosan. Most background data show that chitosan has been evaluated in media such as distilled water or phosphate buffer, and the control of *S'codes ludwigii* in fruit juices, especially grape-must, has been little studied; therefore, their behavior is not clear in these types of matrices. Besides, an important background is that chitosan has shown activity against *Brettanomyces bruxellensis* [117].

In the same vein, a lower antimicrobial effect on *S. cerevisiae* was seen in grape-must than in a synthetic medium at equal doses of DMDC [31]. The authors considered a possible interaction between DMDC and some grape-must/wine compounds, for example with coloring substances. Previously, a significant decrease in the content of ascorbic acid, amino acids, fructose, glucose, lycopene and α-carotene was observed in the presence of DMDC in tomato juice [118]. These possible interactions between DMDC and grape-must/wine components merit further investigation due to their possible technological consequences for wine.

Regarding the hydrolysis of DMDC, the production of methyl carbamate has been detected as a result of its reaction with ammonium, amino acids, polyphenols and organic acids present in grape-must/wine, as well as the formation of other metabolites due to its reaction with the higher alcohol content of wine [47]. Therefore, these interactions must be studied in more detail to verify their potential impacts on the quality of the treated grape-must/wine.

Moreover, the maximum allowed dose of DMDC is 200 mg/L [46], and its complete hydrolysis yields approximately 96 mg of methanol. Although this concentration of methanol is lower than the maximum allowed, 400 mg/L for red wines and 250 mg/L for white and rosé wines [30], the presence of endogenous methanol in wine could increase its concentration to toxic levels [119]. Therefore, the search for alternatives that lower the doses of DMDC becomes of special interest, for example its combination with PEF or gamma radiation, with proven efficacy against *S'codes ludwigii* [44,45].

On the other hand, although the literature does not report previous cases with *S'codes ludwigii*, treatments with gaseous ozone [120] have shown effectiveness to reduce the concentration of ethylphenols in the wine and a partial reduction of *B. bruxellensis* cells, considered among the most common spoilage yeasts in winemaking processes [4].

7.4. Selection of S'codes ludwigii Strains with Differentiated Characteristics

Studies of mixed fermentations between *S'codes ludwigii* and *S. cerevisiae* mention the modulation of the fermentative metabolism between both yeasts, which would have advantages such as a decrease in the alcohol content, an increase in aromatic compounds and a greater release of polysaccharides [10,84].

Most strains studied have shown a high production of acetoin and ethyl acetate. This indicates the need to select strains of *S'codes ludwigii* with low production of these metabolites, which also contributes to an increase in the levels of desirable metabolites, such as isoamyl acetate (banana flavor) and 2-phenylethanol (rose flavor). Likewise, it would be interesting to evaluate the impact of these strains in co-cultures with *S. cerevisiae* [10,78] on the aromatic profile of wines.

Other aspects that require further study are related to the application of *S'codes ludwigii* in AOL, for example, the search for strains with low pigment adsorption [95] and low expression of anthocyanin activity (anthocyanin-β-glucosidase) causing the hydrolysis of anthocyanins, given that a high expression of this activity in some strains of *S'codes ludwigii* has been reported [79,80]. It is also necessary to study the capacity of *S'codes ludwigii* to produce pyranoanthocyanins (vitisins and vinylphenols), which are more stable than monomer anthocyanins in facing the degradation caused by anthocyanase activity [81], thus minimizing the loss of color during AOL in red wines.

Other aspects to be addressed in future studies with *S'codes ludwigii*, given their high release of polysaccharides, are:

The impact of mannose, glucose and protein content of polysaccharides on the wine quality, only studied so far in model media [121].

The use of *S'codes ludwigii* for the exogenous production of polysaccharides, which can be added to wine during AOL [122]. Of course, it is necessary to search for suitable strains, for example those with a low contribution to the reduction of aromas [84].

7.5. Production of Other Fermented Beverages

Another potential industrial application of *S'codes ludwigii* is the production of fermented beverages from other fruits, for example drinks with a higher content of acidity for summer and those with more intense fruity profiles, as demonstrated by Romano et al. [11] with the S81 strain.

Likewise, the high β-glucosidase activity shown by some strains of *S'codes ludwigii* [78] can be used to improve the varietal aromatic profile, given that this enzyme releases aromatic compounds from glycosylated non-aromatic precursors [79,80].

8. Conclusions

S'codes ludwigii is a yeast commonly considered as a wine contaminant due to its high production of ethyl acetate, acetoin or acetaldehyde, with negative effects on the sensory profile at levels above its perception threshold. Traditionally, the control of this and other yeasts is carried out with SO_2, which, however, at the high doses often required, causes health problems and defects in wine that lead to rejection by the consumer. Among the possible alternatives to SO_2, most have been studied for the control of total microbial populations, and not specifically for *S'codes ludwigii*. Of the few studies available, most have been conducted at the laboratory level, which include, for example, physical treatments such as with PEFs and gamma radiation, which still need improvement. Of the chemical treatments available, DMDC, despite being authorized by the OIV, is limited by its rapid hydrolysis and its lack of antioxidant activity, which makes its application in combination with SO_2 necessary. Another alternative is chitosan; however, no applications have been reported in grape-musts. Biological control can also be applied, taking advantage of the killer activity of some strains on *S'codes ludwigii*, an alternative that also requires further studies for its possible scaling at an industrial level.

On the other hand, *S'codes ludwigii* has potential applications in winemaking due to the ability of some strains to reduce the alcoholic degree and volatile acidity, as well as the high production of glycerin, isoamyl acetate, 2-phenylethanol and polysaccharides and its β-glucosidase activity to improve the varietal aroma in white wines. These are considerations that open up new research possibilities without forgetting the potential of *S'codes ludwigii* in the cider and beer brewing industries, to which it would bring many benefits; however, this is not the subject of this review.

Funding: This research received no external funding.

Acknowledgments: Thanks to the National Direction for Research and Development of the Universidad Privada del Norte (UPN) for the financial support in the translation of the manuscript.

Conflicts of Interest: The author declares no conflict of interest.

References

1. Ribéreau-Gayon, P.; Dubourdieu, D.; Donèche, B.; Lonvaud, A. The Microbiology of Wine and Vinification. In *Handbook of Enology*, 2nd ed.; John Wiley and Sons Ltd.: Chichester, UK, 2006; Volume 1, pp. 193–221. ISBN 978-0-470-01034-1.

2. Ciani, M.; Maccarelli, F. Oenological properties of non-*Saccharomyces* yeasts associated with wine-making. *World J. Microb. Biot.* **1997**, *14*, 199–203. [CrossRef]

3. Granchi, L.; Ganucci, D.; Messini, A.; Vincenzini, M. Oenological properties of *Hanseniaspora osmophila* and *Kloeckera cortices* from wines produced by spontaneous fermentations of normal and dried grapes. *FEMS Yeast Res.* **2002**, *2*, 403–407. [CrossRef] [PubMed]

4. Loureiro, V.; Malfeito-Ferreira, M. Spoilage yeasts in the wine industry. *Int. J. Food Microbiol.* **2003**, *86*, 23–50. [CrossRef]

5. Lachance, M.A.; Gilbert, G.D.; Starmer, W.T. Yeast communities associated with *Drosophila* species and related flies in an eastern oak-pine forest: A comparison with western communities. *J. Ind. Microbiol.* **1995**, *14*, 484–494. [CrossRef] [PubMed]

6. Boundy-Mills, K.; Stratford, M.; Miller, M.W. *Saccharomycodes* E.C. Hansen (1904). In *The Yeasts, a Taxonomic Study*, 5th ed.; Kurtzman, C.P., Fell, J.W., Boekhout, T., Eds.; Elsevier: London, UK, 2011; pp. 747–750.

7. Malfeito-Ferreira, M.; Tareco, M.; Loureiro, V. Fatty acid profiling: A feasible typing system to trace yeast contamination in wine bottling plants. *Int. J. Food Microbiol.* **1997**, *38*, 143–155. [CrossRef]

8. Miller, M.W.; Phaff, H.J. *Saccharomycodes* E.C. Hansen. In *The Yeasts. A Taxonomic Study*, 4th ed.; Kurtzman, C.P., Fell, J.W., Eds.; Elsevier: New York, NY, USA, 1998; pp. 372–373.

9. Yamazaki, T.; Oshima, Y. *Saccharomycodes ludwigii* has seven chromosomes. *Yeast* **1996**, *12*, 237–240. [CrossRef]

10. Domizio, P.; Romani, C.; Lencioni, L.; Comitini, F.; Gobbi, M.; Mannazzu, I.; Ciani, M. Outlining a future for non-*Saccharomyces* yeasts: Selection of putative spoilage wine strains to be used in association with *Saccharomyces cerevisiae* for grape juice fermentation. *Int. J. Food Microbiol.* **2011**, *147*, 170–180. [CrossRef] [PubMed]

11. Romano, P.; Marchese, R.; Laurita, C.; Saleano, G.; Turbanti, L. Biotechnological suitability of *Saccharomycodes ludwigii* for fermented beverages. *World J. Microb. Biot.* **1999**, *15*, 451–454. [CrossRef]

12. Stratford, M.; Morgan, P.; Rose, A.H. Sulphur dioxide resistance in *Saccharomyces cerevisiae* and *Saccharomycodes ludwigii*. *J. Gen. Microbiol.* **1987**, *133*, 2173–2179. [CrossRef]

13. Thomas, D.S. Yeasts as spoilage organisms in beverages. In *The yeasts. Yeast Technology*, 2nd ed.; Rose, A.H., Harrison, J.S., Eds.; Academic Press: London, UK, 1993; Volume 5, pp. 517–561. ISBN 0-12-596415-3.

14. Warth, A.D. Resistance of yeast species to benzoic and sorbic acid and sulphur dioxide. *J. Food Prot.* **1985**, *48*, 564–569. [CrossRef]

15. Domizio, P.; Liu, Y.; Bisson, L.F.; Barile, D. Use of non-*Saccharomyces* wine yeasts as novel sources of mannoproteins in wine. *Food Microbiol.* **2014**, *43*, 5–15. [CrossRef] [PubMed]

16. Romano, P.; Fiore, C.; Paraggio, M.; Caruso, M.; Cepece, A. Function of yeast species and strains in wine flavour. *Int. J. Food Microbiol.* **2003**, *86*, 169–180. [CrossRef]

17. Fugelsang, K.C.; Edwards, C.G. Yeasts. In *Wine Microbiology Practical Applications and Procedures*, 2nd ed.; Fugelsang, K.C., Edwards, C.G., Eds.; Springer Science+Business Media: New York, NY, USA, 2007; pp. 3–14. ISBN 978-0-387-33349-6.

18. Remize, F.; Roustan, J.L.; Sablayrolles, J.M.; Barre, P.; Dequin, S. Glycerol overproduction by engineered *Saccharomyces cerevisiae* wine yeast strains leads to substantial changes in byproduct formation and to a stimulation of fermentation rate in stationary phase. *Appl. Environ. Microb.* **1999**, *65*, 143–149.

19. Santos, A.; Marquina, D. Ion channel activity by *Pichia membranifaciens* killer toxin. *Yeast* **2004**, *21*, 151–162. [CrossRef] [PubMed]

20. Gschaedler, A. Contribution of non-conventional yeasts in alcoholic beverages. *Curr. Opin. Food Sci.* **2017**, *13*, 73–77. [CrossRef]

21. Costa, A.; Barata, A.; Malfeito-Ferreira, M.; Loureiro, V. Evaluation of the inhibitory effect of dimethyl dicarbonate (DMDC) against wine microorganisms. *Food Microbiol.* **2008**, *25*, 422–427. [CrossRef] [PubMed]

22. Roller, S.; Covill, N. The antifungal properties of chitosan in laboratory media and apple juice. *Int. J. Food Microbiol.* **1999**, *47*, 67–77. [CrossRef]

23. Lachance, M.A. Yeast communities in a natural tequila fermentation. *Antonie van Leeuwenhoek* **1995**, *68*, 151–160. [CrossRef] [PubMed]

24. Barnett, J.A.; Payne, R.W.; Yarrow, D. *Yeasts: Characteristics and Identification*, 3rd ed.; Cambridge University Press: Cambridge, UK, 2000.

25. Ogilvie, L. Observations on the "slime-fluxes" of trees. *Trans. Br. Mycol. Soc.* **1924**, *9*, 167–182. [CrossRef]

26. Stringini, M.; Comitini, F.; Taccari, M.; Ciani, M. Yeast diversity during tapping and fermentation of palm wine from Cameroon. *Food Microbiol.* **2009**, *26*, 415–420. [CrossRef] [PubMed]

27. Beech, F.W.; Carr, J.G. Cider and perry. In *Economic Microbiology, Volume 1, Alcoholic beverages*; Rose, A.H., Ed.; Academic Press: London, UK, 1977; pp. 139–313. ISBN 978-0125965507.

28. Morata, A.; Vejarano, R.; Ridolfi, G.; Benito, S.; Palomero, F.; Uthurry, C.; Tesfaye, W.; González, C.; Suárez-Lepe, J.A. Reduction of 4-ethylphenol production in red wines using HCDC+ yeasts and cinnamyl esterases. *Enzyme Microb. Technol.* **2013**, *52*, 99–104. [CrossRef] [PubMed]

29. Jarvis, B.; Lea, A.G.H. Sulphite binding in ciders. *Int. J. Food Sci. Technol.* **2000**, *35*, 113–127. [CrossRef]

30. OIV. *Compendium of International Methods of Wine and Must Analysis*, 2018 ed.; International Organization of Vine and Wine (OIV): Paris, France, 2018; Volume II, OIV-MA-C1-01: R2011; ISBN 979-10-91799-79-9. Available online: http://www.oiv.int/public/medias/5773/compendium-2018-en-vol2.pdf (accessed on 25 July 2018).

31. Delfini, C.; Gaia, P.; Schellino, R.; Strano, M.; Pagliara, A.; Ambró, S. Fermentability of grape must after inhibition with dimethyl dicarbonate (DMDC). *J. Agric. Food Chem.* **2002**, *50*, 5605–5611. [CrossRef] [PubMed]

32. Santos, M.C.; Nunes, C.; Saraiva, J.A.; Coimbra, M.A. Chemical and physical methodologies for the replacement/reduction of sulfur dioxide use during winemaking: Review of their potentialities and limitations. *Eur. Food Res. Technol.* **2012**, *234*, 1–12. [CrossRef]

33. Vally, H.; Misso, N.L.A.; Madan, V. Clinical effects of sulphite additives. *Clin. Exp. Allergy* **2009**, *39*, 1643–1651. [CrossRef] [PubMed]

34. Bech-Larsen, T.; Scholderer, J. Functional foods in Europe: Consumer research, market experiences and regulatory aspects. *Trends Food Sci. Technol.* **2007**, *18*, 231–234. [CrossRef]

35. Cravero, F.; Englezos, V.; Torchio, F.; Giacosa, S.; Río Segade, S.; Gerbi, V.; Rantsiou, K.; Rolle, L.; Cocolin, L. Post-harvest control of wine-grape mycobiota using electrolyzed water. *Innov. Food Sci. Emerg.* **2016**, *35*, 21–28. [CrossRef]

36. Hinze, H.; Holzer, H. Analysis of the energy metabolism after incubation of *Saccharomyces cerevisiae* with sulfite or nitrite. *Arch. Microbiol.* **1986**, *145*, 27–31. [CrossRef] [PubMed]

37. Stratford, M.; Rose, A.H. Transport of sulphide dioxide by *Saccharomyces cerevisiae*. *J. Gen. Microbiol.* **1986**, *132*, 1–6. [CrossRef]

38. Kaneko, H.; Hosahara, M.; Tanaka, M.; Itoh, T. Lipid composition of 30 species of yeast. *Lipids* **1976**, *11*, 837–844. [CrossRef] [PubMed]

39. Terrell, F.R.; Morris, J.R.; Johnson, M.G.; Gbur, E.E.; Makus, D.J. Yeast inhibition in grape juice containing sulfur dioxide, sorbic acid, and dimethyldicarbonate. *J. Food. Sci.* **1993**, *58*, 1132–1134. [CrossRef]

40. Threlfall, R.T.; Morris, J.R. Using dimethyldicarbonate to minimize sulfur dioxide for prevention of fermentation from excessive yeast contamination in juice and semi-sweet wine. *J. Food Sci.* **2002**, *67*, 2758–2762. [CrossRef]

41. Sawant, A.D.; Abdelal, A.T.; Ahearn, D.G. Purification and characterization of the anti-Candida toxin of *Pichia anomala* WC 65. *Antimicrob. Agents Chemother.* **1989**, *33*, 48–52. [CrossRef] [PubMed]

42. Palpacelli, V.; Ciani, M.; Rosini, G. Activity of different 'killer' yeasts on strains of yeast species undesirable in the food industry. *FEMS Microbiol. Lett.* **1991**, *84*, 75–78. [CrossRef]

43. Oro, L.; Ciani, M.; Comitini, F. Antimicrobial activity of *Metschnikowia pulcherrima* on wine yeasts. *J. Appl. Microbiol.* **2014**, *116*, 1209–1217. [CrossRef] [PubMed]

44. Youssef, B.M.; Asker, A.A.; El-Samahy, S.K.; Swailam, H.M. Combined effect of steaming and gamma irradiation on the quality of mango pulp stored at refrigerated temperature. *Food Res. Int.* **2002**, *35*, 1–13. [CrossRef]

45. González-Arenzana, L.; Portua, J.; López, R.; López, N.; Santamaría, P.; Garde-Cerdán, T.; López-Alfaro, I. Inactivation of wine-associated microbiota by continuous pulsed electric field treatments. *Innov. Food Sci. Emerg.* **2015**, *29*, 187–192. [CrossRef]

46. OIV. *International Code of Oenological Practices*; Issue 2018; International Organization of Vine and Wine (OIV): Paris, France, 2018; OENO 5/01, OENO 421-2011; ISBN 979-10-91799-88-1.

47. Peterson, T.W.; Ough, C.S. Dimethyldicarbonate reaction with higher alcohols. *Am. J. Enol. Viticult.* **1979**, *30*, 119–123.

48. Suárez-Lepe, J.A.; Morata, A. New trends in yeast selection for winemaking. *Trends Food Sci. Technol.* **2012**, *23*, 39–50. [CrossRef]

49. Porter, L.T.; Ough, C.S. The effects of ethanol, temperature and dimethyldicarbonate on viability of *Saccharomyces cerevisiae* Montrachet No 522 in wine. *Am. J. Enol. Viticult.* **1982**, *33*, 222–225.

50. Divol, B.; Strehaiano, P.; Lonvaud-Funel, A. Effectiveness of dimethyldicarbonate to stop alcoholic fermentation in wine. *Food Microbiol.* **2005**, *22*, 169–178. [CrossRef]

51. Feng, M.; Lalor, B.; Hu, S.; Mei, J.; Huber, A.; Kidby, D.; Holbein, B. Inhibition of yeast growth in grape juice through removal of iron and other metals. *Int. J. Food Sci. Technol.* **1997**, *32*, 21–28. [CrossRef]

52. Jackson, S.L.; Heath, I.B. Roles of calcium ions in hyphal tip growth. *Microbiol. Mol. Biol. R.* **1993**, *57*, 367–382.

53. Fang, S.W.; Li, C.F.; Shih, D.Y.C. Antifungal activity of chitosan and its preservative effect on low-sugar candied kumquat. *J. Food Protect.* **1994**, *56*, 136–140. [CrossRef]

54. Leuba, J.L.; Stossel, P. Chitosan and other polyamines: Antifungal activity and interaction with biological membranes. In *Chitin in Nature and Technology*; Muzzarelli, R., Jeuniaux, C., Gooday, G.W., Eds.; Springer: Boston, MA, USA, 1986; pp. 215–222.

55. Muzzarelli, R.A.A. Chitosan-based dietary foods. *Carbohydr. Polym.* **1996**, *29*, 309–316. [CrossRef]

56. Ren, J.; Liu, J.; Li, R.; Dong, F.; Guo, Z. Antifungal properties of chitosan salts in laboratory media. *J. Appl. Polym. Sci.* **2012**, *124*, 2501–2507. [CrossRef]

57. Papineau, A.M.; Hoover, D.G.; Knorr, D.; Farkas, D.F. Antimicrobial effect of water-soluble chitosans with high hydrostatic pressure. *Food Biotechnol.* **1991**, *5*, 45–57. [CrossRef]

58. Schmitt, M.; Radler, F. Molecular structure of the cell wall receptor for killer toxin KT28 in *Saccharomyces cerevisiae*. *J. Bacteriol.* **1988**, *170*, 2192–2196. [CrossRef] [PubMed]

59. Saravanakumar, D.; Ciavorella, A.; Spadaro, D.; Garibaldi, A.; Gullino, M.L. *Metschnikowia pulcherrima* strain MACH1 outcompetes *Botrytis cinerea*, *Alternaria alternata* and *Penicillium expansum* in apples through iron depletion. *Postharvest Biol. Technol.* **2008**, *49*, 121–128. [CrossRef]

60. Combina, M.; Elia, A.; Mercado, L.; Catania, C.; Ganga, A.; Martinez, C. Dynamics of indigenous yeast populations during spontaneous fermentation of wine from Mendoza, Argentina. *Int. J. Food Microbiol.* **2005**, *99*, 237–243. [CrossRef] [PubMed]

61. Lopes, C.A.; Sangorrín, M.P. Optimization of killer assays for yeast selection protocols. *Rev. Argent. Microbiol.* **2010**, *42*, 298–306. [CrossRef] [PubMed]

62. Türkel, S.; Ener, B. Isolation and characterization of new *Metschnikowia pulcherrima* strains as producers of the antimicrobial pigment pulcherrimin. *Z. Naturforsch. C* **2009**, *64*, 405–410. [CrossRef] [PubMed]

63. Holmes-Hampton, G.P.; Jhurry, N.D.; McCormick, S.P.; Lindahl, P.A. Iron content of *Saccharomyces cerevisiae* cells grown under iron-deficient and iron-overload conditions. *Biochemistry* **2013**, *52*, 105–114. [CrossRef] [PubMed]

64. Comitini, F.; Gobbi, M.; Domizio, P.; Romani, C.; Lencioni, L.; Mannazzu, I.; Ciani, M. Selected non-*Saccharomyces* wine yeasts in controlled multistarter fermentations with *Saccharomyces cerevisiae*. *Food Microbiol.* **2011**, *28*, 873–882. [CrossRef] [PubMed]

65. Mehlomakulu, N.N.; Setati, M.E.; Divol, B. Characterization of novel killer toxins secreted by wine-related non-*Saccharomyces* yeasts and their action on *Brettanomyces* spp. *Int. J. Food Microbiol.* **2014**, *188*, 83–91. [CrossRef] [PubMed]

66. Santos, A.; Navascués, E.; Bravo, E.; Marquina, D. *Ustilago maydis* killer toxin as a new tool for the biocontrol of the wine spoilage yeast *Brettanomyces bruxellensis*. *Int. J. Food Microbiol.* **2011**, *145*, 147–154. [CrossRef] [PubMed]

67. Santos, A.; Alonso, A.; Belda, I.; Marquina, D. Cell cycle arrest and apoptosis, two alternative mechanisms for PMKT2 killer activity. *Fungal Genet. Biol.* **2013**, *50*, 44–54. [CrossRef] [PubMed]

68. Alonso, A.; Belda, I.; Santos, A.; Navascués, E.; Marquina, D. Advances in the control of the spoilage caused by *Zygosaccharomyces* species on sweet wines and concentrated grape musts. *Food Control.* **2015**, *51*, 129–134. [CrossRef]

69. Berbegal, C.; Garofalo, C.; Russo, P.; Pati, S.; Capozzi, V.; Spano, G. Use of autochthonous yeasts and bacteria in order to control *Brettanomyces bruxellensis* in wine. *Fermentation* **2017**, *3*, 65. [CrossRef]

70. Morata, A.; Loira, I.; Vejarano, R.; González, C.; Callejo, M.J.; Suárez-Lepe, J.A. Emerging preservation technologies in grapes for winemaking. *Trends Food Sci. Technol.* **2017**, *67*, 36–43. [CrossRef]

71. Golberg, A. The impact of pulsed electric fields on cells and biomolecules: Comment on "Lightning-triggered electroporation and electrofusion as possible contributors to natural horizontal gene transfer" by Tadej Kotnik. *Phys. Life Rev.* **2013**, *10*, 382–383. [CrossRef] [PubMed]

72. Puértolas, E.; López, N.; Condón, S.; Raso, J.; Álvarez, I. Pulsed electric fields inactivation of wine spoilage yeast and bacteria. *Int. J. Food Microbiol.* **2009**, *130*, 49–55. [CrossRef] [PubMed]

73. Timmermans, R.; Nederhoff, A.; Groot, M.N.; van Boekel, M.; Mastwijk, H. Effect of electrical field strength applied by PEF processing and storage temperature on the outgrowth of yeasts and moulds naturally present in a fresh fruit smoothie. *Int. J. Food Microbiol.* **2016**, *230*, 21–30. [CrossRef] [PubMed]

74. Garde-Cerdán, T.; Arias-Gil, M.; Marsellés-Fontanet, A.R.; Ancín-Azpilicueta, C.; Martín-Belloso, O. Effects of thermal and non-thermal processing treatment on fatty acids and free amino acids of grape juice. *Food Control* **2007**, *18*, 473–479. [CrossRef]

75. Beveridge, J.R.; Wall, K.; MacGregor, S.J.; Anderson, J.G.; Rowan, N.J. Pulsed electric field inactivation of spoilage microorganisms in alcoholic beverages. In Proceedings of the 14th IEEE International Pulsed Power Conference, Dallas, TX, USA, 15–18 June 2003; pp. 1138–1143.

76. Milani, E.A.; Alkhafaji, S.; Silva, F.V.M. Pulsed electric field continuous pasteurization of different types of beers. *Food Control* **2015**, *50*, 223–229. [CrossRef]

77. Bely, M.; Stoeckle, P.; Masnuef-Pomarède, I.; Dubourdieu, D. Impact of mixed *Torulaspora delbrueckii*–*Saccharomyces cerevisiae* culture on high-sugar fermentation. *Int. J. Food Microbiol.* **2008**, *122*, 312–320. [CrossRef] [PubMed]

78. Bovo, B.; Carlot, M.; Lombardi, A.; Lomolino, G.; Lante, A.; Giacomini, A.; Corich, V. Exploring the use of *Saccharomyces cerevisiae* commercial strain and *Saccharomycodes ludwigii* natural isolate for grape marc fermentation to improve sensory properties of spirits. *Food Microbiol.* **2014**, *41*, 33–41. [CrossRef] [PubMed]

79. Fia, G.; Giovani, G.; Rosi, I. Study of beta-glucosidase production by wine-related yeasts during alcoholic fermentation. A new rapid fluorimetric method to determine enzymatic activity. *J. Appl. Microbiol.* **2005**, *99*, 509–517. [CrossRef] [PubMed]

80. Ugliano, M.; Bartowsky, E.J.; McCarthy, J.; Moio, L.; Henschke, P.A. Hydrolysis and transformation of grape glycosidically bound volatile compounds during fermentation with three *Saccharomyces* yeast strains. *J. Agric. Food Chem.* **2006**, *54*, 6322–6331. [CrossRef] [PubMed]

81. Wightman, J.D.; Wrolstad, R.E. β-glucosidase activity in juice-processing enzymes based on anthocyanin analysis. *J. Food Sci.* **1996**, *61*, 544–548. [CrossRef]

82. Vejarano, R.; Morata, A.; Loira, I.; González, M.C.; Suárez-Lepe, J.A. Theoretical considerations about usage of metabolic inhibitors as possible alternative to reduce alcohol content of wines from hot areas. *Eur Food Res. Technol.* **2013**, *237*, 281–290. [CrossRef]

83. Kulkarni, P.; Loira, I.; Morata, A.; Tesfaye, W.; González, M.C.; Suárez-Lepe, J.A. Use of non-Saccharomyces yeast strains coupled with ultrasound treatment as a novel technique to accelerate ageing on lees of red wines and its repercussion in sensorial parameters. *LWT-Food Sci. Technol.* **2015**, *64*, 1255–1262. [CrossRef]

84. Palomero, F.; Morata, A.; Benito, S.; Calderón, F.; Suárez-Lepe, J.A. New genera of yeasts for over-lees aging of red wine. *Food Chem.* **2009**, *112*, 432–441. [CrossRef]

85. Giovani, G.; Rosi, I.; Bertuccioli, M. Quantification and characterization of cell wall polysaccharides released by non-*Saccharomyces* yeast strains during alcoholic fermentation. *Int. J. Food Microbiol.* **2012**, *160*, 113–118. [CrossRef] [PubMed]

86. Charpentier, C.; N'guyen Van Long, T.; Bonaly, R.C.; Feuillat, M. Alteration of cell wall structure in *Saccharomyces cerevisiae* and *Saccharomyces bayanus* during autolysis. *Appl. Microbiol. Biotechnol.* **1986**, *24*, 405–413. [CrossRef]

87. Klis, F.M.; Boorsma, A.; De Groot, P.W.J. Cell wall construction in *Saccharomyces cerevisiae*. *Yeast* **2006**, *23*, 185–202. [CrossRef] [PubMed]

88. Vidal, S.; Francis, L.; Williams, P.; Kwiatkowski, M.; Gawel, R.; Cheynier, V.; Waters, E. The mouth-feel properties of polysaccharides and anthocyanins in a wine like medium. *Food Chem.* **2004**, *85*, 519–525. [CrossRef]

89. Guadalupe, Z.; Palacios, A.; Ayestarán, B. Maceration enzymes and mannoproteins: A possible strategy to increase colloidal stability and color extraction in red wines. *J. Agric. Food Chem* **2007**, *55*, 4854–4862. [CrossRef] [PubMed]

90. Chalier, P.; Angot, B.; Delteil, D.; Doco, T.; Gunata, Z. Interactions between aroma compounds and whole mannoprotein isolated from *Saccharomyces cerevisiae* strains. *Food Chem.* **2007**, *100*, 22–30. [CrossRef]

91. Gonzalez-Ramos, D.; Cebollero, E.; Gonzalez, R. A recombinant *Saccharomyces cerevisiae* strain overproducing mannoproteins stabilizes wine against protein haze. *Appl. Environ. Microbiol.* **2008**, *74*, 5533–5540. [CrossRef] [PubMed]

92. Lubbers, S.; Léger, B.; Charpentier, C.; Feuillat, M. Effet colloïdes protecteurs d'extraits de parois de levures sur la stabilité tartrique d'un vin modèle. *J. Int. Sci. Vigne. Vin.* **1993**, *27*, 13–22.

93. Rodrigues, A.; Ricardo-Da-Silva, J.M.; Lucas, C.; Laureano, O. Effect of commercial mannoproteins on wine colour and tannins stability. *Food Chem.* **2012**, *131*, 907–914. [CrossRef]

94. Iriti, M.; Varoni, E.M. Cardioprotective effects of moderate red wine consumption: Polyphenols vs. Etanol. Review. *J. Appl. Biomed.* **2014**, *12*, 193–202. [CrossRef]

95. Loira, I.; Vejarano, R.; Morata, A.; Ricardo-da-Silva, J.M.; Laureano, O.; González, M.C.; Suárez-Lepe, J.A. Effect of *Saccharomyces* strains on the quality of red wines aged on lees. *Food Chem.* **2013**, *139*, 1044–1051. [CrossRef] [PubMed]

96. Rosi, I.; Gheri, A.; Domizio, P.; Fia, G. Production de macromolecules parietals de *Saccharomyces cerevisiae* au cours de la fermentation et leur influence sur la fermentation malolactique. *Revue des Œnologues* **2000**, *94*, 18–20.

97. Moreno-Arribas, V.; Pueyo, E.; Nieto, F.J.; Martín-Álvarez, P.J.; Polo, M.C. Influence of the polysaccharides and the nitrogen compounds on foaming properties of sparkling wines. *Food Chem.* **2000**, *70*, 309–317. [CrossRef]

98. Moruno, E.G.; Sanlorenzo, C.; Boccaccino, B.; Di Stefano, R. Treatment with yeast to reduce the concentration of ochratoxin A. in red wine. *Am. J. Enol. Viticult.* **2005**, *56*, 73–76.

99. Vejarano, R.; Siche, R.; Tesfaye, W. Evaluation of biological contaminants in foods by hyperspectral imaging (HSI): A. review. *Int. J. Food Prop.* **2017**, *20*, 1264–1297. [CrossRef]

100. Morata, A.; Gómez-Cordovés, M.C.; Suberviola, J.; Bartolomé, B.; Colomo, B.; Suárez-Lepe, J.A. Adsorption of anthocyanins by yeast cell walls during the fermentation of red wines. *J. Agric. Food Chem.* **2003**, *51*, 4084–4088. [CrossRef] [PubMed]

101. Morata, A.; Gómez-Cordovés, M.C.; Calderón, F.; Suárez-Lepe, J.A. Effects of pH, temperature and SO_2 on the formation of pyranoanthocyanins during red wine fermentation with two species of *Saccharomyces*. *Int. J. Food Microbiol.* **2006**, *106*, 123–129. [CrossRef] [PubMed]

102. Rokhina, E.V.; Piet, L.; Virkutyte, J. Low-frequency ultrasound in biotechnology: State of the art. *Trends Biotechnol.* **2009**, *27*, 298–306. [CrossRef] [PubMed]

103. Koda, S.; Miyamoto, M.; Toma, M.; Matsuoka, T.; Maebayashi, M. Inactivation of *Escherichia coli* and *Streptococcus mutants* by ultrasound at 500 kHz. *Ultrason. Sonochem.* **2009**, *16*, 655–659. [CrossRef] [PubMed]

104. Zhou, C.; Ma, H. Ultrasonic degradation of polysaccharide from a red algae (*Porphyra yezoensis*). *J. Agric. Food Chem.* **2006**, *54*, 2223–2228. [CrossRef] [PubMed]

105. Liu, S.Q.; Aung, M.T.; Lee, P.R.; Yu, B. Yeast and volatile evolution in cider co-fermentation with *Saccharomyces cerevisiae* and *Williopsis saturnus*. *Ann. Microbiol.* **2016**, *66*, 307–315. [CrossRef]

106. Lu, Y.; Huang, D.; Lee, P.R.; Liu, S.Q. Assessment of volatile and non-volatile compounds in durian wines fermented with four commercial non-*Saccharomyces* yeasts. *J. Sci. Food Agric.* **2016**, *96*, 1511–1521. [CrossRef] [PubMed]

107. Fugelsang, K. Population dynamics and effects of *Brettanomyces bruxellensis* strains on Pinot noir (*Vitis vinifera* L.) wines. *Am. J. Enol. Viticult.* **2003**, *54*, 294–300.

108. Crauwels, S.; Van Opstaele, F.; Jaskula-Goiris, B.; Steensels, J.; Verreth, C.; Bosmans, L.; Paulussen, C.; Herrera-Malaver, B.; de Jonge, R.; De Clippeleer, J.; et al. Fermentation assays reveal differences in sugar and (off-) flavor metabolism across different *Brettanomyces bruxellensis* strains. *FEMS Yeast Res.* **2017**, *17*, fow105. [CrossRef] [PubMed]

109. Smith, B.D.; Divol, B. *Brettanomyces bruxellensis*, a survivalist prepared for the wine apocalypse and other beverages. *Food Microbiol.* **2016**, *59*, 161–175. [CrossRef] [PubMed]

110. Nardi, T.; Remize, F.; Alexandre, H. Adaptation of yeasts *Saccharomyces cerevisiae* and *Brettanomyces bruxellensis* to winemaking conditions: A comparative study of stress genes expression. *Appl. Microbiol. Biotechnol.* **2010**, *88*, 925–937. [CrossRef] [PubMed]

111. Avramova, M.; Vallet-Courbin, A.; Maupeu, J.; Masneuf-Pomarede, I.; Albertin, W. Molecular diagnosis of *Brettanomyces bruxellensis'* sulfur dioxide sensitivity through genotype specific method. *Front Microbiol.* **2018**, *9*, 1260. [CrossRef] [PubMed]

112. Capozzi, V.; Di Toro, M.R.; Grieco, F.; Michelotti, V.; Salma, M.; Lamontanara, A.; Russo, P.; Orrù, L.; Alexandre, H.; Spano, G. Viable but not culturable (VBNC) state of *Brettanomyces bruxellensis* in wine: New insights on molecular basis of VBNC behaviour using a transcriptomic approach. *Food Microbiol.* **2016**, *59*, 196–204. [CrossRef] [PubMed]

113. Avramova, M.; Cibrario, A.; Peltier, E.; Coton, M.; Coton, E.; Schacherer, J.; Spano, G.; Capozzi, V.; Blaiotta, G.; Salin, F.; et al. *Brettanomyces bruxellensis* population survey reveals a diploid-triploid complex structured according to substrate of isolation and geographical distribution. *Sci. Rep.* **2018**, *8*, 4136. [CrossRef] [PubMed]

114. Guerrero, R.F.; Cantos-Villar, E. Demonstrating the efficiency of sulphur dioxide replacements in wine: A parameter review. *Trends Food Sci. Technol.* **2015**, *42*, 27–43. [CrossRef]

115. Saldaña, G.; Puértolas, E.; Álvarez, I.; Meneses, N.; Knorr, D.; Raso, J. Evaluation of a static treatment chamber to investigate kinetics of microbial inactivation by pulsed electric fields at different temperatures at quasi-isothermal conditions. *J. Food Eng.* **2010**, *100*, 349–356. [CrossRef]

116. Marsellés-Fontanet, À.R.; Puig, A.; Olmos, P.; Mínguez-Sanz, S.; Martín-Belloso, O. Optimising the inactivation of grape juice spoilage organisms by pulse electric fields. *Int. J. Food Microbiol.* **2009**, *130*, 159–165. [CrossRef] [PubMed]

117. Portugal, C.; Sáenz, Y.; Rojo-Bezares, B.; Zarazaga, M.; Torres, C.; Cacho, J.; Ruiz-Larrea, F. *Brettanomyces* susceptibility to antimicrobial agents used in winemaking: In vitro and practical approaches. *Eur. Food Res. Technol.* **2014**, *238*, 641–652. [CrossRef]

118. Bizri, J.N.; Wahem, I.A. Citric acid and antimicrobial affect microbiological stability and quality of tomato juice. *J. Food Sci.* **1994**, *59*, 130–134. [CrossRef]

119. Barceloux, D.G.; Bond, G.R.; Krenzelok, E.P.; Cooper, H.; Vale, J.A. American Academy of Clinical Toxicology practice guidelines on treatment of methanol poisoning. *J. Toxicol. Clin. Toxicol.* **2002**, *40*, 415–446. [CrossRef] [PubMed]

120. Cravero, F.; Englezos, V.; Rantsiou, K.; Torchio, F.; Giacosa, S.; Río Segade, S.; Gerbi, V.; Rolle, L.; Cocolin, L. Control of *Brettanomyces bruxellensis* on wine grapes by post-harvest treatments with electrolyzed water, ozonated water and gaseous ozone. *Innov. Food Sci. Emerg. Technol.* **2018**, *47*, 309–316. [CrossRef]

121. Charpentier, C.; Escot, S.; González, E.; Dulau, L.; Feuillat, M. The influence of yeast glycosylated proteins on tannins aggregation in model solution. *Int. J. Vine Wine Sci.* **2004**, *38*, 209–218. [CrossRef]

122. Suárez-Lepe, J.A.; Morata, A. Nuevo Método de Crianza Sobre Lías. Patente P200602423, 25 September 2006.

fermentation

MDPI

Review

A Control Alternative for the Hidden Enemy in the Wine Cellar

Rubén Peña [1], Renato Chávez [2], Arturo Rodríguez [3] and María Angélica Ganga [1,*]

[1] Departamento en Ciencia y Tecnología de los Alimentos, Facultad Tecnológica, Universidad de Santiago de Chile, Santiago 9170201, Chile; ruben.pena@usach.cl

[2] Departamento de Biología, Facultad de Química y Biología, Universidad de Santiago de Chile, Santiago 9170201, Chile; renato.chavez@usach.cl

[3] Departamento de Tecnologías Industriales, Facultad Tecnológica, Universidad de Santiago de Chile, Alameda 3363, Estación Central, Santiago 9170201, Chile; arturo.rodriguez@usach.cl

* Correspondence: angelica.ganga@usach.cl; Tel.: +56-2-2718-4509

Received: 21 December 2018; Accepted: 28 February 2019; Published: 6 March 2019

Abstract: *Brettanomyces bruxellensis* has been described as the principal spoilage yeast in the winemaking industry. To avoid its growth, wine is supplemented with SO_2, which has been questioned due to its potential harm to health. For this reason, studies are being focused on searching for, ideally, natural new antifungals. On the other hand, it is known that in wine production there are a variety of microorganisms, such as yeasts and bacteria, that are possible biological controls. Thus, it has been described that some microorganisms produce antimicrobial peptides, which might control yeast and bacteria populations. Our laboratory has described the *Candida intermedia* LAMAP1790 strain as a natural producer of antimicrobial compounds against food spoilage microorganisms, as is *B. bruxellensis*, without affecting the growth of *S. cerevisiae*. We have demonstrated the proteinaceous nature of the antimicrobial compound and its low molecular mass (under 10 kDa). This is the first step to the possible use of *C. intermedia* as a selective bio-controller of the contaminant yeast in the winemaking industry.

Keywords: antimicrobial peptides; biocontrol; *Brettanomyces bruxellensis*; *Candida intermedia*; wine; off-flavors

1. Introduction

Phenol derivatives have been identified as one of the volatile components which provide a pleasant aroma to wine when produced in low concentration [1,2]. The most important molecules that belong to this group are 4-vinylphenol, 4-vinylguaiacol, 4-ethylphenol, and 4-ethylguaiacol [1,3]. Nevertheless, there are threshold values for these components; thus, an increase of the concentration produces an off-flavor in wine. Some authors have established that concentrations over 620 µg/L of 4-ethylphenol produced aromas related to "phenol", "barn", "horse sweat", "leather", "varnish", among others [2,3], which causes important economic losses for the industry [4,5]. However, concentrations under 400 µg/L, 4-ethylphenol contribute to the aromatic complexity of the product, providing notes of "spices", "leather", and "smoke" which are valued by most wine consumers [2].

2. Production of Phenolic Derivatives

The precursors of the phenol derivatives are the phenolic acids or hydroxycinnamic acids (*p*-coumaric, ferulic, caffeic, and sinapinic acids). These compounds are naturally found in grapes and in vegetal tissues conjugated with tartaric acid as a natural part of the grape peel [2,6]. The hydroxycinnamic acids can be released during winemaking [2,6]. Besides, some microorganisms that are in the must release enzymes which would also help to release these acids [2]. It has been described that hydroxycinnamic acids would exert an inhibitory effect on the growth of microorganisms, due to imbalances produced in the cell medium. In this context, ferulic and *p*-coumaric acids would exert the most inhibitory effects in yeasts [7]. For this, microorganisms that are able to ferment vegetable products show enzymatic activity, which would allow these compounds to metabolize into less toxic ones [1,8]. The most studied and characterized pathway to transform these hydroxycinnamic acids into volatile phenols corresponds to the sequential action of two enzymes; first, the action of phenolic acid decarboxylase transforms hydroxycinnamic acids into vinyl derivatives and, posteriorly, these compounds are reduced to ethyl derivatives by the action of a vinyl reductase [1,8,9].

The presence of phenolic acid decarboxylase activity has been related to *Bacillus* and *Lactobacillus* bacteria and, *Saccharomyces* and no-*Saccharomyces* yeasts [1,3]. On the other hand, it has been described that only yeasts from the genus *Brettanomyces/Dekkera*, *Kluyveromyces*, *Candida*, and *Pichia* would be able to generate ethyl derivatives, even though only *Pichia* and *Brettanomyces/Dekkera* species could produce important quantities that surpass the sensorial threshold [10,11]. The presence of acids in the must, which are mainly esterified with tartaric acid, has been described. Thus, the generation of their free forms is dependent on the presence of microorganisms, which have enzymes with cinnamoyl esterase activities [2]. Among hydroxycinnamic acids present in the must, there are the *p*-coumaric, ferulic, and caffeic acids; being *p*-coumaric found in greater quantity [2,3].

3. *Brettanomyces/Dekkera* as a Wine Spoilage Yeast

In this genus has been described the anamorphs *B. bruxellensis*, *B. anomalus*, *B. custersianus*, *B. naardenensis*, and *B. nanus*, with teleomorphs existing for the first two species, *Dekkera bruxellensis* and *Dekkera anomala* [12]. *B. bruxellensis* has been described as being mainly responsible for off-flavor production in wine worldwide. In the case of *B. bruxellensis*, it presents slow growth, fermentative and oxidative metabolism, consumption of several sugars, production of acetic acid under aerobic conditions, and natural resistance to the antifungal compound cycloheximide [2,6,13]. From the enological point of view, *B. bruxellensis* is recognized for its high tolerance to ethanol and its capacity of surviving in environments which lack nutrients and have a low pH, allowing persistent proliferation in winemaking processes [6,12]. Nevertheless, the distinct characteristic of this species is the capacity of transforming hydroxycinnamic acids present in the must into phenolic derivatives, which affect the organoleptic quality of wine [1,3,6,8–11].

Contamination by *B. bruxellensis* can occur during the winemaking process. Nevertheless, its proliferation is favored during the ageing period in barrels [2]. In this context, *B. bruxellensis* can settle in the pore microstructure of the wood [13]. Besides, it has been described that this yeast can decompose cellobiose, so it can supply its nutritional necessities from wood and keep metabolically active in this structure during several generations [2,13]. These yeast features increase the risk to transform the barrels into carriers and transmitters of contamination. Therefore, its proliferation is very difficult to eradicate.

4. Control of *B. bruxellensis*

For many years the wine industry has looked for tools to eradicate contaminant microorganisms in the fermentation and ageing processes of wine in barrels. This problem has been studied chemically, through anhydrous sulfide addition (SO_2), as a potassium metabisulfite form [14]. This compound is frequently used as a preservative, due to its antimicrobial, antioxidant and stabilizing properties to the final product [4]. SO_2 supplement is carried out over must to decrease its natural microbiological charge. However, it is common to repeat its addition after alcoholic fermentation and during the ageing in barrels, with the aim of avoiding the growth of spoilage microorganisms during this process [4,14]. The anhydrous sulfide is found normally in chemical equilibrium between its molecular form ($SO_2 \cdot H_2O$) and the bisulfite anion ($HSO^{3-} + H^+$). The molecular SO_2 can diffuse into the cell cytoplasm and dissociate now between bisulfite ($HSO^{3-} + H^+$) and sulfite ($SO_3^{2-} + H^+$) [14]. This dissociation produces a sustaining increase of the concentration protons in the cytoplasm, which generates a rapid acidification and an abrupt redox imbalance. Additionally, it has been determined that the sulfite anion (SO_3^{2-}) is highly reactive and produces the inactivation of several metabolites and cell enzymes. Besides the penetration of molecular SO_2 to yeasts, cytoplasm produces the immediate inhibition of the glyceraldehyde-3-phosphate dehydrogenase enzyme, interrupting glycolysis and NADH regeneration, allowing ATP depletion [14]. In the case of *S. cerevisiae*, the presence of *SSU1* gene has been described, which codifies for a SO_2 efflux pump, making this yeast resistant to this compound and able to survive to generate the alcoholic fermentation [15,16]. Among the physiological and molecular studies carried out on *B. bruxellensis*, it has been determined that this yeast shows strain-dependent resistance to SO_2 [4,17]. This phenomenon has been related to the presence of an ortholog gene to *SSU1* in the genome of *B. bruxellensis* AWRI1499, which may affect this strain tolerance to SO_2 [16]. Another study has shown the tolerance profiles to this compound in 108 *B. bruxellensis* strains, obtained from different geographical origins. The results showed that 19 strains do not tolerate 0.1 mg/L SO_2, 29 grow with 0.1 mg/L, 42 tolerate 0.2 mg/L, 16 tolerate 0.4 mg/L, and two tolerate over 0.6 mg/L SO_2 [4]. This phenomenon is relevant in the industry because it has been reported that SO_2 can be a potentially harmful agent for human consumption, due to it producing irritation of the gastric mucosa, dizziness, headache and, in susceptible individuals, it can cause allergic and severe asthmatic crisis [18].

The use of food industry sanitizers, as alkaline detergents and iodophors, has low use in the wine industry due to the complex geometry of their machines (bottling machines, valves, etc.) or to the low access for cleaning of the superficies [13]. On the other hand, dimethyl dicarbonate (DMDC) is a preservative authorized to be used in winemaking in some countries; its efficiency depends on the strain, temperature, ethanol concentration, and pH. DMDC is rapidly hydrolyzed, the effect done instantaneously in must or wine; however, its use in large volumes has low effectiveness. For this reason, DMDC is recommended to be used in the presence of molecular sulfur dioxide [19]. Another compound studied to reduce the *B. bruxellensis* population in wine has been chitosan, a natural polymer obtained from the exoskeletons of crustaceans. At laboratory level, the studies show a control on the growth of *Brettanomyces*; however, at industrial level there is not complete eradication, with there being the fungistatic effect limited in time [20].

So, the wine industry is looking for new technological solutions, which allow eradication of this yeast in the fermentation and ageing processes of wine in barrels. Biotechnological investigation has provided several physics strategies to avoid contamination by *B. bruxellensis*. Thus, some works study the exposition of contaminated must and wine to pulses of a defined electric field. It was reported that the application of a pulse of 29 kV/cm (186 kJ/kg) reduces the viability of contaminant bacteria and yeasts by 99.9%, such as *Lactobacillus hilgardii*, *Lactobacillus plantarum*, *D. anomala*, and *B. bruxellensis* [21]. On the other hand, the study of the treatment of contaminated barrels with deionized water at different temperatures determines that submerging barrels during 19 min in water at 60 °C reduces the growth of four *B. bruxellensis* strains in eight logarithmic cycles [22]. Furthermore, the application of hydrostatic pressure on the growth of strains *B. bruxellensis* in synthetic must, at different pH and ethanol concentrations, was studied. The results showed that one minute treatment at a pressure of 300 MPa totally reduces the viability of contaminant yeast [23]. Nevertheless, these physics strategies have not been effectively incorporated by the industry due to their low technical and economic feasibility for implementation. From a biological point of view, several authors have focused on the identification and characterization of natural killer toxins with antifungal properties. In this context, the first report of a killer toxin that had an effect on the growth of *B. bruxellensis* was found in the non-*Saccharomyces* yeasts, *Pichia anomala* (Pikt) and *Kluyveromyces wickerhamii* (Kwkt). In this study, it was determined that both toxins showed the capacity of modulating the proliferation of a contaminant yeast in wine for 10 days [24]. However, after that period, the bio-controlled efficiency of toxins on the yeast proliferation is not described. Posteriorly, the use of PMKT2 toxin of *Pichia membranefaciens* on *S. cerevisiae* and *B. bruxellensis* in must was described. Here it was determined that PMKT2 is an effective bio-controller of the contaminant yeast, but it also affects the growth of the fermentative yeast [25]. Therefore, it is not considered a good tool for the industry. Later, the production of a killer toxin secreted *Ustilago maydis* has been described [26]. This work demonstrated, using vinification assays, that the toxin affects the growth of *B. bruxellensis*, while *S. cerevisiae* shows complete resistance. Besides, it was observed that supplementing toxins in the fermentation and ageing conditions of wine in barrels reduces the content of 4-ethylphenol produced by the contaminant yeast significantly. This result demonstrates its effective reduction of volatile phenols that cause the aromatic default. Nevertheless, 17 *B. bruxellensis* strains used in the study show relative sensitivities, with 10 of them being low sensitivity [26]. This phenomenon constrains the use of the toxin killer *U. maydis* as a general bio-control strategy against *B. bruxellensis*, since not all strains would be susceptible, so that the accumulation of volatile phenols in wine would be at random.

Additionally, the production of toxins CpKT1 and CpKT2 for *Candida pyralidae* YWBT Y1140 strain has been described. In this study, it was demonstrated that toxins present a molecular mass over 50 kDa and stability in an acidic pH, high alcoholic degree, and different sugar concentrations. Nevertheless, its effectiveness was studied in a laboratory medium. In white and red wine, the toxins showed that only seven strains, out of 15 *B. bruxellensis* strains studied, present sensitivity in the wine matrix [27]. Later, the same authors demonstrated that toxins CpKT1 and CpKT2 can cause damage to cell walls in *B. bruxellensis* sensitive strains [28].

Several investigation groups have focused on the search of antimicrobial peptides (AMPs), which have been studied as different microorganism agent controllers at clinical or industrial importance [27].

5. Antimicrobial Peptides and their Antifungal Action Mechanisms

Antimicrobial peptides (AMPs) are molecules which are found in a broad range of organisms (from prokaryotic cells to human beings) and constitute the first line of defense against potentially pathogenic organisms in multicellular organisms. However, some microorganisms are able to produce AMPs with the purpose of ensuring survival [29]. Generally, AMPs show relative length (below 100 amino acid residues) and can differ in sequence. Nevertheless, they share, as a distinctive characteristic, the presence of amino acid residues charged to physiological pH and non-polar residues, which determine its amphipathic nature [30–32]. The analysis of the tridimensional structure of different AMPs has shown that these can be linear or adopt α-helical or β-laminar conformation, in which charged and hydrophobic residues are aligned in the opposed faces, allowing its water solubility [30,31]. Due to the high number of identified peptides "The Antimicrobial Peptide Database (APD3)" was generated (http://aps.unmc.edu/AP/main.php). This database is a library in which peptides are classified according to origin, sequence, activity, and structural or physiochemical properties [33]. To date, APD3 contains peptides obtained from different kingdoms, from which 13 are produced by species corresponding to *Fungi*. From them, six peptides were identified in *Fungi* such as *Aspergillus giganteus*, *Aspergillus clavatus*, *Aspergillus niger*, *Penicillium chrysogenum*, and *Pseudoplectania nigrella* [34–39]. Other authors have described the genus *Trichoderm* can produced antimicrobial peptides of no-ribosomal synthesis named peptaibols. They are characterized by a length between seven and 20 amino acid residues, from which a high proportion corresponds to no-proteinogenic amino acids, such as isovaline and α–aminoisobutyric. Furthermore, they show acetylation at the N-terminal and amino alcohol at the C-terminal [40]. To date, 317 peptaibols have been described, which have been stored according to origin, sequence, and crystallographic structure at the Peptaibol database (http://peptaibol.cryst.bbk.ac.uk/home.shtml) [41].

One of the main aspects in the study of AMPs and peptaibols has been the determination of their antifungal action mechanism. When the antifungal effect of a peptide produced by *Aspergillus giganteus* was studied, it was determined that their action mechanism is related to cell wall permeabilization [36]. Further, through immunofluorescence experiments, it was determined that this AFP (antifungal protein) is exclusively located in the plasmatic membrane, so that its inhibitory effect would be related to the joining and destabilization of this structure [36]. This mechanism would be similar to what was described for identified peptaibols in different species of genus *Trichoderma* [41]. The peptaibol alamethicin (isolated from culture medium of *Trichoderma viride*) showed amphiphilic characteristics and it is strongly absorbed by natural and synthetic membranes and, consequently, generated cell lysis [42]. On the other hand, different computing simulations were carried out by using the sequence, tridimensional structure, physicochemical properties of several AMPs, and alamethicin peptaibol to potentially determine its action mechanism. It was determined that these attach to the outer face of the plasmatic membrane and its accumulation produces disorganization of the phospholipid bilayer, favoring emerging pores. These pores would allow ion efflux (mainly potassium) from intracellular, which causes an ionic gradient imbalance, oxygen reactive species production (ROS), and the subsequent cell death [43]. In this context, cell membrane permeabilization is not the only action mechanism related to AMPs. When an antifungal peptide secreted by *P. chrisogenum* (PgAF) was characterized, it was determined that it corresponds to a peptide whose length is 55 amino acid residues, rich in cysteine, and 25% hydrophobic amino acid residues [35]. Likewise, by sequence alignment, it has been determined that PgAF presents a 42% sequence identity to an antifungal peptide from *A. giganteus* [36], 37% to a novel antifungal peptide from *A. niger* (AnAFP) [34], and 100% to *Penicillium nalgiovense* antifungal protein (NAF) [35]. Using proteins with antifungal activity from *Penicillium chrysogenum*, it was determined that these proteins produce cell wall disorganization and membrane permeabilization. This produces a great loss of turgidity and ionic gradient due to the rapid potassium efflux. The authors also reported an effect on the growth tips of hyphae and generation of reactive oxygen species (ROS) [35]. This observation was the first evidence of antimicrobial peptides and intracellular toxicity by ROS generation. Another piece of research has studied the antifungal effect of a hexapeptide derived from peptide PAF of *P. chrysogenum*, named PAF26. This lineal hexapeptide (sequence RKKWFW) presents two well defined functional motives. The first, located at the N-terminal (RKK), corresponds to a cationic domain (net charge +3), while the second, located at the C-terminal (WFW), corresponds to a hydrophobic domain [44]. By fluorescence microscopy, it was determined that PAF26 is internalized by *Aspergillus fumigatus*, *Neurospora crassa*, and *S. cerevisiae*, exerting its antifungal action in the intracellular space [31]. Later, these authors proposed an action mechanism for PAF26 using, as models, *N. crassa* and *S. cerevisiae*. They observed two particular sub-mechanisms, which are directly related to the concentration of the hexapeptide in the medium. When concentrations between 2.5–5.0 μM PAF26 are applied, it was determined that the peptide interacts with natural negative charges of the cell wall. Once in contact with the membrane, PAF26 is internalized by the cell via generation of endosomes, producing the accumulation and expansion of vacuoles. After, by active transport of vacuoles, PAF26 is released into the cytoplasm where it produces permeabilization of membranes, allowing the release of mitochondrial ROS. Thus, cell death would be related to the oxidative stress of DNA, oxidative damage to the membrane level, and homeostatic intracellular imbalance [31]. In the second mechanism, when PAF26 concentration exceeds 20 μM, the hexapeptide translocation occurs through the cell membrane, followed by ROS generation and the subsequent oxidative damage on membranes and DNA. This effect produces cell death similar to what was described for low concentrations of PAF26 [31]. These mechanisms demonstrate that cell wall disorganization, membrane permeabilization, and cell internalization of AMPs produced by *Fungi* produce a redox imbalance in the target cell, whose final consequence is cell death.

6. Antimicrobial Peptides as a Contaminant Bio-Control Tool in the Winemaking Industry

Regarding the study of antimicrobial peptides as biocontrol contaminant microorganisms in the winemaking industry, the effect of synthetic fragments built from antimicrobial peptides produced by *P. chrysogenum* [35] and from the antimicrobial bovine peptide named Lactoferricin [45] has been analyzed. All peptides affected the growth of *B. bruxellensis*, *Cryptococcus albidus*, *Pichia membranifaciens*, *Zygosaccharomyces bailii*, and *Zygosaccharomyces* in a laboratory medium. Nevertheless, these peptides also affected (in less proportion) the growth of the fermentative yeast *S. cerevisiae* [46]. Posteriorly, the antifungal effect of the synthetic peptide LfcinB$_{17-31}$ on the growth of *B. bruxellensis* in a laboratory medium, must, and white wine was studied, determining that this affected the growth in all media, and its action mechanism is related to the interaction and penetration of LfcinB$_{17-31}$ into the cell cytoplasm [46].

Actually, *S. cerevisiae* CCMI885 has been the only yeast in which the production of AMPs has been described. The production of antifungal compounds in a protein fraction between 2–10 kDa affected the growth of native wine yeast isolates of *B. bruxellensis*, *Hanseniaspora uvarum*, *Hanseniaspora guilliermondii*, *Candida stellata*, *Kluyveromyces thermotolerans*, *Kluyveromyces marxianus*, and *Torulaspora delbrueckii* [47]. Later, by characterization of these peptides through an Electrospray Ionizacion Mass Spectrometry (ESI-MS), it was determined that it produces two peptides of molecular mass close to 1.6 kDa, which show a high sequence identity with isoforms one and two/three of the enzyme GAPDH [48]. Nevertheless, these peptides do not produce the complete inhibition of *B. bruxellensis* in a laboratory medium. Therefore, the surviving yeasts may continue producing the aroma default in a wine matrix [49]. Posteriorly, the authors studied the antifungal action mechanism of the characterized peptides, demonstrating that they produce disruption in the cell wall integrity in *H. guilliermondii* [50]. Finally, when synthetic isoforms of this AMP were produced, it was observed that they are not as effective as natural peptides [51]. By assessing the biochemical characteristics of AMPs and their antifungal action mechanisms, it can be said that the use of antimicrobial peptides for biocontrol of spoilage yeasts in the winemaking industry could be an effective tool. Our work group described the antimicrobial activity from several strains of yeasts isolated from winemaking environments, among them was *C. intermedia* LAMAP1790, which has antibacterial activity against the food pathogens *Escherichia coli*, *Listeria monocytogenes*, and *Salmonella typhimurium* [52]. Posteriorly, it was demonstrated that the *C. intermedia* LAMAP1790 strain affects the growth of the *B. bruxellensis* LAMAP 2480 strain, determining that the antifungal is released to the culture medium (Figure 1). Moreover, it was demonstrated that this antifungal does not affect the growth of the fermentative yeast *S. cerevisiae*, it being the most important species from the enological point of view [53].

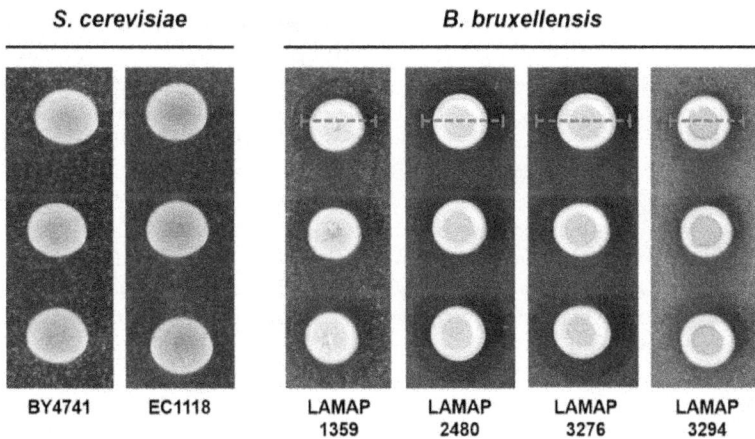

Figure 1. Semi-quantitative assessment of the antifungal action of *C. intermedia* LAMAP1790 on *S. cerevisiae* and *B. bruxellensis*. Each column corresponds to an inoculated strain in agar, while *C. intermedia* was inoculated as a drop in the layer three times. The antifungal capacity was quantified by measuring the diameter of the inhibition halo generated around *C. intermedia* LAMAP1790 (represented with red dotted lines). Left to right columns: *S. cerevisiae* BY4741, *S. cerevisiae* EC1118, *B. bruxellensis* LAMAP1359, *B. bruxellensis* LAMAP2480, *B. bruxellensis* LAMAP3276, and *B. bruxellensis* LAMAP3294 [51].

To determine whether the antifungal compound showed antifungal properties in a liquid medium, a viability assay was carried out by the exposition of the studied strains on culture sterile supernatant of *C. intermedia*. By comparing the counts of both strains of *S. cerevisiae*, it was observed that there are not statistically significant differences in their growth and cell viability, the conclusion being it was not affected by the exposition to supernatant [53]. Nevertheless, by assessing the counts of *B. bruxellensis*, it was determined that there was not growth of *B. bruxellensis* LAMAP1359, *B. bruxellensis* LAMAP2480, and *B. bruxellensis* LAMAP3276 strains, and the *B. bruxellensis* LAMAP3294 strain showed statistically low viability. These results confirmed what was observed in the semi-quantitative assay (Figure 1), it being possible to demonstrate that the released compound to the culture medium, by *C. intermedia*, shows fungicidal activity against *B. bruxellensis*, without affecting *S. cerevisiae* [53].

To define the nature of the antifungal compound, an assay was carried out, in which a fraction of the supernatant was treated at 100 °C for 10 min and its antifungal capacity through viability of *B. bruxellensis* LAMAP2480 post-exposition was analyzed. Culture medium was used as a control, from which a fraction was submitted to thermic treatment [53]. These results show that when supernatant is kept at 4 °C, the antifungal activity was not affected, since the growth of *B. bruxellensis* was decreased by four logarithmic orders. Nevertheless, when supernatant is kept at 100 °C, its antifungal activity decreased significantly [53]. Later, supernatant was concentrated 100 times, and the total proteins were fractionated by ultrafiltration according to their molecular mass. Thus, two fractions were obtained, which contained proteins upper and lower than 10 kDa, respectively. It was demonstrated that the fraction lower than 10 kDa had antifungal activity and it disappeared in the presence of protease, showing the proteinaceous nature of the compound with antifungal activity. So, this is the first report of the production of peptides with antifungal capacity in a yeast different from *S. cerevisiae* [53].

7. Action Mechanisms of AMPs

Despite the knowledge obtained from the use of AMPs and their antimicrobial action mechanisms, to date it is not clear why some yeasts are resistant to AMPs and/or which mechanisms are used by the yeast producer to protect itself from its own antifungal action mechanism. For this reason, resistant mechanisms to AMPs in bacteria have been studied, determining that these can modify certain characteristics of their wall cell and membrane to protect themselves from the action of peptides [29]. Thus, it has been determined that *Staphylococcus aureus* transports alanine and lysine towards its wall to reduce negative net charge of teichoic acids; therefore, generating electrostatic repulsion with cationic AMPs. Additionally, it was determined that there is a positive correlation between the increase of membrane proteins with the resistant antimicrobial peptides in *Yersinia enterocolitica* [54].

To date, the only known resistant mechanisms to AMPs have been described in the pathogen yeast *C. albicans*. This yeast can defend itself from the attack of the antimicrobial peptide salivary histatin-5, which has been described as a relevant part of innate immunity in humans [55]. It has been described that *C. albicans* can defend against this peptide through three different mechanisms. In the first mechanism, this yeast is able to inactivate histatin-5 through secretion of proteases Sap9 and Sap10, which are able to digest antimicrobial peptide, besides degrading tissue of the host, which allows it to avoid the innate defense system and colonize the human oral cavity [56,57]. On the second mechanism, *C. albicans* secretes a glycoprotein named Msb2, which is able to join free AMPs in its glycoside realm, reducing its effective concentration and detoxifying the medium [57]. Finally, in the third mechanism, it has been described that *C. albicans* is able to expel the histatin-5 peptide from its cytoplasm, using as efflux pump Flu1, which is part of the protein resistant family to multi drugs MDR [57]. So far, these mechanisms would be exclusively of *C. albicans*; therefore, it cannot be extrapolated to other organisms. Nevertheless, it is necessary to carry out more studies on other yeasts, with other peptides and other media to determine the existence of similar mechanisms and/or new resistant mechanisms to AMPs.

8. Conclusions

B. bruxellensis is the most important spoilage yeast in wine at a world level, due to the negative sensorial effect when it is present. It has been studied using different methodologies to eradicate its presence in wine, but the cost and efficiencies of these has meant that, actually, in the wine industry, SO_2 is still used as a microbial controller. However, the SO_2 presence in wine can bring health problems; so, it is necessary to search for natural products that allow for the control of microbial growth, especially spoilage microorganisms. The AMPs are peptides of low molecular mass that are secreted by microorganisms ensuring survival. Our work group identified that *C. intermedia* LAMAP1790 secretes peptides with antimicrobial activity, which have a molecular mass lower than 10 kDa. These peptides control the *B. bruxellensis* growth without affecting *S. cerevisiae*. However, it is necessary to determine their action at pH variation, residual sugar concentration, the increase of the ethanol concentration, and proliferation of other yeasts and/or related bacteria in winemaking to define the industrial potential of these AMPs. Likewise, the low production of these peptides is a problem, it being necessary to develop biotechnological tools that allow larger production of these peptides and thus enable the carrying out of assays at an industrial scale.

Author Contributions: Conceptualization, R.P. and M.A.G.; methodology, R.P, R.C., and M.A.G.; writing—review and editing, R.C, A.R, R.P., and M.A.G.

Funding: This research was funded by grants Dicyt 081871GM_DAS, Proyecto Fortalecimiento Usach USA1398_GM181622 and the Comisión Nacional de Investigación Científica y Tecnológica (CONICYT) PCHA/DoctoradoNacional/2013-21130439 doctoral fellowship.

Conflicts of Interest: The authors declare no conflicts of interest.

References

1. Chatonnet, P.; Dubourdie, D.; Boidron, J.-N.; Pons, M. The Origin of Ethylphenols in Wines. *J. Sci. Food Agric.* **1992**, *60*, 165–178. [CrossRef]
2. Oelofse, A.; Pretorius, I.S.; du Toit, M. Significance of *Brettanomyces* and *Dekkera* during Winemaking: A Synoptic Review. *S. Afr. J. Enol. Vitic.* **2008**, *29*, 128–144. [CrossRef]
3. Chatonnet, P.; Dubourdieu, D.; Boidron, J. The Influence of *Brettanomyces/Dekkera* Sp. Yeasts and Lactic Acid Bacteria on the Ethylphenol Content of Red Wines. *Am. J. Enol. Vitic.* **1995**, *46*, 463–468.
4. Vigentini, I.; Lucy Joseph, C.M.; Picozzi, C.; Foschino, R.; Bisson, L.F. Assessment of the *Brettanomyces bruxellensis* Metabolome during Sulphur Dioxide Exposure. *FEMS Yeast Res.* **2013**, *13*, 597–608. [CrossRef] [PubMed]
5. Avramova, M.; Cibrario, A.; Peltier, E.; Coton, M.; Coton, E.; Schacherer, J.; Spano, G.; Capozzi, V.; Blaiotta, G.; Salin, F.; et al. *Brettanomyces bruxellensis* Population Survey Reveals a Diploid-Triploid Complex Structured According to Substrate of Isolation and Geographical Distribution. *Sci. Rep.* **2018**, *8*, 1–13. [CrossRef] [PubMed]
6. Wedral, D.; Shewfelt, R.; Frank, J. The Challenge of *Brettanomyces* in Wine. *LWT-Food Sci. Technol.* **2010**, *43*, 1474–1479. [CrossRef]
7. Harris, V.; Jiranek, V.; Ford, C.M.; Grbin, P.R. Inhibitory Effect of Hydroxycinnamic Acids on *Dekkera* spp. *Appl. Microbiol. Biotechnol.* **2010**, *86*, 721–729. [CrossRef] [PubMed]
8. Godoy, L.; Martínez, C.; Carrasco, N.; Ganga, M.A. Purification and Characterization of a p-Coumarate Decarboxylase and a Vinylphenol Reductase from *Brettanomyces Bruxellensis*. *Int. J. Food Microbiol.* **2008**, *127*, 6–11. [CrossRef] [PubMed]
9. Godoy, L.; Garrido, D.; Martínez, C.; Saavedra, J.; Combina, M.; Ganga, M.A. Study of the Coumarate Decarboxylase and Vinylphenol Reductase Activities of *Dekkera bruxellensis* (Anamorph *Brettanomyces bruxellensis*) Isolates. *Lett. Appl. Microbiol.* **2009**, *48*, 452–457. [CrossRef] [PubMed]
10. Loureiro, V.; Malfeito-Ferreira, M. Spoilage Yeasts in the Wine Industry. *Int. J. Food Microbiol.* **2003**, *86*, 23–50. [CrossRef]
11. Dias, L.; Dias, S.; Sancho, T.; Stender, H.; Querol, A.; Malfeito-Ferreira, M.; Loureiro, V. Identification of Yeasts Isolated from Wine-Related Environments and Capable of Producing 4-Ethylphenol. *Food Microbiol.* **2003**, *20*, 567–574. [CrossRef]
12. Ciani, M.C.F. Brettanomyces. *Encycl. Food Microbiol.* **2014**, *1*, 316–323. [CrossRef]
13. Malfeito-Ferreria, M. Two Decades of "Horse Sweat" Taint and Brettanomyces Yeast in Winw: Where do We Stand Now? *Beverages* **2018**, *4*, 32. [CrossRef]
14. Divol, B.; Du Toit, M.; Duckitt, E. Surviving in the Presence of Sulphur Dioxide: Strategies Developed by Wine Yeasts. *Appl. Microbiol. Biotechnol.* **2012**, *95*, 601–613. [CrossRef] [PubMed]

15. Avram, D. SSU1 Encodes a Plasma Membrane Protein with a Central Role in a Network of Proteins Conferring Sulfite Tolerance in *Saccharomyces* SSU1 Encodes a Plasma Membrane Protein with a Central Role in a Network of Proteins Conferring Sulfite Tolerance in Saccharom. *J. Bacterial.* **1997**, *179*, 5971–5974. [CrossRef]

16. Curtin, C.D.; Borneman, A.R.; Chambers, P.J.; Pretorius, I.S. De-Novo Assembly and Analysis of the Heterozygous Triploid Genome of the Wine Spoilage Yeast *Dekkera bruxellensis* AWRI1499. *PLoS ONE* **2012**, *7*, 1–10. [CrossRef] [PubMed]

17. Curtin, C.; Kennedy, E.; Henschke, P.A. Genotype-Dependent Sulphite Tolerance of Australian *Dekkera* (*Brettanomyces*) *bruxellensis* Wine Isolates. *Lett. Appl. Microbiol.* **2012**, *55*, 56–61. [CrossRef] [PubMed]

18. Devalia, J.L.; Rusznak, C.; Herdman, M.J.; Trigg, C.J.; Davies, R.J.; Tarraf, H. Effect of Nitrogen Dioxide and Sulphur Dioxide on Airway Response of Mild Asthmatic Patients to Allergen Inhalation. *Lancet* **1994**, *344*, 1668–1671. [CrossRef]

19. Costa, A.; Barata, A.; Malfeito-Ferrerira, M.; Loureiro, V. Evaluation of the inhibitory effect of dimethyl dicarbonate (DMDC) against wine microorganisms. *Food Microbiol.* **2008**, *25*, 422–427. [CrossRef] [PubMed]

20. Giacosa, S.; Rio Segade, S.; Cagnasso, E.; Caudana, A.; Rolle, L.; Gerbi, V. SO$_2$ in wine. *Red Wine Technol.* **2019**, 309–321. [CrossRef]

21. Puértolas, E.; López, N.; Condón, S.; Raso, J.; Álvarez, I. Pulsed Electric Fields Inactivation of Wine Spoilage Yeast and Bacteria. *Int. J. Food Microbiol.* **2009**, *130*, 49–55. [CrossRef] [PubMed]

22. Fabrizio, V.; Vigentini, I.; Parisi, N.; Picozzi, C.; Compagno, C.; Foschino, R. Heat Inactivation of Wine Spoilage Yeast *Dekkera bruxellensis* by Hot Water Treatment. *Lett. Appl. Microbiol.* **2015**, *61*, 186–191. [CrossRef] [PubMed]

23. González-Arenzana, L.; Sevenich, R.; Rauh, C.; López, R.; Knorr, D.; López-Alfaro, I. Inactivation of *Brettanomyces bruxellensis* by High Hydrostatic Pressure Technology. *Food Control* **2016**, *59*, 188–195. [CrossRef]

24. Comitini, F.; Ingeniis De, J.; Pepe, L.; Mannazzu, I.; Ciani, M. *Pichia anomala* and *Kluyveromyces wickerhamii* Killer Toxins as New Tools against *Dekkera/Brettanomyces* Spoilage Yeasts. *FEMS Microbiol. Lett.* **2004**, *238*, 235–240. [CrossRef] [PubMed]

25. Santos, A.; San Mauro, M.; Bravo, E.; Marquina, D. PMKT2, a New Killer Toxin from Pichia Membranifaciens, and Its Promising Biotechnological Properties for Control of the Spoilage Yeast *Brettanomyces bruxellensis*. *Microbiology* **2009**, *155*, 624–634. [CrossRef] [PubMed]

26. Santos, A.; Navascués, E.; Bravo, E.; Marquina, D. *Ustilago maydis* Killer Toxin as a New Tool for the Biocontrol of the Wine Spoilage Yeast *Brettanomyces bruxellensis*. *Int. J. Food Microbiol.* **2011**, *145*, 147–154. [CrossRef] [PubMed]

27. Mehlomakulu, N.N.; Prior, K.J.; Setati, M.E.; Divol, B. Candida Pyralidae Killer Toxin Disrupts the Cell Wall of *Brettanomyces bruxellensis* in Red Grape Juice. *J. Appl. Microbiol.* **2017**, *122*, 747–758. [CrossRef] [PubMed]

28. Mehlomakulu, N.N.; Setati, M.E.; Divol, B. Characterization of Novel Killer Toxins Secreted by Wine-Related Non-*Saccharomyces* Yeasts and Their Action on *Brettanomyces* spp. *Int. J. Food Microbiol.* **2014**, *188*, 83–91. [CrossRef] [PubMed]

29. Mahlapuu, M.; Håkansson, J.; Ringstad, L.; Björn, C. Antimicrobial Peptides: An Emerging Category of Therapeutic Agents. *Front. Cell. Infect. Microbiol.* **2016**, *6*, 1–12. [CrossRef] [PubMed]

30. Zhang, L.; Gallo, R.L. Antimicrobial Peptides. *Curr. Biol.* **2016**, *26*, R14–R19. [CrossRef] [PubMed]

31. Muñoz, A.; Gandía, M.; Harries, E.; Carmona, L.; Read, N.D.; Marcos, J.F. Understanding the Mechanism of Action of Cell-Penetrating Antifungal Peptides Using the Rationally Designed Hexapeptide PAF26 as a Model. *Fungal Biol. Rev.* **2013**, *26*, 146–155. [CrossRef]

32. López-García, B.; Harries, E.; Carmona, L.; Campos-Soriano, L.; López, J.J.; Manzanares, P.; Gandía, M.; Coca, M.; Marcos, J.F. Concatemerization Increases the Inhibitory Activity of Short, Cell-Penetrating, Cationic and Tryptophan-Rich Antifungal Peptides. *Appl. Microbiol. Biotechnol.* **2015**, *99*, 8011–8021. [CrossRef] [PubMed]

33. Wang, G.; Li, X.; Wang, Z. APD3: The Antimicrobial Peptide Database as a Tool for Research and Education. *Nucleic Acids Res.* **2016**, *44*, D1087–D1093. [CrossRef] [PubMed]

34. Gun Lee, D.; Yub Shin, S.; Maeng, C.-Y.; Zhu Jin, Z.; Lyong Kim, K.; Hahm, K.-S. Isolation and Characterization of a Novel Antifungal Peptide from *Aspergillus niger*. *Biochem. Biophys. Res. Commun.* **1999**, *263*, 646–651. [CrossRef] [PubMed]

35. Kaiserer, L.; Oberparleiter, C.; Weiler-Görz, R.; Burgstaller, W.; Leiter, E.; Marx, F. Characterization of the Penicillium Chrysogenum Antifungal Protein PAF. *Arch. Microbiol.* **2003**, *180*, 204–210. [CrossRef] [PubMed]
36. Theis, T.; Wedde, M.; Meyer, V.; Stahl, U. The Antifungal Protein from *Aspergillus giganteus* Causes Membrane Permeabilization. *Antimicrob. Agents Chemother.* **2003**, *47*, 588–593. [CrossRef] [PubMed]
37. Mygind, P.H.; Fischer, R.L.; Schnorr, K.M.; Hansen, M.T.; Sönksen, C.P.; Ludvigsen, S.; Raventós, D.; Buskov, S.; Christensen, B.; De Maria, L.; et al. Plectasin Is a Peptide Antibiotic with Therapeutic Potential from a Saprophytic Fungus. *Nature* **2005**, *437*, 975–980. [CrossRef] [PubMed]
38. Hajji, M.; Jellouli, K.; Hmidet, N.; Balti, R.; Sellami-Kamoun, A.; Nasri, M. A Highly Thermostable Antimicrobial Peptide from *Aspergillus clavatus* ES1: Biochemical and Molecular Characterization. *J. Ind. Microbiol. Biotechnol.* **2010**, *37*, 805–813. [CrossRef] [PubMed]
39. Rodríguez-Martín, A.; Acosta, R.; Liddell, S.; Núñez, F.; Benito, M.J.; Asensio, M.A. Characterization of the Novel Antifungal Protein PgAFP and the Encoding Gene of *Penicillium chrysogenum*. *Peptides* **2010**, *31*, 541–547. [CrossRef] [PubMed]
40. Kubicek, C.P.; Komoń-Zelazowska, M.; Sándor, E.; Druzhinina, I.S. Facts and Challenges in the Understanding of the Biosynthesis of Peptaibols by *Trichoderma*. *Chem. Biodivers.* **2007**, *4*, 1068–1082. [CrossRef] [PubMed]
41. Whitmore, L.; Wallce, B. The Peptaibol Database: A Database for Sequences and Structures of Naturally Occurring Peptaibols. *Nucleic Acids Res.* **2004**, *32*, D593–D594. [CrossRef] [PubMed]
42. Gisin, B.F.; Kobayashi, S.; Hall, J.E. Synthesis of a 19-Residue Peptide with Alamethicin-like Activity. *Proc. Natl. Acad. Sci. USA* **1977**, *74*, 115–119. [CrossRef] [PubMed]
43. La Rocca, P.; Biggin, P.C.; Tieleman, D.P.; Sansom, M.S.P. Simulation Studies of the Interaction of Antimicrobial Peptides and Lipid Bilayers. *Biochim. Biophys. Acta* **1999**, *1462*, 185–200. [CrossRef]
44. Muñoz, A.; Harries, E.; Contreras-Valenzuela, A.; Carmona, L.; Read, N.D.; Marcos, J.F. Two Functional Motifs Define the Interaction, Internalization and Toxicity of the Cell-Penetrating Antifungal Peptide PAF26 on Fungal Cells. *PLoS ONE* **2013**, *8*, 1–11. [CrossRef] [PubMed]
45. Enrique, M.; Marcos, J.F.; Yuste, M.; Martínez, M.; Vallés, S.; Manzanares, P. Antimicrobial Action of Synthetic Peptides towards Wine Spoilage Yeasts. *Int. J. Food Microbiol.* **2007**, *118*, 318–325. [CrossRef] [PubMed]
46. Enrique, M.; Marcos, J.F.; Yuste, M.; Martínez, M.; Vallés, S.; Manzanares, P. Inhibition of the Wine Spoilage Yeast *Dekkera bruxellensis* by Bovine Lactoferrin-Derived Peptides. *Int. J. Food Microbiol.* **2008**, *127*, 229–234. [CrossRef] [PubMed]
47. Albergaria, H.; Francisco, D.; Gori, K.; Arneborg, N.; Gírio, F. *Saccharomyces cerevisiae* CCMI 885 Secretes Peptides That Inhibit the Growth of Some Non-Saccharomyces Wine-Related Strains. *Appl. Microbiol. Biotechnol.* **2010**, *86*, 965–972. [CrossRef] [PubMed]
48. Branco, P.; Francisco, D.; Chambon, C.; Hébraud, M.; Arneborg, N.; Almeida, M.G.; Caldeira, J.; Albergaria, H. Identification of Novel GAPDH-Derived Antimicrobial Peptides Secreted by *Saccharomyces cerevisiae* and Involved in Wine Microbial Interactions. *Appl. Microbiol. Biotechnol.* **2014**, *98*, 843–853. [CrossRef] [PubMed]
49. Branco, P.; Albergaria, H.; Arneborg, N.; Prista, C. Effect of GAPDH-Derived Antimicrobial Peptides on Sensitive Yeasts Cells: Membrane Permeability, Intracellular PH and H+-Influx/-Efflux Rates. *FEMS Yeast Res.* **2018**, *18*. [CrossRef] [PubMed]
50. Branco, P.; Viana, T.; Albergaria, H.; Arneborg, N. Antimicrobial Peptides (AMPs) Produced by Saccharomyces cerevisiae Induce Alterations in the Intracellular PH, Membrane Permeability and Culturability of *Hanseniaspora guilliermondii* Cells. *Int. J. Food Microbiol.* **2015**, *205*, 112–118. [CrossRef] [PubMed]
51. Branco, P.; Francisco, D.; Monteiro, M.; Almeida, M.G.; Caldeira, J.; Arneborg, N.; Prista, C.; Albergaria, H. Antimicrobial Properties and Death-Inducing Mechanisms of Saccharomycin, a Biocide Secreted by *Saccharomyces cerevisiae*. *Appl. Microbiol. Biotechnol.* **2017**, *101*, 159–171. [CrossRef] [PubMed]
52. Acuña-Fontecilla, A.; Silva-Moreno, E.; Ganga, M.A.; Godoy, L. Evaluación de La Actividad Antimicrobiana de Levaduras Vínicas Nativas Contra Microorganismos Patógenos de La Industria Alimentaria. *CYTA-J. Food* **2017**, *15*, 457–465. [CrossRef]
53. Peña, R.; Ganga, M.A. Novel Antimicrobial Peptides Produced by *Candida intermedia* LAMAP1790 Active against the Wine-Spoilage Yeast *Brettanomyces bruxellensis*. *Antonie Van Leeuwenhoek* **2018**, *9*. [CrossRef] [PubMed]
54. Brogden, K.A. Antimicrobial Peptides: Pore Formers or Metabolic Inhibitors in Bacteria? *Nat. Rev. Microbiol.* **2005**, *3*, 238–250. [CrossRef] [PubMed]

55. Ahmad, M.; Piludu, M.; Oppenheim, F.G.; Helmerhorst, E.J.; Hand, A.R. Immunocytochemical Localization of Histatins in Human Salivary Glands. *J. Histochem. Cytochem.* **2004**, *52*, 361–370. [CrossRef] [PubMed]
56. Peters, B.M.; Shirtliff, M.E.; Jabra-Rizk, M.A. Antimicrobial Peptides: Primeval Molecules or Future Drugs? *PLoS Pathog.* **2010**, *6*, 4–7. [CrossRef] [PubMed]
57. Swidergall, M.; Ernst, J.F. Interplay between *Candida albicans* and the Antimicrobial Peptide Armory. *Eukaryot. Cell* **2014**, *13*, 950–957. [CrossRef] [PubMed]

fermentation

MDPI

Comment

Enological Repercussions of Non-*Saccharomyces* Species

Agustín Aranda

Institute for Integrative Systems Biology, I2SysBio, University of Valencia-CSIC, 46980 Paterna, Spain;
agustin.aranda@csic.es

Received: 13 July 2019; Accepted: 18 July 2019; Published: 24 July 2019

The bulk of the sugar fermentation in grape juice, in order to produce wine is carried out by yeasts of the genus *Saccharomyces*, mainly *S. cerevisiae*. However, *S. cerevisiae* is not the only wine yeast, as spontaneous grape juice fermentation involves a complex succession of growth and death of different yeasts [1,2], and each of them contribute to the organoleptic properties of the final product. *Saccharomyces* are not usually found in the epiphytic yeasts present on the surface of grapes, where *Hanseniaspora*, *Candida*, *Pichia*, and *Hansenula* are dominant [3]. However, *Saccharomyces* imposes itself due to its higher tolerance to the stressful conditions of fermentation, due to its resistance to the addition of sulfite and to the ethanol that it itself produces [4]. Therefore, most non-*Saccharomyces* species relevant to winemaking have been traditionally overlooked, except when they act as spoilage agents [5]. However, the use of selected non-conventional yeasts to improve the organoleptic properties of wine is probably the most exciting trend in modern enological microbiology [6–8]. This Special Issue gives a complete picture of the most promising non-*Saccharomyces* strains and their contributions towards more complex and balanced wines.

Spoilage yeasts are still the subject of much interest. One of the most resilient contaminants in the cellar are *Brettanomyces/Dekkera* yeasts. *B. bruxellensis* has been described as being mainly responsible for worldwide off-flavor wine production, due to its ability to transform hydroxycinnamic acids present in the grape juice into phenolic derivatives [9]. *Brettamomyces* has been controlled traditionally by the addition of sulfur dioxide, but the current trend is to reduce the use of such chemical agents due to its negative impact on human health. Therefore, alternatives are being explored. The use of killer toxins produced by different non-*Saccharomyces* yeasts have been described, and the most promising one is the use of antimicrobial peptides. Interestingly, a peptide derived from a strain of *Candida intermedia* isolated form wine fermentations proved useful against *B. bruxellensis*, which leads to the possibility of this and other yeasts to be used as agents for biological control. *Zygosaccharomyces* is also a well-known wine spoilage yeast, causing re-fermentation in sweet wines and producing not just CO_2 but also undesired compounds such as some esters [10]. However, *Z. rouxii* has been used to produce low-alcohol beer, and co-cultivation of. *S. cerevisiae* and *Z. bailii* produces wine with lower ethanol content. High stress tolerance and low oxygen requirements would make this genus fit to produce sparkling wines, but its production of off-flavors, such as acetoin, might be detrimental. Another spoilage yeast with a high stress tolerance is *Saccharomycodes ludwigii* [11]. Due to its high tolerance to sulfur dioxide it can become a serious problem in the winery. However, it has some positive abilities, like reducing the alcoholic content in mixed fermentations, the high release of polysaccharides in wine (during aging-on-lees, as a result of cellular autolysis, but also during growth and alcoholic fermentation) and the aromatic profile improvement of mixed fermentations.

Fission yeast *Schizosaccharomyces pombe* is one of the non-conventional yeasts that has raised more early interest [12]. *S. pombe* is naturally found in grape juice and has the ability to reduce the L-malic in wine, through maloalcoholic fermentation into ethanol and CO_2. That is a way to control the acidity of wines, producing more balanced products when acidity is high. However, due to its slow growth and the fact that it produces wines with a less fruity tone, its use as unique fermenting yeasts is ruled out.

Fermentation **2019**, *5*, 68

However, the use of immobilized *S. pombe* has proved to be an alternative to reduce malic acid in a controlled way. Additionally, this yeast helps to stabilize color and to release polysaccharides during ageing on lees. *Torulaspora delbrueckii* is probably the most suitable of non-*Saccharomyces* yeasts for use in winemaking [13]. It has a better fermentative performance that any of the other non-conventional yeasts mentioned here, and it has some positive aspects that might improve wine fermentation when compared to *S. cerevisiae*—low acetic acid and ethanol production, high glycerol production and mannoprotein and polysaccharide release—along with some interesting contributions to the final aroma. However, it is more sensitive to stress than *S. cerevisiae*, it dies faster, and also its metabolic activity declines markedly as a result of the environmental stress.

The genus of apiculate yeast *Hanseniaspora* and its asexual anamorph *Kloeckera* is the most abundant yeast associated with grapes. Its tolerance to ethanol and sulfite is low, so their contribution during a spontaneous fermentation is restricted to the first stages [14]. However, its contribution to the organoleptic properties of the final product produced by *S. cerevisiae* is remarkable. Some specific species that are relatively well-adapted to wine fermentation have attracted interest. For instance, strains of *H. vineae* has been demonstrated to increase fruity aromas by increasing 2-phenylethyl acetate and ethyl acetate. All *Hanseniaspora* species increase the level of almost all acetate esters. Some species, such *H. clermontiae*, *H. opuntiae*, *H. guilliermondii*, and *H. vineae* are also able to stabilize the color of wines. Other genera that are easily found in grapes and wineries are *Candida*. Some strains of *C. stellata* have been associated with food production for a long time [15]. It has a positive impact in winemaking for its fructophilic character, its high glycerol production, and its efficient production of extracellular enzymes, such as pectinases, cellulases, proteases, glycosidases, and so on. *Candida* is a complex genus from the genetic point of view, and for instance a *C. stellata* strain of enological interest has been renamed *Starmerella bombicola*, while other food related species *C. zemplinina* has been renamed *Starmerella bacillaris*, so this variety of yeasts with interesting industrial properties have to be carefully studied. A similar change of name has suffered *Wickerhamomyces anomalus*, which was previously known as *Pichia anomala*, *Hansenula anomala* or *Candida pelliculosa* [16]. This yeast is present in grape juice fermentations, and while it was traditionally associated with a high ethyl acetate production it has been proved to be a good enzyme producer. For instance, it produces high amounts of glycosidases that contribute to the release of primary aromas from the grape. It also produces a high level of proteases that prevent wine haze.

Most non-*Saccharomyces* yeasts change the wine properties with a variety of small changes in the final product, while others have a very distinct advantage. The main advantage of *Lachancea thermotolerans* (previously known as *Kluyveromyces thermotolerans*) is its strong production of lactic acid, a fact that can lower the wine pH by 0.5 units or even more [17]. This prevents the addition of tartaric acid in cases where the acidity is sub-optimal. Additionally, it helps to reduce the volatile acidity when mixed with *S. cerevisiae*. *Metschnikowia (Candida) pulcherrima* can be used as a biological control agent due to its production of a natural antimicrobial compound, pulcherrimin, which has antifungal properties [18]. It also helps stabilize the wine's color due to its low anthocyanin absorption rates and the formation of stable pigments (pyranoanthocyanins and polymers). The saprophytic yeast-like fungus *Aureobasidium pullulans* shows antagonistic activity against plant pathogens and it has the ability to decrease ochratoxin A (OTA) production and OTA biosynthetic gene expression of the contaminating fungus [19].

Due to their lower fermentative potential, compared to *S. cerevisiae*, the non-*Saccharomyces* species are used in mixed or sequential fermentation with an *S. cerevisiae* strain with good fermentative power to achieve the complete consumption of sugars from grape juice. Non-conventional yeasts might help to improve the most undesired aspect of a given fermentation. With regards to wine acidity, up to seven non-*Saccharomyces* species can increase or decrease acidity [20]. By doing so they can be useful to prevent the unbalances produced through the increasing levels of global warming. Finally, non-*Saccharomyces* yeasts can be useful in the production of wines involving specific post-fermentation processes, such as the case of sparkling wine [21]. Secondary fermentation can be carried out by

Fermentation **2019**, *5*, 68

T. delbrueckii giving positive effects on the overall aroma and sensory characteristics. Presence of *S. pombe* and *Saccharomycodes ludwigii* influence the acidity and color of the final product, and of course, the presence of non-*Saccharomyces* in the primary fermentation has its fingerprint in the sensory quality of sparkling wines. For all these reasons there is an increasingly longer list of non-*Saccharomyces* species used as commercial starters, particularly of *T. delbrueckii*, where five distinct commercial brands are now available.

Funding: This work has been funded by a grant from the Spanish Ministry of Economy and Competiveness (AGL2017-83254-R).

Conflicts of Interest: The authors declare no conflict of interest.

References

1. Fleet, G.H. The microorganisms of wine making. In *Wine Microbiology and Biotechnology*; CRC Press: Boca Raton, FL, USA, 1993; pp. 1–27.
2. Ribéreau-Gayon, P.; Dubourdieu, D.; Donèche, B. *Handbook of Enology*, 2nd ed.; John Wiley: Chichester, UK; Hoboken, NJ, USA, 2006.
3. Jackson, R.S. *Wine Science: Principles, Practice, Perception*, 2nd ed.; Academic Press: San Diego, Spain, 2000; p. 648.
4. Matallana, E.; Aranda, A. Biotechnological impact of stress response on wine yeast. *Lett. Appl. Microbiol.* **2017**, *64*, 103–110. [CrossRef] [PubMed]
5. Loureiro, V.; Malfeito-Ferreira, M. Spoilage yeasts in the wine industry. *Int. J. Food Microbiol.* **2003**, *86*, 23–50. [CrossRef]
6. Fleet, G.H. Wine yeasts for the future. *FEMS Yeast Res.* **2008**, *8*, 979–995. [CrossRef] [PubMed]
7. Jolly, N.P.; Varela, C.; Pretorius, I.S. Not your ordinary yeast: Non-*Saccharomyces* yeasts in wine production uncovered. *FEMS Yeast Res.* **2014**, *14*, 215–237. [CrossRef] [PubMed]
8. Padilla, B.; Gil, J.V.; Manzanares, P. Past and Future of Non-*Saccharomyces* Yeasts: From Spoilage Microorganisms to Biotechnological Tools for Improving Wine Aroma Complexity. *Front. Microbiol.* **2016**, *7*, 411. [CrossRef] [PubMed]
9. Peña, R.; Chávez, R.; Rodríguez, A.; Ganga, M.A. A Control Alternative for the Hidden Enemy in the Wine Cellar. *Fermentation* **2019**, *5*, 25. [CrossRef]
10. Escott, C.; Del Fresno, J.M.; Loira, I.; Morata, A.; Suárez-Lepe, J.A. *Zygosaccharomyces rouxii*: Control Strategies and Applications in Food and Winemaking. *Fermentation* **2018**, *4*, 69. [CrossRef]
11. Vejarano, R. *Saccharomycodes ludwigii*, Control and Potential Uses in Winemaking Processes. *Fermentation* **2018**, *4*, 71. [CrossRef]
12. Loira, I.; Morata, A.; Palomero, F.; González, C.; Suárez-Lepe, J.A. *Schizosaccharomyces pombe*: A Promising Biotechnology for Modulating Wine Composition. *Fermentation* **2018**, *4*, 70. [CrossRef]
13. Ramírez, M.; Velázquez, R. The Yeast *Torulaspora delbrueckii*: An Interesting But Difficult-To-Use Tool for Winemaking. *Fermentation* **2018**, *4*, 94. [CrossRef]
14. Martin, V.; Valera, M.J.; Medina, K.; Boido, E.; Carrau, F. Oenological Impact of the *Hanseniaspora/Kloeckera* Yeast Genus on Wines—A Review. *Fermentation* **2018**, *4*, 76. [CrossRef]
15. García, M.; Esteve-Zarzoso, B.; Cabellos, J.M.; Arroyo, T. Advances in the Study of *Candida stellata*. *Fermentation* **2018**, *4*, 74. [CrossRef]
16. Padilla, B.; Gil, J.V.; Manzanares, P. Challenges of the Non-Conventional Yeast *Wickerhamomyces anomalus* in Winemaking. *Fermentation* **2018**, *4*, 68. [CrossRef]
17. Morata, A.; Loira, I.; Tesfaye, W.; Bañuelos, M.A.; González, C.; Suárez Lepe, J.A. *Lachancea thermotolerans* Applications in Wine Technology. *Fermentation* **2018**, *4*, 53. [CrossRef]
18. Morata, A.; Loira, I.; Escott, C.; del Fresno, J.M.; Bañuelos, M.A.; Suárez-Lepe, J.A. Applications of *Metschnikowia pulcherrima* in Wine Biotechnology. *Fermentation* **2019**, *5*, 63. [CrossRef]

19. Bozoudi, D.; Tsaltas, D. The Multiple and Versatile Roles of *Aureobasidium pullulans* in the Vitivinicultural Sector. *Fermentation* **2018**, *4*, 85. [CrossRef]
20. Vilela, A. Use of Nonconventional Yeasts for Modulating Wine Acidity. *Fermentation* **2019**, *5*, 27. [CrossRef]
21. Ivit, N.N.; Kemp, B. The Impact of Non-*Saccharomyces* Yeast on Traditional Method Sparkling Wine. *Fermentation* **2018**, *4*, 73. [CrossRef]

MDPI

St. Alban-Anlage 66

4052 Basel

Switzerland

Tel. +41 61 683 77 34

Fax +41 61 302 89 18

www.mdpi.com

Fermentation Editorial Office

E-mail: fermentation@mdpi.com

www.mdpi.com/journal/fermentation

www.ingramcontent.com/pod-product-compliance
Lightning Source LLC
Chambersburg PA
CBHW051753200326
41597CB00025B/4543